2372

THE
LINGUISTICS
WARS

RANDY ALLEN HARRIS

THE
LINGUISTICS
WARS

New York   Oxford
OXFORD UNIVERSITY PRESS
1993

Oxford University Press

Oxford   New York   Toronto
Delhi   Bombay   Calcutta   Madras   Karachi
Kuala Lumpur   Singapore   Hong Kong   Tokyo
Nairobi   Dar es Salaam   Cape Town
Melbourne   Auckland   Madrid

and associated companies in
Berlin   Ibadan

Copyright © 1993 by Randy Allen Harris

Published by Oxford University Press, Inc.
200 Madison Avenue, New York, New York 10016

Library of Congress Cataloging-in-Publication Data
Harris, Randy Allen.
The linguistics wars / Randy Allen Harris.
p.   cm.   Includes bibliographical references and index.
ISBN 0-19-507256-1
1. Chomsky, Noam.   2. Linguistics—History—20th century.
3. Generative grammar.   I. Title.
P85.C47H34   1993   410'.904—dc20   92-34789

1 3 5 7 9 8 6 4 2

Printed in the United States of America
on acid-free paper

For Tom and Dot Harris,
for life, the universe, and everything

Preface

Approach

There is, most unfortunately, a widespread ignorance and trepidation about linguistics—peculiarly so, since language is unutterably fundamental to our humanhood. Look at this odd comparison Herbert Spencer offered in the nineteenth century (when linguistics was making spectacular advances):

> Astonished at the performances of the English plough, the Hindoos paint it, set it up, and worship it; thus turning a tool into an idol. Linguists do the same with language. (1865 [1852]:33)

The comparison, in addition to inventing some mythically idiotic Hindoos, has it exactly backwards. Linguists would take out their wrenches and screwdrivers, pull the plow apart, and try to figure out how it works. The ignorant worship and fear. Scientists worship and investigate. But the attitude Spencer displays—something we might call *linguiphobia* if that term didn't conjure up a fear of certain sexual practices or pasta cuts—has not abated. Continuing to pick on the British, we can cite one Auberon Waugh, who finds contemporary linguists to be so evil as to play into the hands of Neanderthal conservatives:

> Linguistics [has been] reduced by Chomsky and his disciples to a positively mind-boggling level of stupidity and insignificance. If ever [the Prime Minister] wants an excuse to close down a university, she has only to look at its department of linguistics. (1988)

Ignorance is a kind word for Waugh, who is in need of kind words, but his isn't a unique ignorance.

This book—a "popular science" look at linguistics by way of narrating an influential dispute in the sixties and seventies—attempts to clarify what linguists do, why they do it, and why everyone else should care about what they do.

My hope is that linguists will find this book useful, since many of them have a shaky or partisan view of their own recent history, but my greater hope is that nonlinguists will find an entertaining and informative account of the science of our most profound and pervasive human attribute, language.

Technical Notes

The bibliographical references in the text are quite standard. Citations are given by author, year of publication, and page number—usually in the main text—with the Works Cited accordingly organized primarily by author and year of publication. (A lowercase letter is suffixed to the year of publication, in the text and the Works Cited, when there is more than one bibliographical entry in the same year by the same author.) But there are two wrinkles.

First, there was a great deal of underground literature circulating during the period of interest, only some of which eventually surfaced, so that time-of-composition is often more important than time-of-publication, and I have tried to provide a bit of a road map here by including year-of-composition, in brackets, after year-of-publication. For instance, George Lakoff's claim that "a generative semantic theory may well be simpler and more economical than an interpretive theory" shows up as "Lakoff (1976a [1963]:50)." Consistency called for me to maintain this style for most other authors as well, leading at times to some awkward looking citations like "James (1981 [1907])," for his *Pragmatism* lectures. The only authors for whom I have avoided this style are the ancients—preferring the conventional "Aristotle (*Rhetoric* 1358^b)" to the ugly and overly specific "Aristotle (1954 [c355 B.C.]:32)."

Second, the research for this book blurred at times into an oral-history project, generating hundreds of pages of interview transcripts, hundreds of pages more of letters and e-mail printouts, and almost as many pages of telephone bills. When quoting from this material—which I do quite extensively in the later chapters—there is no citation at all. An embedded quotation, then, like "In Ross's terms, Lakoff 'is fearless, absolutely fearless'," without further attribution, is by default a remark made directly to me by Ross.

Acknowledgments

I would like, very gratefully, to acknowledge the contributions and assistance of a great many people to this book.

First, I thank the many scholars who responded to my letters, e-mail, phone calls, and hallway ambushes with (mostly) good will, candor, and generosity: Dwight Bolinger, Guy Carden, Wallace Chafe, Bruce Derwing, Matthew Dryer, Joseph Emonds, Charles Fillmore, Susan Fischer, Bruce Fraser, Gerald Gazdar, Allan Gleason, John Goldsmith, Joseph Greenberg, Geoffrey Huck, Robert Kirsner, E. F. Konrad Koerner, Yuki Kuroda, Robin Lakoff, Judith Levi, George Miller, Stephen Murray, Greg Myers, Margaret Nizhnikov, Gary Prideaux, Geoffrey Pullum, Wallace Reid, Sally Rice, Sarah Grey Thomason, Don Walker, and Anna Wierzbicka.

Second, and with even more gratitude, I thank the scholars who set aside valuable time to speak with me at length, also with (mostly) good will, candor, and generosity: Thomas Bever, Noam Chomsky, Jerry Fodor, Morris Halle, Ray Jackendoff, Jerrold Katz, Jay Keyser, Susumu Kuno, George Lakoff, Howard Lasnik, Robert Lees, James McCawley, Frederick Newmeyer, Paul Postal, Háj Ross, Jerrold Sadock, and Arnold Zwicky.

Bolinger, Chafe, Chomsky, Gleason, Jackendoff, Lakoff, Lasnik, McCawley,

Murray, Myers, Newmeyer, Postal, Prideaux, and Ross deserve still more gratitude, for reading and commenting on various earlier bits and pieces of this book, as do Victoria Bergval, Allen Fitchen, Michael Halloran, Lou Hammer, Leo Mos, Terry Nearey, Cynthia Read, Nicolette Saina, and Jim Zappen. Cynthia, my editor, gets another large measure of gratitude for cheerfulness beyond the call of duty. Many thanks also to Susan Hannan, Kim Honeyford, Sylvia Moon, Jean Mulder, and Philip Reynolds.

Not all of these people agree with what this book says, of course, or how it says what it says, and most will happily insist that any flaws are mine alone, but two people disagree so violently with the substance of this book as to require special notice, Chomsky and Lakoff. Both had very extensive comments on the same previous incarnation of the book, comments which I found mostly very profitable, and for which I remain extremely grateful, but both had very strong negative responses to the overall arrangement and orientation. Their responses were essentially inverse, Lakoff finding me to have sided with the interpretive semanticists, Chomsky finding me to have told the generative semantics version, both feeling that I slighted or ignored their own impressions or interpretations of the dispute. I should stress that the version they saw is very different in many ways from the one in your hands, but I have reason to believe that neither of them will be much more pleased with this version (their displeasure, in fact, may very well increase, since some of the latent elements that they found objectionable in the earlier version are stated a little more directly here; my correspondence with them sharpened my judgments on several matters, sometimes in directions neither of them would have preferred). Indeed, Chomsky even objects strenuously to my characterization of him in this preface; he sees no "symmetry" between his and Lakoff's opposition to the book. I am naturally distressed by their negative reactions, but it would have unquestionably been impossible to satisfy both; perhaps by satisfying neither, I am closer to neutrality than either of them believe.

I also thank the interlibrary loan people at the University of Alberta, and especially, at Rensselaer Polytechnic Institute, for above-and-beyond-the-call assistance; Madam X, for diverting to higher education a little time, money, and photocopying resources that the State of New York had not allocated to higher education; the many people at Bell-Northern Research who supported my work, especially Bill Hosier, John Phillips, and Andy Sutcliffe; the Heritage Trust Fund, Rensselaer Polytechnic Institute, the Social Sciences and Humanities Research Council of Canada, and, particularly, the estate of Dorothy and Izaak Walton Killam, for financial assistance; the editors and publishers of *Historiographia Linguistica, Rhetoric Review, Rhetoric Society Quarterly, College English, Word,* and *Rhetoric and Ideology* (Kneupper, 1989) for permission to reprint the portions of this book that have appeared elsewhere; and the staff and students of the department of linguistics, at the University of Alberta, for stimulating discussions of my work and theirs.

Lastly, and therefore most importantly, I thank my wife and guiding light, Indira Naidoo-Harris.

Waterloo, Ontario R.A.H.

Contents

THE
LINGUISTICS
WARS

Language, Thought, and the Linguistics Wars

Language and our thought-grooves are inextricably related, are, in a sense, one and the same.

Edward Sapir

Every utterance is deficient—it says less than it wishes to say.
Every utterance is exuberant—it conveys more than it plans.

José Ortega y Gasset

"Never eat more than you can lift," advises Miss Piggy; or, to quote the somewhat less voracious Ms. Anonymous, "Don't bite off more than you can chew." Putting aside the differences of scale which give these cautions their flavor, we get a very clear warning, one whose kernel is so apparent that paraphrase only mangles it. But it's not an especially easy warning to follow, especially for juicy topics. And we are pulling up to a spread of the juiciest topics associated with the human mouth: the commingle of meaning, noise, and power bundled into the word *language*. Every morsel on the table savours of faraway regions, deep, mysterious, compelling; even the gustatory metaphor we currently find closing in on us. Perhaps another paragraph will help.

Language is the subject and the object of this book. It is the method and the material, the product and the process, the chisel and the stone—points which, language being what it is, often slip noiselessly away while more immediate matters occupy us, but which, language being what it is, also lurch crashing from the shadows when we least expect it. Sometimes we seem to look right through language, hear right past it, and apprehend directly the ideas beneath the writing on the page, behind the words in the air; sometimes we can't get it out of the way. Sometimes another paragraph helps.

Sometimes not.

But always weaving in and out, off and on, through and through the discourse, are the infinite, indescribably subtle sinews that bind language and thought. As Miss

3

Piggy and Ms. Anonymous demonstrate, separating the idea from the vehicle is not a job for the faint; nor, in fact, for the sturdy. We can say that Miss Piggy's and Ms. Anonymous's expressions mean the same thing, despite the significant difference in their specific words. We can even say they both mean the same thing as the Scottish proverb, "If ye canna see the bottom, dinna wade." Moreover, we can state that 'same thing' as "Don't tackle a job beyond your capacity," or, more baldly, as "Do not do something which is not within your abilities," or, more baldly,

∀x ∀y ¬((HAVE (YOU (ABILITYx)) & NEED (TASKy (ABILITYx)) ⊃ ¬DO (YOU (TASKy)).

But, of course, we are only moving around in language, trying to hold the thought steady. We haven't peeled away the language to get at the thought; indeed, we've mangled the thought a bit with every translation. We can also move around in thought, trying to hold the language steady—by setting out the meanings of an ambiguous word, like *bank* or the meanings of an ambiguous sentence, like "Roberto saw the man from the library" (Roberto could be looking out the library window and see the man, or he could be walking down the street and see a man he knows from the library), or an ambiguous discourse, like *Hamlet*. The rub here is that, despite important similarities, *bank* is not the same word when it refers to a place where you keep your money and when it refers to land next to a river; the two Roberto-sentences are not the same; your *Hamlet* is not my *Hamlet*. The corresponding rub for Miss Piggy and Ms. Anonymous is that, despite important similarities, their thoughts are not the same.

Language and thought are not identical, since each can be partially manipulated independently of the other; but only partially, and only by willfully ignoring infinite, indescribably subtle sinews.

Something is always lost. Which brings us to linguistics, the science with the unenviable task of disentangling language and thought.

Not all linguists would agree that their science charts the sinuous relations of language to thought, thought to language, nor even that linguistics is a science, nor, if it is, about what sort of science it is. And these disagreements are crucial themes in much of what follows, as is the unavoidable conclusion that linguists are a contentious lot. Take the *dramatis personae* from the story at the heart of this book, the wars fought among (one-time or still) adherents to the principles of the first man in the list:

Noam Chomsky
Ray Jackendoff
Jerrold Katz
George Lakoff
James McCawley
Paul Postal
Háj Ross

The definition for *linguistics* we just gave runs afoul of several of them. Katz and Postal, for instance, regard linguistics as something very much like mathematics, a pristine formal science without connection to anything as messy as thought. Lakoff and Chomsky both agree that linguistics is very much concerned with mind, and that it is an empirical science, but disagree severely on many specifics, including what it is to be an empirical science. Ross, McCawley, and Jackendoff are in the empirical science camp, but fall between Lakoff and Chomsky on various specifics, depending on the issues. All of these people and issues show up recurrently in the story of the linguistics wars. For now, we will alleviate the sense of discord over fundamental issues by offering a more conventional definition of *linguistics,* one that virtually all linguists would agree to (although with linguists, as with most reflective humans, we can't do without that *virtually*): the study of the links between sound and meaning.

Two qualifications, though, are immediately necessary. First, *sound* is something of a short-hand here for the most accessible elements of language; *meaning,* for the most elusive. That is, *sound* in this definition includes the noises we make, but also stands in for the letters of written languages (like English), the characters of pictographic languages (like Chinese), the gestures of signing languages (like Ameslan). *Meaning* runs the gamut from logical and grammatical concepts (like negation and subject/predicate relations) to the nebulous domains of implication and nuance (like getting someone to close the window by snarling "It's cold in here" at her, enforcing social relations to boot). Sound is the hard currency; meaning is the network of cultural and formal conventions that turns it into a stick of gum at the candy store.

Second, the idea of standing-in is a critical, but implicit, part of the definition of linguistics, so much so that the definition would be more accurately rendered as "the study of the links between *symbolic* sound and meaning." The clatter of a train is a sound that means you should clear off the tracks, but sound and meaning are causally related here, the way a thermometer reading is linked to heat. Symbols— like "Watch out for the train!"—carry their meaning more tenuously, more subtly, more inscrutably.

Such is the tremendous mystery linguists plumb. It can look pretty mundane at times—when the phenomena under analysis are as familiar and vacuous as "Hello" or "Please pass the salt" or "Hot enough for you?"—but it is every fathom as deep as the search for the fundamental bits and pieces of the physical universe or for the guiding principles of life, and it is far more intimately connected with what it means (there's that word again) to be human.

Linguists examine language in a variety of largely opportunistic ways, as physicists examine matter, biologists life, but among their primary methods are those of the surveyor. They carve up the vast territory between sound and meaning into more manageable provinces. The borders between these provinces are frequently in dispute and hang on some very technical issues, only some of which play a role in the linguistics wars, but their existence and their primary concerns are well established. Moving in the conventional direction, *phonetics* concerns the acoustic dimensions of linguistic sound. *Phonology* studies the clustering of those acoustic properties into significant cues. *Morphology* studies the clustering of those cues into meaningful units. *Syntax* studies the arrangement of those meaningful units into expressive sequences. *Semantics* studies the composite meaning of those sequences.

For anyone unfamiliar with linguistics, those definitions are sure to constitute a stew of alien and undigestible terms. As they become relevant to our story, they become clear. But, as a crash course, consider the sentence, "Fideau chased the cat." Phonetics concerns the acoustic waveform itself, the systematic disruptions of air molecules that occur whenever someone utters the expression. Phonology concerns the elements of that waveform which recognizably punctuate the sonic flow—consonants, vowels, and syllables, represented on this page by letters. Morphology concerns the words and meaningful subwords constructed out of the phonological elements—that *Fideau* is a noun, naming some mongrel, that *chase* is a verb signifying a specific action which calls for both a chaser and a chasee, that *-ed* is a suffix indicating past action, and so on. Syntax concerns the arrangement of those morphological elements into phrases and sentences—that *chased the cat* is a verb phrase, that *the cat* is its noun phrase (the chasee), that *Fideau* is another noun phrase (the chaser), that the whole thing is a sentence. Semantics concerns the proposition expressed by that sentence—in particular, that it is true if and only if some mutt named *Fideau* has chased some definite cat.

These details of the linguistic land grants are not especially important in and of themselves, beyond illustrating one of the key uses to which linguists put the divide-and-conquer approach endemic to science, but a trend should be very clear: their direction is from sound to meaning, from accessible to elusive. We start with the observable—clacking tongues, disturbed air molecules, vibrating ear drums—and move toward significance—meaning, content, sense. Phonetics tells us such things as the amplitude, duration, and component frequencies of the speech signal; semantics tells us people use that speech signal to make assertions about a dog and a cat; the intermediate branches chart the growth of meaning. We also move, then, despite the reservations of some linguists, unmistakably toward thought. Indeed, *meaning* is in many of its uses just an alias for *thought;* more specifically, many of its uses tag certain important subsets of thought, the ones which have the most to do with being human. When I say "I mean X" to you, I am saying that "X" is in my head and, by way of my clacking tongue or clacking keyboard, I want it to end up in your head too.

The events at the heart of this book—the work of Noam Chomsky, the semantic rebellion it sparked, and the impact of both on modern linguistics—have everything to do with thought and being human. The story begins with Chomsky's compelling arguments that fundamental aspects of human behavior (linguistic creativity, for instance, and language acquisition) are inaccessible without his innovations. It develops further when his followers, principally Lakoff and McCawley, extend this work much deeper into the territory of thought than Chomsky intended. And it erupts into open warfare when Chomsky, soon abetted by the work of other followers, most notably Jackendoff, retrenches aspects of his work to banish such extensions, repudiating the work of Lakoff, McCawley, and their compatriots. How it ends, even *if* it ends, is controversial, but the received view is that Lakoff and McCawley were routed for irrationality and error, and that linguistics is much the better for their defeat. Perhaps. But, although the name for their movement, *generative semantics,* has become something of a snide joke in linguistic orthodoxy, one of the aims of this book is to help it regain a bit of its lost virtue—keeping in mind, however, that it deserves some of its shame; Chomsky's camp, some of its glory.

The events at the heart of this book also have everything to do with borders; more specifically, with border disputes. The ones involving phonetics and phonology saw very little action in the debate, but those among morphology, syntax, and semantics—the provinces more directly involved in meaning—were all flash points, and the closer the territory was to the holy land of meaning, the hotter the battles. In extreme, generative semanticists argued that language was one big shmoosh, with no place at all for borders, even in principle; sound was at one end of the linguistic continuum, meaning at the other, and a small group of uniform rules, untagged as to traditional linguistic subdiscipline, mapped one into the other. In extreme, Chomsky's camp, the interpretive semanticists, were demarcation fetishists, redrawing their borders daily; one day a given phenomenon was syntactic, the next day morphological; one day it was semantic, the next syntactic. Each saw the other side as perverse, and said so in graphic, uncompromising terms.

Such internal border disputes are largely a matter of one theory against another, much the same as a dispute between a cloud-like subatomic model and a mini-solar-system model, between fixed continents and drifting ones, between Darwinian and Lamarckian evolution. Generative semantics wanted to leave the language pie pretty much as a whole, describing its shape and texture noninvasively. Interpretive semantics wanted to slice it into more manageable pieces. But as the battle became more fierce another border dispute arose, an extra-theoretical one, concerning the definition of the entire field, the scope of language study, the answer to the question, *What is linguistics?*

Every science needs to rope off those phenomena for which it can reasonably generate explanatory theories. Nature, it has been clear since at least Heraclitus, is in dizzying flux, abuzz with colliding, chaotic, blurred events; it is a universe of infinitesimal detail and immeasurable vastness. Our senses have adapted to this by tuning to only a tiny range of those events, the ones most relevant to our survival and propagation. We see only a certain narrow band of light frequencies, hear only a small range of sound, smell and taste and feel only the grossest of data. Everything else we filter off, ignore. Sciences do exactly the same thing. Collectively they have overcome many corporeal limitations, augmenting our senses astonishingly well, but they also make even more exclusive choices than our senses. Even in the outlandishly general schemes of some physicists, currently working on a Theory of Everything, only the narrowest of phenomena would be covered; a Theory of Everything would not explain, for instance, a moth drawn to a flame, a wolf baying at the moon, a physicist writing a grant proposal. Nor should it. Science, like any other form of apprehending the world, would be impossible without its self-imposed limits.

Chomsky argued forcefully that in linguistics such limits should be drawn between the knowledge of language and the use of language. Consider the difference between knowing how to play chess and making a specific move. The first is relatively tidy—the rook goes horizontally and vertically, the bishop goes diagonally, the knight does a buttonhook. The second depends on a welter of ephemeral conditions—past moves, adversarial skill, emotional state, even the amount of light on the board or the cough of a spectator. The first can be described comprehensively by a body of rules. The second can only be described broadly, and never predicted with anything approaching certainty by an observer. In language, knowledge is relatively stable after childhood acquisition, though vocablary and conceptual knowledge grow and decay, while use is subject to all the vicissitudes of life—stress, distraction, altered states of consciousness. A speaker who knows the pronunciation of *two* and *martinis* might still claim to have had only "tee martoonies" if pulled over for erratic driving, especially in a 1950s sitcom. Chomsky and the interpretivists felt the only way to isolate tractable problems for linguistics was to focus on knowledge and filter off the ephemera of use.

Generative semanticists found this approach absurd and arbitrary, regarding accounts of linguistic knowledge to be completely artifactual when separated from the application of that knowledge, its use; McCawley's analogy for the interpretivist separation of form and function was to a theory of the stomach which ignored digestion. And, of course, however worthy the metaphors, language is neither chess nor digestion. It is far messier and far less exact than chess, far more ramified than digestion, though perhaps not so messy: separating knowledge from use is not easy. In extreme, generative semantics said there was no defensible separation. Responding appropriately to "Hot enough for you?" was the same for them as a rule for making sure pronouns matched their antecedents. In extreme, interpretive semanticists shifted their definitions daily. Yesterday's knowledge was today's use; today's use, tomorrow's knowledge. Again, each side saw little more than perversion in the other's methodological proclivities.

The story, then, is in large measure about how much is too much, about how big a bite of language is more than linguistics can chew. Chomsky charged the generative semanticists with gluttony beyond even Miss Piggy's broad constraints, of trying to swallow every conceivable thing with the most oblique relation to language. The return accusation was that interpretive semanticists took only conservative, tasteless, nutritionless little nibbles from the immense, and immensely challenging, human phenomenon, language.

The data of this dispute included such things as sentences and their meanings. So, for instance, sentences like 1a and 1b were important in the germinal stages of the debate; sentences like 2a, in its death throes.

1 a Everyone on Cormorant Island speaks two languages.
   b Two languages are spoken by everyone on Cormorant Island.

2 a Spiro conjectures Ex-Lax.

The issue and appeals surrounding 1a and 1b are very narrow, highly technical, and revolve exclusively around the formal machinery required by the competing theories to explain their different implications for Cormorant Islanders: 1a implies a world where they are all bilingual, but the languages they speak might be quite diverse; 1b implies a world where they all speak the same two languages (say, Kwakwala and English). Sentence 1a could be true in circumstances where 1b was false, and vice versa. The issues and appeals surrounding 2a are very wide, relatively informal, and revolve around much bigger questions than which theory is better; they ask, What is language? and What is linguistics?

By the time these questions surfaced, the interpretive-generative semantics differences had outgrown and exhausted the term *debate*. What began as a compact, in-house disagreement over a single hypothesis within Chomskyan linguistics had mushroomed into foundational proportions. Both sides saw the relevance of 1a and 1b very clearly, and both sides saw a resolution within reach. But the years of acrimony and diverging arguments between those sentences and 2a had altered their vision. Interpretive semanticists didn't even see 2a as data, and regarded its invocation by the other camp as clear and damning evidence they were no longer doing linguistics; generative semanticists saw 2a as the crux of an *experimentum crucis,* and saw its dismissal by the other camp as clear and damning evidence that they were practicing a brand of linguistics so sterile and navel-contemplative that their work was completely hollow. Even the political and whimsical elements of 2a chart the chasm that had grown between the erstwhile companions.

Simply put, the chasm stretched between consensus and dissensus, although these terms are not particularly simple. When 1a and 1b were relevant to our story, all the arguers agreed closely about their implications, and about what sort of enterprise linguistics should be; with 2a, there was so little agreement that *arguers* hardly applies. But this picture only catches the grossest image of the conflict, the shadows on the wall. In the mid-sixties, with the two-languages sentences, interpretive and generative semanticists agreed with one another about how to study language, certainly, but they disagreed collectively with their immediate predecessors. By the mid-seventies, with Spiro's laxative conjecture, they disagreed with one another, but now the generative semanticists began to find points of agreement with pre-Chomskyan linguists. This shifting ground of agreements—that is, history—forms not only the defining backdrop for the interpretive-generative semantics dispute, but for all the *whys, whats,* and *hows* of language study. The issues which crystallized in the divergences of Chomsky and his former disciples echo back through the centuries to other controversies, other clusters of assent and dissent, back to the earliest investigations of language, back to the birth of linguistics, and science, all of which we will get to anon.

Before we do, though, Spiro is still on the table, and we should clear him away: the nub of 2a is that it is hopelessly nonsensical in isolation (the way interpretive semanticists always preferred their sentences), but is perfectly fine in context (the way generative semanticists grew to prefer their sentences); namely, as a response to the question in 2b.

2 b Does anyone know what Pat Nixon frosts her cakes with?

Linguistics

Linguistics, in the widest sense, is that branch of science which contains all empirical investigations concerning languages.

Rudolph Carnap

To put it briefly, in human speech, different sounds have different meanings. To study this co-ordination of certain sounds with certain meanings is to study language.

Leonard Bloomfield

The Science of Language

Linguistics is, concisely but not uncontroversially, the science of language. There are various circumlocutions available, if necessary, but language is unquestionably the object of study, and *scientific* best captures the spirit of investigation common to almost everyone who has examined that object in a way that (currently or retrospectively) fits the term, *linguistic.* Other approaches to studying language, and there are many, go by names like *poetics, philology,* and *rhetoric,* but as long as we have had the word in English, *linguistics* has been associated with the methods, goals, and results of science.[1] When William Whewell (who is also responsible for the coinage, *scientist*) first proposed the term, it was in his *History of the Inductive Sciences* (1837.1:cxiv; he was borrowing it from the Germans, who, Teutonically enough, later came to prefer *Sprachwissenschaft*).

Ultimately, the matter of linguistics' fit to the category of science (or, in terms more befitting the charismatic power of *science* in the twentieth century, the matter of linguistics' merit for the status of science) is a pretty trivial one. Clearly there are compelling reasons for linguists to emulate workers in disciplines like physics, chemistry, and biology—the prototypical sciences. Physicists, chemists, and biologists have been immensely successful, producing vast quantities of results about the natural world.

There are also some striking parallels between linguistics and these other sciences, and the stronger those parallels are—the closer linguistics is to these pursuits

10

in methods, goals, and results—the more confidence we have in giving it the label, *science.*

But, more crucially, each intellectual domain requires a certain measure of integrity, and there are equally compelling reasons not to emulate these fields too closely or too blindly. The object under investigation must be allowed to guide the analysis, and a syllable is not a quark. A meaning is not a molecule. A sentence is not a liver.

Nor does linguistics need the nominal blessing of *science.* It is some sort of systematic, truth-seeking, knowledge-making enterprise, and as long as it brings home the epistemic bacon by turning up results about language, the label isn't terribly important. Etymology is helpful in this regard: *science* is a descendant of a Latin word for knowledge, and it is only the knowledge that matters.

Having said all that, however, there is certainly a range of methods, goals, and results that places such pursuits as literary criticism, philosophy, and history at one end of a continuum of knowledge-making pursuits; physics, chemistry, and biology at the other. For lack of a better term, we can call the criticism and philosophy end *humanities.* For lack of a better term, we can call the physics and chemistry end *sciences.* And defining linguistics as "the science of language" acknowledges that it falls much closer to the physics end than the criticism end. Its methods, goals, and resilient results come from a long tradition of treating language as a natural object—sometimes a social object, sometimes a mental object, sometimes both, but always as something which could be observed, like the stars and the rocks, and sometimes poked, like the animals and the plants.

Sound and Meaning

> Speech is meaning—an incorporeal thing—expressed in sounds, which are material things.
>
> Ernst Cassirer

Although the formal study of language dates at least back to the Akkadians, and there was surely campfire linguistics—Fred and Barney must have had some way of talking about talking, or what they were using wouldn't have been language—the winds of time have erased all but a very few vestiges of pre-Hellenic work. We can start with the Stoics, who, among their other activities, systematically investigated language as an object in the natural world. They were philosophers, and rhetoricians, and political scientists, and proverbial tough-guy fatalists, but they were also linguists.

Linguists *qua* linguists are interested in language in and of itself, the way a physicist is interested in matter, or a biologist in life. This statement, as simple as it is, actually conflicts with the stated goals of a great many linguists, including several who take center stage in our story. Noam Chomsky, in particular, says flatly and often that he has very little concern for language in and of itself; never has, never will. His driving concern is with mental structure, and language is the most revealing tool he has for getting at the mind. Most linguists these days follow Chomsky's lead here. The subtitle of George Lakoff's major book, for instance, is *What Categories Reveal about the Mind,* and Ray Jackendoff, who works in a department of cognitive science, has one entitled *Semantics and Cognition;* in general, linguists regard their discipline now as a branch of psychology. For most of this century, though, linguists had quite different allegiances, seeing their discipline as a branch of cultural anthropology. Earlier yet many linguists had frankly theological goals— historical linguists in the nineteenth century were after the one "pure" Adamic language, spoken from Eden to the collapse of the Nimrod's tower, and the Medieval Modistae used language to map the hidden structures of creation.

But, of course, scientists almost always hold distant goals while they work on more immediate data and theories, especially religious goals. Astrophysicists like Kepler and Newton and Einstein were trying to uncover the workings of God in nature, as are more recent physicists with much different notions of God and nature, like Capra and Zukav; even the church's biggest bogey men—Galileo and Darwin—portrayed their research as branches of natural theology, revealing the subtlety and beauty of God's handiwork. Quasi-secular motives are also popular with scientists, particularly in this century, like high-energy physicists looking for the beginning of time or the tiniest bits of matter, or molecular biologists looking for the secret of life. Whatever their ultimate motives, though, physicists look at matter, biologists look at organisms, geologists look at rocks. That is where they go for their data, what they seek to explain with their theories. Linguists look at language. That is where they go for their data, what they seek to explain with their theories.

The most frequently invoked definition of *linguistics,* a version of which begins this section, calls language a path running from sound to meaning, and calls linguistics the exploration of that path. The Stoics were the first to formalize the two end points of this path, "distinguishing between 'the signifier' and 'the signified'" (Robins, 1967:16), an utterly fundamental insight, the first principle of linguistics. The scientific approach to language has uniformly proved more valuable for exploring the sonic side of the split (the signifier), including the arrangement of sounds into words and sentences. The meaning side of the divide (the signified) has remained shrouded in speculation, and many of the most substantial contributions have come from philosophers, but linguists have always found the prospect of getting at the signifieds very compelling. The Stoics were also the first to identify distinct areas within the study of signification—phonetics, morphology, and syntax.

These were major advances, establishing the parameters of linguistics as an autonomous pursuit, and the key to these advances was clearly the same as the key to Greek advances in cosmology and mechanics: abstraction. Language is so intimately tied to consciousness, reason, and being human, that it is difficult for many thinkers to detach themselves to the point where they can look at it in general rather than specific terms. But the Stoics flourished at a time when contacts between Greek speakers and non-Greeks were on the rise; indeed, the head Stoic's (Zeno) first language was Semitic. This exposure forced the Stoics to realize that there was nothing inherent to the sound of a word or the pattern of a sentence which carried the meaning. There is nothing inherent in the sound of *chien,* or *Hund,* or *dog,* that evokes a loyal, barking quadruped; rather, as the Stoics found, the links between signifier and signified are the product of convention, consensus, and reason.

The Stoics also participated in an important controversy about language, which historians tag with the words *analogy* and *anomaly*. The analogists saw language in terms of order and regularity; the anomalists saw it as far more haphazard, particularly in the domain of meaning. The participants in the debate did not cut the pie this cleanly, and the issues were not even delineated very precisely until Varro reexamined them in the first century B.C. History has not treated this dispute with much sympathy, and it is easy to see why. "The business of science," as Russell tells us, "is to find uniformities" (1967 [1912]:35), so the position that language is fundamentally haphazard is tantamount to abandoning science. If order is illusory or superficial, there is no point in looking for patterns—in systematizing, or classifying, or abstracting. Indeed, abstraction is unthinkable in a world of totally unique objects; more importantly, such a world is itself unthinkable, since our brains are fundamentally pattern detectors.

The Stoics, curiously enough, were pretty much in the anomalist camp. But the positions were neither rigid nor absolute—rather, they were "two attitudes to language, each in itself reasonably justified by part of the evidence" (Robins, 1967:19)—and the Stoics were reacting to analogists who over-generalized, ignored data, and attempted to prescribe usage. The Stoics were empirical, with a healthy respect for the complexity of language—an important cornerstone of their advances was rejecting the simple equation of one word with one meaning. They were also less concerned than the analogists with issues of linguistic "purity," and correspondingly more tolerant of dialectal variation. The dispute subsided with the discovery of more regularities in language, such as the critical distinction between inflectional morphemes and the semantically heavier, more idiosyncratic, derivational morphemes, and with the general neglect of meaning. In short, it was settled, quietly, in favor of the analogists, though it has flared up consistently in virtually every other divisional debate in linguistics, and it plays an especially critical role in the generative-interpretive schism, when one camp became consumed with semantic questions and pursued language deep into irregularity and chaos while the other stayed safely near the surface.

A crucial term—*formal*—has snuck into the discussion in several places, and it signals the last criterial lesson we need to take from the Greeks. *Formal* has a nasty ring about it for some linguists (mostly linguists opposed to Chomsky's program, though others attack him for not being formal enough), but it is absolutely essential to linguistics, as it is to any science, and means nothing more than codified abstraction. For instance, /str/ is a representation of an abstract sound string, an instance of which occurs in the pronunciation of *string*. *String* is an expression in the formal system of English orthography. "NP + VP" is a formal expression which represents the syntactic structure of the previous sentence (since it contains a Noun Phrase followed by a Verb Phrase). And so on. The Greeks explored the abstract codification of language, adapting the Phonecian alphabet and using it to carve up the relatively continuous acoustic waveforms of speech into discrete sentences, phrases, words, morphemes, and phonemes.

The Greeks stayed pretty close to the sonic (and graphic) aspects of language, as did their Roman and early Medieval grammatical descendants, but the study of language veered sharply off toward more obscure matters when classical grammar met up with the unique brand of Aristotelian thought in the high Middle Ages known as *Scholasticism.* In modern terms, the resulting synthesis is probably closer to one of the humanities, philosophy, than to natural science, but in the terms of the period the Modistae were rigorously scientific, and, also like philosophy, had significant ties to a formal science, logic. They got their name from a collection of representative writings entitled *De Modis Significandi,* and significance was their defining concern, but they were more generally interested in the threads weaving among *modi essendi* (the ways things are), *modi intellegendi* (the ways we conceive them), and the titular *modi significandi* (the ways we express them): in short, among reality, thought, and language.

"No idea is older in the history of linguistics," Pieter Seuren writes, "than the thought that there is, somehow hidden underneath the surface of sentences, a form or a structure which provides a semantic analysis and lays bare their logical structure" (1973 [1971]:528); with the Modistae, this thought became the driving concern. Modistic grammar is best characterized by the systematic extension of formal logic to the study of language, and by the adoption of Aristotle's preoccupation for causation. In a mood swing typical of most intellectual pursuits, the Modistae jumped all over their predecessors for not looking deeply enough into causes, with settling for mere taxonomy when explanation was required.

The general explanation to which their rigid deductive methodology led strikes moderns as somewhat mystical—that there is a universal grammar underlying language which is "dependant on the structure of reality" (Bursill-Hall, 1971:35)—but it is the consequences of this position that are relevant. The Modistae were far more concerned with abstracting general principles of language than the ancients (who tended to look for general principles of individual languages, particularly Greek). Roger Bacon, for instance, said that there were problems specific to a given language, and problems common to all languages, and only the latter were of scientific interest. As a natural extension of this approach, they came to the position that all languages were in essence the same, and "that surface differences between them are merely accidental variations" (Robins, 1967:77), a position we will see again. In the standard definitional schema of the field, which sees linguistics as the investigation of links between the signifier and the signified, the Modistae were a great deal more interested in the links at the signified end of the chain than in the accidental variations of the signifiers. Indeed, they ruled all matters directly concerning sound completely out of the realm of grammatical study.

The scholastics were the victims of a rather violent mood swing themselves. They were driven from the intellectual scene by the increased concern for empirical research and mathematical modeling that marks the beginnings of modern science, and the work of the Modistae was largely forgotten. Jespersen's survey of linguistic history, for instance, dismisses the entire Middle Ages in two sentences (1922:21), and Modistic grammar had very little direct influence on modern linguistics, aside from some terminological remnants. But its indirect influence is substantial: Chomsky studied the Modistae as a young man, and it shows. Modistic grammar also had an impact on Renaissance philosophers of language, especially the Port-Royal school that Chomsky has warmly acknowledged as an intellectual forerunner of his program.

The next critical step in the history of linguistics, and the one generally taken to mark the emergence of "modern linguistics" comes with the famous chief justice of Bengal, William Jones, in his 1786 Third Annual Discourse to the Royal Asiatic Society, in which he suggested that Sanskrit, Latin, and Greek were all the descendants of "some common source, which, perhaps, no longer exists," and that Gothic and Celtic might have similar roots. There had already been substantial work done on Sanskrit; there had been debate about classification, genetic relations, and hypothetical sources, and the proposal of a common source for Sanskrit, Latin, and Greek had even been advanced. But Jones's paper is a convenient crystallization for historical purposes, because it draws all these threads into a succinct discussion; it was even commonplace for quite some time to view Jones's paper as the dividing point between the pre-scientific and scientific periods of language study.[2] The work which came to a head in Jones's address rapidly hardened into the paradigm known as *comparative linguistics.*

The comparative method was extremely simple, though its results frequently depended on staggering diligence and an astonishing breadth of knowledge. Linguists just looked closely for packages of sound and meaning in one language which were similar to packages of sound and meaning in another language and worked out explanations for the similarities. The process is exactly parallel to that of other observational sciences, like astronomy and paleontology; indeed, Kiparsky (1974) calls its practitioners *paleogrammarians.*

In some cases, the explanation of similarity the comparativists came up with might be that a word was adopted by neighboring language groups, in other cases the correspondences could only be explained as coincidences, but it became very clear that many of the European and West Asian languages were "related," descendants of the same parent. The most famous demonstration of these relations is Grimm's law (which, however, Jacob Grimm simply called a "sound shift," not a law, and which Rasmus Rask had observed before him), accounting for the parallels among, for instance, Latin *pater,* German *Vater,* and English *father,* and among Latin *piscis,* German *Fisch,* and English *fish.* The beauty of Grimm's law is that it very neatly identified a major branch of the Indo-European family tree, the Germanic languages, by way of a few simple articulatory similarities (such as the fact that both *p* and *f* are pronounced using the lips), and within a few intense decades, similar insights had established the present configuration of the Indo-European family as a hard scientific fact—solidly among the chief intellectual accomplishments of the nineteenth century.

The comparativist results have withstood the corrosive passage of time remarkably well, but the comparativists themselves were viciously attacked by the self-styled neogrammarians toward the end of the century, in a power shift that many linguists regard as a "false revolution"—in fact, as the prototypical false revolution, all heat, no light—but which is best regarded as a demi-revolution. It affected the data and the scope of the field substantially. The neogrammarians (the most famous being Karl Brugmann and Hermann Paul) attended more widely to contemporary languages and dialects as valuable in their own rights, where the comparativists had focused largely on dead languages, looking to contemporary languages primarily for the light they could throw on the past. This shift also affected the goals and argument lines of linguistics, by turning toward psychological questions and generating new classes of warrants and appeals. The neogrammarians, for instance, looked for laws rather than regularities, and banned speculation on nonverifiable matters, like the origins of language. They fancied themselves much like physicists; the comparativists' favorite analogy was to naturalists. What this shift did not alter in any interesting way was the bulwark of comparative linguistics' great success, its methodology—the neogrammarian codifications of scientific principles "were largely drawing out what had been implied by [the comparativists' work]" (Robins, 1967:187)—and it had no effect on comparativists' results, except perhaps to strengthen some of them. The other revolutionary shoe fell with Ferdinand de Saussure's monumental *Course in General Linguistics* (1966 [1915]), which initiated the linguistic strain commonly known as *structuralism* (and which, incidentally, is another point that marks, some say, the beginning of modern, scientific linguistics; just as the middle class is always rising, linguistics is always becoming a science).

Structuralism

> The first thing that strikes us when we study the facts of language is that their succession in time does not exist insofar as the speaker is concerned. He is confronted with a state.
>
> Ferdinand de Saussure

Saussure's influence was vast, but somewhat indirect, since his *Course* is a posthumous reconstruction of some of his late lectures by two of his colleagues (Charles Bally and Albert Sechehaye, in collaboration with one of Saussure's better note-taking students, Albert Riedlinger). For the purposes of our Grand Prix review of linguistic history, though, we need to consider only two of Saussure's most ramified conceptual impacts, both idealizations which help to isolate the object of linguistics.

Before Saussure, many people cared passionately about the object of linguistics—language—but no one was particularly concerned about defining it in a rigorous way. Language was just that thing that happened when you opened your mouth at the table, squeezed a few noises out of your vocal chords, and induced Socrates thereby to pass the salt. The Stoics wanted to see what its bits and pieces were—sounds, morphemes, words. The Modistae took some of these discoveries (and ignored others), along with many of their own, and sifted through them for the structure of reality (or, what was the same, the mind of God). The comparativists added time, huge stretches of time, to linguistics, trying to reel it back to the starting point. And all of them had some background notion of what language "really was"—the Adamic tongue, of which only degenerate scraps remained; the blueprint of the universe; or, for the deeply chauvinistic Greeks, Greek. But they weren't especially concerned with defining the perfectly obvious, language. Saussure was.

He was so concerned that he felt almost paralyzed in the face of the neogrammarian continuation of this disregard, telling one of his friends that he couldn't write anything on language because no one in the field knew what they were doing. First he left Leipzig, the center of the neogrammarian universe, for a chair in Paris; then he left Paris, still too close to the misguided mainstream, for the relative obscurity of Geneva; then, before he died, he destroyed most of the lecture notes articulating his notions of language and language study. But, teaching a general course in the linguistic outback of Geneva, he was free of suppositional constraints, and redefined the field.

The first idealization to this end was to separate language from the weight of the centuries the comparativists had laid on it, a weight which pressed heavily on linguists but was wholly unnoticed by speakers in their daily trade of meanings. Saussure distinguished sharply between diachronic linguistics and synchronic linguistics. *Diachronic* literally means *across-time,* and it describes any work which maps the shifts and fractures and mutations of languages over the centuries. In gross outline, it is similar to evolutionary biology, which maps the shifts and fractures and mutations of species over time, and to geology, which maps the shifts and transformations of rocks. *Synchronic* literally means *with-time,* though etymology is misleading here, since Saussure's term describes an atemporal linguistics, linguistics which proceeds without time, which abstracts away from the effects of the ages and studies language at a given, frozen moment. Two other words he used in this regard—*evolutionary* and *static* linguistics—help make the distinction clearer, but they also draw attention to the peculiarity of studying language as if time didn't matter.

Static linguistics is a pretty baffling notion, to which there are no clear analogies in other natural sciences. Ecological biology is similar, in that it looks at the interactions of species at a given time, without too much regard for the selective pressures that gave rise to them, and so is chemistry, in that it looks at the interactions of chemicals, irrespective of their history, but both of these sciences have definite temporal dimensions. The closest analogies, in fact, are to formal sciences, like most branches of mathematics and logic; triangles and existential quantifiers are outside of time. But how can language, an inescapably empirical phenomenon, be the object of a formal science? How can a word be like a triangle? The answers are as problematic as the questions, and we will see a good deal of this issue before we are through, but whatever the in-principle complications are, in practice Saussure's distinction is very workable. In practice, *synchronic* means something like "within a generation," since it is only through the innovations and misunderstandings of sons and daughters, grandsons and granddaughters, that languages change, and Saussure asked his students for a thought-experiment to make this point. "Imagine an isolated individual living for several centuries," he asks. "We would probably notice no change; time would not influence language" (1966 [1916]:78).[3]

Synchronic and *diachronic,* then, refer not to aspects of language so much as perspectives on language.

The key term in Saussure's thought experiment is clearly *isolated.* Even an ageless speaker—Dick Clark, for instance—has to change his speech to keep up with the generational tide. That is, language is a social product, which brings us to Saussure's second idealization, another sharp division, this one between language when it is put to use, hawking records on television, and the system that makes hawking possible. The first, language in use, Saussure called *parole;* the second, the system behind language use, he called *langue.* The difference is roughly the one between the ordinary parlance terms, *speech* and *language,* words which are pretty loose in their own right, but which are, respectively, still the two best English translations for Saussure's terms; *speech* is more closely associated with talking and listening, *language* with the principles and rules which make the trade of meanings possible when we talk and listen. More abstractly, we might identify Saussure's terms, respectively, with behavior and grammar. *Parole* is verbal activity: speaking, writing, listening, reading. *Langue* is the background system that makes linguistic behavior possible.

The scientific approach to language means, in large measure, taking it to be a natural object, something which exists in nature, and this notion clearly lies in back of Saussure's thinking—in the *Course* language on a few occasions is even called a "concrete object," though there is nothing concrete about it at all. The most concrete aspects of language—acoustic disturbances in the air or characters on the page—are only reflexes, virtually accidental. Certainly the dancing air molecules, the ink and the page, are not what we mean when we talk about language. It is the patterns in the air and on the page, and the network of relations which link those patterns to actions and beliefs. The patterns and their network constitute Saussure's *langue*. *Parole* is largely a filter for his approach, to screen out the variable, vulnerable, ephemeral echoes of those patterns. *Parole* is, he says, outside the scope and capabilities of linguistics. Saussurean linguistics studies the system, the rules of the game, not the individual moves of a specific contest. (Chess, by the way, was a favorite Saussurean analogy for language.)

There are, it is easy to see, some daunting complications to this style of reasoning. The data must come from *parole,* from people opening their mouths and blurting out significant sounds, but the theories concern *langue,* the system that links those signifiers to signifieds. More troublesome, the signifiers are public items, sensible only in concert with a notion of community; the signifieds are private items, sensible only in concert with a notion of individual cognition. Language is a "social product deposited in the brain of each individual" (Saussure, 1966 [1916]:23). To the extent that language is a natural object, then, there are only two conceivable locations for it to reside in nature, both of them necessary but both of them very amorphous and poorly understood themselves, society and mind. This situation makes linguistics a very Januslike profession, one head facing toward anthropology and sociology, the other toward psychology. (Saussure's thought, in fact, accommodates both heads, but he was strongly influenced by Durkheim, and his overwhelming tendency is to face toward sociology.)

Linguistic theory in Saussure's mode—that is, structuralism—charts the system underlying speech, not speech itself. This system, best known in linguistic circles as *grammar,* now takes center stage.

Sapir, and, Especially, Bloomfield

Very roughly, the first half of the twentieth century saw the following major theoretical developments in [linguistics]: (1) the confluence, with all appropriate turbulence, of the two relatively independent nineteenth century traditions, the historical-comparative and the philosophical-descriptive, the practical descriptivism of missionaries and anthropologists coming in as an important tributary. (2) serious efforts by Saussure, Sapir, and especially Bloomfield, not only to integrate the positive findings of these traditions into a single discipline but, even more, to establish that discipline as a respectable branch of science with the proper degree of autonomy from other branches. (3) The discovery and development of the phonemic principle.

Charles Hockett

In North America, where our story now takes us, structuralism took very firm root in the twenties and thirties, and continues to flourish (though the word, *structuralism,* is actually in some disrepute).[4] But it was a home-grown structuralism. As happens so often at critical junctures in the history of science, structuralism was in the air. It was, hindsight reveals, incipient in the neogrammarian moves to introduce rigor and systematicity into comparativist approaches, but several important threads are also noticeable in a number of independent scholars—in particular, in the linguistic work of three guys named Will: the philosopher, Wilhelm von Humboldt; the psychologist, Wilhelm Wundt; and the only American in the group, linguist William Dwight Whitney. Humboldt was one of the few nineteenth-century scholars of language not primarily concerned with its historical aspects, and he was (in a way that partially recalls the Modistae) far more interested in the general properties of language, its system, than his contemporaries. Wundt, who was strongly influenced by Humboldt, wove linguistic interests into his *Völkerpsychologie*— roughly, "cultural psychology"—and *völkerpsychologischen* interests into his linguistics. Whitney, who was trained among the German neogrammarians, also had a solid concern for the social-psychological dimensions of language, and, most importantly, argued for a more systematic and independent approach to language (Bloomfield credits him with helping to banish the "mystic vagueness and haphazard theory" of earlier approaches—1914:312). None of these Wills could be called a structuralist, and their contributions to linguistics are quite varied, but they all contributed substantially to the climate which gave rise to Saussure's views and their North American cognates.

The most important figures in the development of American structuralism, far and away, are Edward Sapir and Leonard Bloomfield; and, given the subsequent direction of the field, the most important of these two, far and away, is Bloomfield.[5]

Sapir—and, to a lesser extent, the early Bloomfield—had the cultural-psychological interests of Wundt and Whitney, and he had Humboldt's concern for the general, systematic properties of language, for what he called, after Humboldt, its *inner form.* Without the explicit here-a-distinction-there-a-distinction theorizing of Saussure, he wove from these strands a remarkably parallel approach to linguistic analysis, the specifics of which (in both Saussure and Sapir) would take us too far afield. But there was something else in the weave as well, the most important characteristic separating American linguists from their European cousins, a defining trait best termed "the Amerindian imperative." Sapir's teacher was the intellectual and political juggernaut of U.S. language studies at the turn of the century, Franz Boas, a.k.a. Papa Franz, a.k.a. The Father of American Linguistics. Boas recognized both the opportunity and the obligation that came with the rich, diverse, challenging languages of the Americas—languages very different from the Indo-European tongues which dominated Old World linguistics.

(As a very superficial example of these differences, take the verbs of Kwakwala, a language native to the damp western reaches of Canada, including Cormorant Island. Kwakwala verbs are wholly indifferent to time of occurrence, and needn't be marked for tense in the way most Indo-European verbs are. But they are highly concerned about the authority of the speaker, and have to be marked to indicate the speaker's justification for making a statement about the described action—marked to indicate whether the speaker saw the action, just heard about it from someone else, or experienced it in a dream—a notion highly alien to the languages, and the speakers, of the Indo-European families.)[6]

Much of the earliest research into non-European languages had one or the other, or both, of two straightforwardly rapacious motives: conquest and conversion. Diversity was therefore a problem, something which impeded "the advance of civilization and the labours of the missionary" (Lyell, 1870:461). Grammatical research primarily looked for ways of forcing the concepts of Christianity or of European administration into the native language, so they could be served up later from pulpit or page. This goal, along with haphazard training, a general belief in the racial, cultural, and linguistic inferiority of "primitives," and a warping streak of chauvinism which held Latin to be Pope of all Languages, led to treatments of Amerindian languages almost as barbarous as the treatment of their speakers. Algonquin and Mohawk and Delaware expressions were pounded into categories like dative and subjunctive and partitive-genitive, and what couldn't be pounded into these slots was ignored. Boas and his students had nothing but contempt for this bungling and mangling. Sapir put it this way:

> A linguist who insists on talking about the Latin type of morphology as though it were necessarily the high-water mark of linguistic development is like the zoölogist that sees in the organic world a huge conspiracy to evolve the race-horse or the Jersey cow. (1922:124)

The reference to zoology is not accidental. Boas recognized and enforced the integrity of Amerindian languages, prizing the collection of textual specimens above all, and steered his students, along with (through his influence at such institutes as the American Bureau of Ethnology) most of the available funds, in a primarily descriptive, data-driven direction. Though other attitudes and other approaches continued, under Boas the sanctioned mainstream of linguistics was what he called the "analytic technique"—to describe languages in their native habitat, extracting the regularities that presented themselves, imposing none from without. (Humboldt, incidentally, was also influential here; he had argued, for instance, that certain Malayo-Polynesian words which looked superficially like European verbs were in fact better analyzed, within their own linguistic systems, as nouns; see Koerner, 1990.) Variety for Boas and his students was not a hindrance, but a cause for celebration, and they also came to have a healthy respect for the various world views bundled up in the diverse Amerindian languages. Boas certainly had, like most scientists, interests beyond the brute facts. He called language a "window on the soul," which was not so much a spiritual definition as a cultural and psychological one. But the overwhelming impact of Papa Franz was to focus closely on languages in and of themselves; this emphasis made him, for many, "the father of the authentically scientific study of language in North America" (Anderson, 1985:198).

Sapir—isolated, like Saussure, in a cold intellectual backwater, Ottawa—augmented Boas's data-driven program with a theoretically richer, philosophically deeper, but somewhat eclectic approach, developing a uniquely American structuralism.[7] He wrote about the dangers of succumbing wholly to the "evolutionary prejudice" of historical linguistics (1949a[1921]:123), for instance, and he articulated a notion closely parallel to Saussure's *langue,* saying that the defining aspects of language lie "in the formal patterning and in the relating of concepts," and that "it is this abstracted language, rather more than the physical facts of speech" which forms the subject matter of linguistics (1949a[1921]:22). Where he departs most clearly from Saussure is in the explicit appreciation of variety which grew out of the Amerindian imperative. (Notice, incidentally, that this imperative in and of itself was enough to determine a strong synchronic bent to American linguistics, since there was virtually no written records with which to plumb linguistic history; too, Boas—who, in any event, had little historical training—actively discouraged his students from comparativist work.) Sapir's work is remarkable for penetrating insights, brilliant leaps, and a careful balancing of the tension between the general properties of language and the astonishing range of concepts and categories employed by languages; between uniformity and diversity; between, in Varro's somewhat stilted terms, analogy and anomaly.

He writes eloquently about the "deep, controlling impulse to form" and "the great underlying ground-plans," and (in a phrase particularly evocative of Saussure) argues for "an ideal linguistic entity dominating the speech habits" of language users (1949a[1921]:144, 148). But he is equally eloquent, and more voluble, about variety, about the defining traits that keep speakers of different languages from truly understanding one another, even in translation, because each lacks "the necessary form-grooves in which to run" one another's thoughts (1949a[1921]:106).

Sapir's structuralism, then, was more thoroughly psychological than Saussure's, and it was—thanks to the wealth of native data that kept American linguists skeptical of general claims about language—much more aware of the diversity and volatility in the human trade of meanings. Sapir was ingenious, and very influential. But he was not, even though there were linguists sometimes known as *Sapirians* into the forties and fifties, the sort to sponsor a school; Joos (1957:25) cites him not for "the developing of any method, but rather the establishing of a charter for the free intellectual play of personalities more or less akin to his own," and, in fact, Joos wags his finger a bit at "the essential irresponsibility of what has been called Sapir's 'method'." Sapirians (almost entirely made up of Sapir's students) were distinguished mostly by their unorthodox interest in the mental life of language, for a certain methodological elasticism, and for their occasional critiques of the orthodoxy, not for a specific body of unique postulates and principles.

The same might have been said of Sapir's colleague at the University of Chicago, and his successor to the Sterling Professorship in Linguistics at Yale, and the definer of the orthodoxy in the forties and fifties, Leonard Bloomfield. The same might have been said of Bloomfield, but for two things. He found behaviorism and he found logical positivism, for both of which he is now widely snickered at; behaviorism is an outmoded brand of psychology, positivism an outmoded brand of philosophy. So, Bloomfield's name shows up frequently as little more than a cipher in the linguistics of the last few decades, a foil to another name we have already seen a good deal of, and will see much more of, *Chomsky*. In part, the role of foil is natural, since understanding Chomsky's impact comes most easily when it is viewed as a reaction, if not a corrective, to certain Bloomfieldian trends. In part, the role of foil is imposed, since the victors write the history, and Chomsky's rise came at the expense of a generation inspired and strongly influenced by Bloomfield.

The word which best captures Bloomfield, especially in distinction to his partial rival, Sapir, is *methodical*. (Chomsky was never Bloomfield's rival except in the abstract; Bloomfield died before Chomsky came on the scene.) They both wrote books entitled *Language,* for instance, and the differences are telling. Sapir's (1949a[1921]) is a rich, invigorating essay—certainly not without structure and theoretical import, but heaped high with brilliant insights and imaginative leaps. Bloomfield's (1933) is a cookbook—certainly not without brilliance and imagination, but far more systematic, and far more careful about giving its readers recipes with which to obtain similar results, leading them to their own insights, guiding their imagination. The comparison may be less than flattering to Bloomfield, and it caricatures two books which hold up astonishingly well, despite more than sixty intervening years of feverish linguistic activity, but it catches the primary difference between the books, the linguists, and their respective impacts on the field. Sapir's book is more enjoyable, and perhaps more passionate, but it is also less practical, less useful. Bloomfield gave a generation of linguists a handbook. He gave them something to do (and, of course, many said, he made linguistics a science).[8] Even Sapir's most devout students had to admit Bloomfield's impact on the discipline was far more comprehensive:

> Although Sapir used linguistic methods and procedures with consummate skill, he was an artist rather than a scientist in this regard. It was Bloomfield who formulated the methods of linguistic science into a clearly defined and tightly coherent body of doctrine. (Newman, 1951:86)

Little more than a decade separates Bloomfield's *Language* from Sapir's, but it was an important one for American linguistics and goes almost as far toward explaining the differences between those two books as does the difference in their authors' temperaments. The defining event of that decade was the formation of the Linguistic Society of America, whose name proclaims the success of the independence movement early in the century and declares another one just under way; the modifiers on either side of *Society* say it all.[9] The prepositional phrase, *of America,* codifies the developments separating its members from their European relatives. The adjective, *Linguistic,* signals a separation from their academic relatives studying language in parallel disciplines. Appropriately, Boas, Sapir, and Bloomfield were all instrumental in forming the society: Boas and Sapir were the main forces in cutting the umbilical cord to Europe; Bloomfield was rapidly becoming the main force in cutting the apron strings to psychology and ethnology. He wrote the LSA's manifesto, calling for an organization distinct from "the existing societies, Philological, Oriental, Modern Language, Anthropological, Psychological, and what not," most of whose members "[do] not know that there is a science of language" (1925:1; 1970:109). Most people who called themselves *linguists* were in fact still housed in language or literature or anthropology departments—only three of the 264 Foundational Members of the LSA listed linguistics among the courses they taught—but they were beginning to feel more kinship with others who called themselves linguists than with their immediate colleagues, and Bloomfield articulated that kinship. Even linguists who maintained strong interests in literature or philology, for instance, took their papers in these areas to other forums. (Hill writes that he "felt forced to present" his literary analyses elsewhere—1991:14.)

The LSA soon fired up what has become a prominent feature of the field's landscape ever since, its summer Linguistic Institute. The Institute was (and remains) a very important training and indoctrination ground for scholars who saw themselves, or thereafter came to see themselves, as scientists of language first, scholars of culture or mind or French, second. Bloomfield was a regular and inspiring teacher at the Institute for most of its first decade (Sapir taught there only once). With the LSA also came a publishing organ—taking the common, omnivorous, but apropos title, *Language*—which soon became hugely influential to the profession and practice of linguistics, and no article was more influential in both regards than Bloomfield's contribution to the second issue, "A Set of Postulates for the Science of Language" (1926; 1970:128–40)—three decades later still being called "the Charter of contemporary descriptive linguistics" (Joos, 1957:31). The postulates take up a now-familiar topic, the object of linguistics (Saussure's *Course* and Sapir's *Language* are both cited as inspirations), but with considerably more rigor than they had been tackled by any of Bloomfield's predecessors. Here is a sample, kept mercifully brief:

8. Def. A *minimum* X is an X which does not consist entirely of lesser X's.
Thus, if X_1 consists of $X_2X_3X_4$, then X_1 is not a minimum X. But if X_1 consists of X_2X_3A, or of A_1A_2, or is unanalyzable, then X_1 is a minimum X.

9. Def. A minimum form is a *morpheme;* its meaning a *sememe.*
Thus a morpheme is a recurrent (meaningful) form which cannot in turn be analyzed into smaller recurrent (meaningful) forms. Hence any unanalyzable word or formative is a morpheme.

10. Def. A form which may be an utterance is *free.* A form which is not free is *bound.*
Thus, *book, the man* are free forms; *-ing* (as in *writing*), *-er* (as in *writer*) are bound forms, the last-named differing in meaning from the free form *err.*

11. Def. A minimum free form is a *word.* (1926:155–56; 1970:130)

There are seventy-three more—fifty in all for synchronic linguistics, seventeen for diachronic linguistics (historical studies being still very much alive, but no longer in the driver's seat). All seventy-seven look equally pedantic. But only to someone unwilling to grant the need for precision in the study of language. They were necessary to give linguistics a formal backbone. Newton's *Opticks* may have looked pedantic to some of his contemporaries, Euclid's *Principles* to some of his;[10] certainly Bloomfield had such contemporaries. Sapir, for one. Sapir was no enemy of precision or of rigor, but his view of language was far too ramified for a neat natural science approach, and he surely had Bloomfield in his mind, if not his sights, when he argued a few years later for a linguistics "which does not ape nor attempt to adopt unrevised the concepts of the natural sciences." Too, he was clearly worried about Bloomfieldian scissors at the apron strings when he followed that argument with a plea for linguists to "become increasingly concerned with the many anthropological, sociological, and psychological problems which invade the field of language" (1929:214), to no avail.[11] The strings were cut, at least far as the majority of linguists was concerned, especially the younger ones, who took the antiseptic postulates to heart and their fullest exposition, Bloomfield's *Language,* to bed with them at night.

Two collateral developments, outside the field of linguistics, in the *Language*-to-*Language* decade were even more important for Bloomfield's handbook, both apparently crystallizing for him at Ohio State, where he became fast friends with Alfred Weiss (in fact, his postulates were explicitly modeled on Weiss' postulates for psychology—1925). These developments, foreshadowed a few pages back, were the rises of behaviorism and positivism, both of which reared their seductive heads in the twenties.

Behaviorist psychology had been building since Pavlov's famous Nobel-winning, ring-the-dinner-bell-and-watch-the-dog-drool experiments at the turn of the century, but it didn't hit its stride, or get its name, until the work of John Watson and his collaborators in the teens and twenties. In the baldest terms, behaviorism is the position that beliefs, actions, and knowledge are all the products of rewards and punishments. Give a pigeon food every time it sneezes, and it will soon start sneezing whenever it gets hungry. Shock a rat whenever it attacks another rat, and it will soon show less aggression. Smile at a baby and give her extra attention when she calls you "papa" and she will (though all too briefly) say "papa" whenever she wants some extra attention from you. Expose a child to censure or ridicule when she mispronounces "light" or gets an irregular plural wrong or spells a word incorrectly, and her linguistic behavior will converge on the norm; it will become grammatical. Behaviorism is a simple, powerful, compelling theory, especially for simple behavioral phenomena. Its attraction for Bloomfield was not so much that he could put it to work explaining linguistic behavior. Quite the opposite. It was so successful, he felt comfortable leaving the psychological ends of language to the psychologists.[12]

In short, it let him comfortably avoid the messier aspects of language—learning a language, knowing it, using it, understanding it—aspects that nagged earlier linguists, the pre-behaviorist Bloomfield included. His early writings show a concern for the mental tentacles of language, and a dependence on psychology ("linguistics is, of all the mental sciences, in need of guidance at every step by the best psychologic insight available"—1914:323). This concern dropped away, and Bloomfield became profoundly anti-mental. The psychology that entered his early work was pretty muddy stuff, sometimes tentatively offered, and his first text was criticized for it.[13] Far more important than this criticism, though, was that in the two decades between writing *An Introduction to the Study of Language* and what he called its "revised version" (but which everyone else called "a wholly new work"),[14] Bloomfield did serious field research on non-Indo-European languages (Tagalog, Ilocano, Menomini, Fox, Ojibwa, and Cree)—languages he couldn't study at leisure, under a master at university, languages which didn't already have centuries of research on them. He bumped nose-to-pillar into the Amerindian imperative.[15] In the process, he came to view all the mental aspects of language as distractions from the real job, description: getting the phonological and morphological structure right. The relation of mental explanation to linguistic data was for him something like the relation "of the House of Lords to the House of Commons: when it agrees, superfluous; when it disagrees, obnoxious" (Hockett, 1965 [1964]:196).

His principal use of behaviorism, then, was as an appeal to justify cutting linguistics loose from the forbidding complexities of the mind. On this score, too, he took his lead from psychologists like Weiss, whom he praised for banishing "the specters of our tribal animism (mind, consciousness, will, and the like)" (1931:220; 1970:238). These psychologists, like Bloomfield, were motivated by the desire to be rigorous, and therefore scientific. Listen to George Miller's paraphrase of Watson and the early behaviorists:

> Look, introspection is unreliable, different people introspect differently, there's no way I can verify that you really had the experience you told me you had. Let's throw the mind out of psychology—that's all religious superstition. We'll be hard-headed, hard-nosed scientists. (in J. Miller, 1983:21)

With behaviorism you get the curious spectacle of a psychology that throws out the mental in order to talk exclusively about directly observable behavior; a psychology, in a very real sense, not of the mind, but of the body. Banishing everything not directly verifiable, for the behaviorists and for Bloomfield, was the way to be a science. They knew this because the logical positivists told them so.

Positivism has ancient roots, reaching back to the Epicureans and beyond, by way of the powerful British thinkers of the eighteenth century, Locke, Berkeley, and Hume; it is an articulation, somewhat extreme, of the grand philosophical tradition which says that all knowledge comes from the senses—empiricism. But its formal beginnings are with the famous *Wiener Kreiss* of twenties Vienna, a circle of thoroughly empiricist philosophers who took Wittgenstein's insufferably titled *Tractatus Logico-Philosophicus* (1961 [1921]) as their defining document. The short version of positivist thought, particularly as it relates to the enterprise that most concerned the circle, science, comes in the verification principle: The meaning of a proposition is the method of its verification. The meaning of "It's raining" is sticking your hand out the window; if it gets wet, the proposition is true; if not, not. The meaning of Galileo's law of descent is his ball-on-the-inclined-plane experiment, "repeated a full hundred times" and finding each time that "the times of descent, for various inclinations of the plane, bore to one another precisely the ratio" entailed by the law (1954 [1638]:179). The meaning of $E = mc^2$ is the apocalyptic detonation on 16 July 1945, in the New Mexico desert, repeated hundreds of times since. There is more than one method to skin a verification, of course—listen to the rain on the roof, look out the window, rub Fideau's head to see if it is wet as he comes through the doggy door—but the critical point for the positivists is that there be, in principle, *some* empirical method of verification. For Bloomfieldians, this method—called *mechanism,* in contrast to *mentalism*—was to link all explanations to the body. "The mechanist," Bloomfield told his followers, "believes that *mental images, feelings,* and the like are merely popular terms for various bodily movements" (1933:142; Bloomfield's italics).

One of the important contributions of positivism (and empiricism generally) is its insistence on skepticism, and its consequent disavowal of subjects without even the most tenuous possibilities for verification. In particular, positivists continued the rejection of metaphysics begun in the previous century, condemning it as utter nonsense—not just in the informal sense of "silly," but, literally, without sense. Since Plato's Realm of the Forms, for instance, could be verified by no conceivable method, all statements about it were meaningless. The Vienna Circle refracted Wittgenstein's famous "Whereof one cannot speak, thereof one must be silent" (Wittgenstein, 1961 [1921]:150) into the slogan "Metaphysicians: Shut your traps!" The behaviorist and Bloomfieldian move from this slogan to "Mentalists: Shut your traps!" was a short drive. Not even that. A putt.[16]

So, mentalism in psychology and linguistics went the way of vitalism in biology, phlogiston in chemistry, ether in physics, and, also like those other notions, mentalism packed its bags when it left. One of its suitcases was particularly crucial for the discipline, however, the one into which mentalism threw a rather critical part of linguistics' subject matter, meaning. The Amerindian imperative had disposed linguists to concentrate on phonological and morphological description anyway, which kept their attention on the signifiers, away from the signifieds—away from the messier, harder-to-isolate-and-catalog aspects of language, away from meaning. But Bloomfield raised this reluctance from practice to principle. He was certainly well aware of the attractions meaning holds for linguists; if language was just some systematic noises humans made, with no connections to thought or society, it would be of no more interest than coughing, or sneezing, or playing the bagpipes. But he also recognized that linguistics' big successes (principally those of the comparativists) were much nearer the signifier shore of the gulf between sound and meaning; more pointedly, that "the statement of meaning is the weak point of language study" (1933:140).

His aversion of meaning involved some interesting sleight of discipline. One way to get on with the business at hand, Bloomfield held, was to establish your borders firmly on this side of messy data and recalcitrant issues, leaving them in someone else's backyard: "matters which form no real part of the subject should properly be disposed of by merely naming them as belonging to the domain of other sciences" (1926:154; 1970:129). Despite the peculiarity of saying that it formed no real part of language, meaning was one of those disposable matters for Bloomfield; he regularly suggested that it belonged more properly to psychology, sociology, anthropology, anything but linguistics. Linguistics had more immediate concerns, and in order to satisfy those concerns, he confessed of linguists, "we define the meaning of a linguistic form, wherever we can, in terms of some other science. Where this is impossible, we resort to makeshift devices" (1933:140). Bloomfield intended this statement as a description of the way things were generally done when linguists looked at languages, and it was, but for the generation of linguists which learned the field from his *Language,* it also became a prescription.[17]

There is an irony as big as Everest that positivism—a theory of meaning—undergirded the exclusion of meaning from linguistics, but Bloomfield, like most scientists of the period who concerned themselves with philosophy, attended much more to the predicate of the verification principle than its subject, and one word was particularly eye-catching. The positivists cast metaphysics into the darkness by putting their spotlight brightly on the *method* of verification. The important thing about being a science was having a method; better yet, having a methodology. And the home-grown American structuralism that Bloomfield codified in his *Language* was nothing if not rigorously methodological.

Bloomfieldian methodology—and at this point we can safely start using his name as a descriptive adjective, the one which best characterizes American structuralism for, roughly, the three decades following publication of his text—was not, of course, strictly Bloomfield's.[18] It was a Saussurean-Sapirian mélange, strongly influenced by the practical necessities of analyzing the diverse, disappearing aboriginal languages of the Americas; mildly influenced by a few post-Saussurean European linguists; reworked, winnowed, and augmented by Bloomfield; and tied up with antimentalist, meaning-fearing ribbons. From Bloomfield's hands, it passed to several influential successors—most notably, Bernard Bloch, George Trager, Zellig Harris, and Charles Hockett—some of whom were considerably more dogmatic than their inspiring leader. The approach, in an epitome which does some violence to its flexibility, began with a large collection of recorded utterances from some language, a corpus. The corpus was subjected to a clear, stepwise, bottom-up strategy of analysis which began by breaking acoustic streams into discrete sounds, like the puff of air released when the lips are unsprung to form the first sound in *pin* or the stream of air forced through a channel formed by the tongue against the teeth as the last sound in *both*. These sounds were classified into various phonemes, each with a small range of acoustic realizations. Next were the small units like *per-* and *-vert,* classified into various morphemes, each with a small range of realizations. Next came words, classified into such categories as nouns and verbs, each with its realizations. Then there were utterances themselves—like *Oh my!* and *Where did I leave the dental floss?*—and further than that Bloomfieldians did not care to go. The enterprise surely seems dull and tedious to anyone who has not engaged in it, but it is a very demanding, intellectually challenging job to confront an alien stream of noises and uncover the structure in those noises that allows people to get one another to pass the pemmican.

American structuralism, in fact, was more diverse, and more interesting, than we really have time to appreciate here (see, for instance, Hymes and Fought, 1981 [1974]). There were several identifiable strains—including Sapir-tinged approaches to language, which did not outlaw mental considerations, and Christian approaches, which still studied languages primarily as means to missionary ends—along with a smattering of *indépendistes,* pluralists, and cranks. And they were a lively bunch, feuding among themselves, attacking their European counterparts, and pursuing cultural imperialism. But they were also quite cohesive, in methods and beliefs, the bulk of which followed from the theoretical structure erected by Bloomfield. His influence was everywhere, at first in person and through his textbook, then, increasingly, through his students and their students, especially by way of the LSA, its publishing arm, *Language,* and the Linguistic Institute.

Bloomfield's ideas defined the temper of the linguistic times: that it was primarily a descriptive and taxonomic science, like zoology, geology, and astronomy; that mental speculations were tantamount to mysticism, an abandonment of science; that all the relevant psychological questions (learning, knowing, and using a language) would be answered by behaviorism; that meaning was outside the scope of scientific inquiry. This program was methodologically rigorous, and very successful.

There was a good deal of confidence and optimism in the air. Structuralism had proven so successful that linguistics was widely hailed as the most rigorous and fruitful knowledge-gathering activity outside the prototypical sciences, much as comparative linguistics had been hailed in the previous century. Sociologists, anthropologists, even folklorists were explicitly adopting its classificatory methods (the most famous of these adoptions being Levi-Strauss's structuralist anthropology), and the hybrid discipline of psycholinguistics was taking a few promising steps. Linguists had also proven their patriotic mettle during the Second World War (designing courses, writing books, and preparing audio materials to teach soldiers European, Pacific, and Asian languages; designing and analyzing secret codes; working as translators), and the field was therefore partaking in the postwar financial boom. Major projects were under way: language atlases of Canada and the U.S., an American English supplement to the great *Oxford English Dictionary,* and a project to document all the known languages of the world. There was even a popular radio show by one of the leading Bloomfieldian theorists, Henry Lee Smith's "Where Are You From?" With George Trager, Smith had also built a model with the promise of completeness, and the 1953 LSA Annual Meeting saw his extended paper on that model, "the fullest presentation ever made of the upward-looking technique [beginning with sound and moving 'upward' into morphemes and syntax]" (Hill, 1991:34). Linguists had reason to be a little smug.

The bad news amid all this promise, however, was the pronounced gaps in this work—the mind, meaning, thought; in short, the good stuff.

Lo, in the east, Chomsky arose.

Chomskyan Linguistics[19]

> I am interested in meaning and I would like to find a way to get at it.
>
> Noam Chomsky

Chomsky's rapid and radical success in restructuring linguistics—and this is one of those places in science where the unqualified use of the abused term, *revolution,* is wholly appropriate; Chomsky spun linguistics on its axis, if not its head—has almost everything to do with bringing the good stuff back into linguistics, and much of the generative-interpretive debate which flowed out of his revolution hinged on how much of the good stuff linguistics can handle, and in what ways, and still responsibly do its main job, accounting for the structure of language.

The good stuff came slowly, though. Chomsky was quite circumspect about mind and meaning in his early publications, offering his work in the conciliatory tones of measured expansion. There are two general ways scientists can present innovative theoretical proposals to their field: as an extension of existing theory, the way Newton framed his optical proposals; or as a replacement of existing theory, the way Lavoisier framed his chemical proposals. Both are extremes, only partially connected to the distance between those innovations and the prevailing notions of the field, and Chomsky's proposals are best seen, like Newton's and Lavoisier's, as revisions—an extension here, a replacement there, a reinterpretation or a deletion somewhere else. But both are effective, and Chomsky has used both well, starting with the extension strategy.

Bloomfieldians had very little difficulty in seeing Chomsky's revisions as a methodological appendix to their concerns. For one thing, he was known to be the student of Zellig Harris, a brilliant, somewhat eccentric, but thoroughly Bloomfieldian, and very highly regarded linguist—"perhaps the most skillful and imaginative prophet [of the period]" (Bar-Hillel, 1967:537)—and Harris had developed a body of analyses and procedures from which Chomsky borrowed liberally. Harris, in fact, gave the 1955 LSA presidential address, "Transformation in Linguistic Structures," just as Chomsky was breaking onto the Bloomfieldian stage; Chomsky's first important paper, and a very Bloomfieldian one at that, was given the same year at the Georgetown Round Table on Linguistics. Chomsky's first important transformational book, *Syntactic Structures,* also coincided with an important Harris Paper, "Co-occurrence and Transformation in Linguistic Structure" (Chomsky, 1957a, Harris, 1957). The scene couldn't have been better set.

More importantly, though, among the most obvious lacunae in American linguistics of the period was one it had inherited from the neogrammarians, and they from the comparativists, a gaping hole in linguistic coverage which had been ignored since the Modistae and their Renaissance followers, a hole which Harris's research was trying to fill in. Syntax was AWOL.

The absence has many sources. In part, it was inertia. In part, it was Bloomfield's own confusing treatment of syntax; he had wrestled with it valiantly, but left his followers very little into which they could sink their methodical teeth. In part it was a reflection of Saussure's view, which saw the syntactic atom, the sentence, as "the typical unit of *parole*" (Wells, 1947a:15); that is, as outside the true subject matter of linguistics, *langue.* In part, it was the result of a methodological proscription which developed in Bloomfieldian linguistics against "mixing levels"—in effect, the insistence that a linguist first work out the sounds of a language (the level of phonology), then the words (the level of morphology), then the phrases and sentences (the level of syntax). Since the sounds and the words presented so many problems, it was tough to do the syntax justice.

But much of this absence also had to do with the primary data base. The dividing line between any two levels of linguistic analysis is not especially clear, and in many of the Amerindian languages that fueled Bloomfieldian research, the line between morphology and syntax is especially difficult to make out. Here is a standard delineation of the provinces of morphology and syntax, borrowed from an important Bloomfieldian text, Bloch and Trager's *Outline of Linguistic Analysis* (1942:53):[20]

MORPHOLOGY deals with the structure of words; SYNTAX deals with the combinations of words into phrases and sentences.

Now how, in Manitou's name, is a linguist to apply these notions to an utterance like 1?

1   ɑːwlisɑutissʔɑrsiniɑrpuŋɑ

From an Eskimo dialect, this expression translates into English roughly as "I am looking for something suitable to use as a fishing line," but it is neither word, the province of morphology, nor sentence, the province of syntax; neither reptile nor mammal; or, rather, it is both, like a platypus, one webbed foot in the reptilian class, one in the mammalian. But the Bloomfieldians were in the position of biologists who are really good at fish and reptiles, but uncomfortable over milk secretion, fur, and warm-bloodedness, leaving them aside as problems for another day. They could get a long way applying their morphological tools to an utterance like 1, but consider one from a language closer to home:

2   A cat is on the mat.

This utterance is somewhat atypical, as the utterances of philosophers tend to be, but it is unambiguously a sentence, furry and warm-blooded, with an individual slot for everything, and everything in its slot. There are six little words, each pretty much on its own, morphologically speaking. In concert, the words comprise an assertion about the world, and smaller groups of these words form components of the assertion—*a* and *cat* go naturally together, in the same way that *the* and *mat* go together; *is* and *on* don't go naturally together (*a cat,* for instance, can be used to answer a question like "What is sitting on the mat?" but *is on* would only be uttered in very unusual circumstances). In short, the English sentence is full of plums for the syntactic picking, ripe and inviting. The Eskimo sentence, while not immune to syntactic analysis, yields much more revealing fruits to morphological harvesters.[21]

The Bloomfieldians, once they had their morphological crop, usually left the orchard, and therefore had almost nothing of interest to say about linguistic clusters the size of phrases and sentences. It's not that the Bloomfieldians ignored syntax, or, with meaning, defined it as outside the scope of linguistics. Bloomfieldian linguistics was said to concern three core topics—phonology, morphology, and syntax—none of which was ever left out of a textbook, and none of which was ever ignored in the Bloomfieldian mainstay, fieldwork. When linguists wrote textbooks or overviews, there was always a chapter on syntax. When they went out to bag a language, their descriptive grammars always dutifully included a chapter or two on syntax (usually called *grammar*). But such chapters often betray an informality that was anathema to the Bloomfieldian program. The phonological and morphological analyses were rich, detailed, revealing investigations; the methodologies were probing. Syntactic analyses were haphazard, seat-of-the-pants outlines, and contained far more discussion of phenomena linguists would now call morphological than they would call syntactic; the methodologies were limp extensions of slice-and-dice, sounds-and-words procedures.

What was missing was method, and since method was the defining notion of science for Bloomfield, syntactic work usually came with an air of embarrassment. Charles Fries begins his *American English Grammar* with a preface apologizing to those readers "who are well trained in the scientific approach to language" (1940:viii) for the casualness of his treatment. Trager and Smith, not known for humility, note only that "syntax, as yet only begun (as will become evident) is necessarily treated sketchily" (1957 [1951]:8). Trager and Smith's syntactic program— and it *was* a program, going by the suitably sound-based label, phonological syntax—was in fact widely considered to be the most promising approach in the Bloomfieldian tool shed.

In brief, American structuralists' results in syntax are dwarfed by their advances in phonology and morphology, and not even in the same league with Chomsky's first book, *Syntactic Structures,* let alone such post-Chomskyan works as McCawley's (1988) masterful two-volume *Syntactic Phenomena of English.*[22]

Still, syntactic poverty didn't cause too much anxiety. The Bloomfieldian universe was unfolding as it should, and the gap would be filled in due course, once the final intricacies of sounds and words had been worked out. Hockett (1987:81) calls the ten or so years after Bloomfield's *Language,* "the Decade of the Phoneme;" the ten years after that, "the Decade of the Morpheme," and there was reason to believe the Decade of the Sentence was impending. Indeed, Bloomfieldian successes with sounds and words were so impressive that there was a kind of gloomy optimism in the air, at least in Pennsylvania, where Chomsky was working under Harris. As Chomsky recalls,

> In the late 1940s, Harris, like most structural linguists, had concluded that the field was essentially finished, that linguistics was finished. They had already done everything. They had solved all the problems. You maybe had to dot a couple of i's or something, but essentially the field was over. (See Chomsky 1991a [1989]:11 for similar comments.)

The finished-field theme is a common one in the story of scientific shake-ups, signaling the calm before the storm, and Chomsky's story could easily be Max Planck's:

> When he was seventeen years old and ready to enter the university, Planck sought out the head of the physics department and told the professor of his ambition. The response was not encouraging: "Physics is a branch of knowledge that is just about complete," the professor said drearily. "The important discoveries, all of them, have been made. It is hardly worth entering physics anymore." (Cline, 1987:34).[23]

Planck went on to discover his famous constant, the spark that ignited the quantum revolution. Harris developed the transformation, the spark that set Chomsky alight. And, just as Planck was working conscientiously to elaborate the Newtonian paradigm, Harris was working to expand the Bloomfieldian paradigm. He set out to find methods for boiling down syntax to a set of patterns small enough and consistent enough that structuralist methods could go to work on them, and he imported a concept from mathematics to get him there, the transformation. Some Bloomfieldians may have found Harris's work in syntax a little premature, but still a palatable extension of structuralism, and he was not the only theorist beginning to scratch around in syntax. Younger linguists, in particular, found the prospect of syntax inviting; indeed, inevitable. The previous generation had conquered phonology and morphology. The next domain for conquest in the inexorable march toward meaning was syntax, and some small incursions had been made by the early fifties. In particular, an approach called *Immediate Constituent analysis* was generating a fair amount of attention.[24]

Chomsky's approach to syntax has two critical components, both signalled by the most common term for his program, *transformational-generative grammar*. The most wide-reaching of his innovations is in the second half of the compound. A generative grammar is a formal mechanism which generates structural descriptions of the sentences in a language, in the mathematical sense of *generate*. A generative grammar is a collection of rules which define the sounds, words, phrases, and sentences of a language in the same way that geometry is a collection of rules which define circles and squares. And the point of both systems is also the same—geometry represents knowledge about space; generative grammar represents knowledge about language—which is where the transformation comes in.

There are strong empirical constraints in both domains. Certain concepts are necessary to model knowledge of space, and certain concepts are necessary to model knowledge of language. One of these concepts, from Chomsky's perspective, is given in the first half of the compound: Harris's structure-relating device, the transformation. Again, the analogy with mathematics is important, where a wealth of procedures exists for transforming data structures into other data structures, coordinates into shapes, circles into spheres. In fact, Klein has said "geometry is not really about points, lines, angles and parallels; it is about transformations" (Stewart, 1990:31). In linguistics, the transformation adds, deletes, and permutes—for instance, transforming the a-sentences below into the b-sentences.

3   a   A mat is on the cat.
    b   There is a mat on the cat.

4   a   Aardvarks like ants, and they like Guinness too.
    b   Aardvarks like ants, and Guinness too.

5   a   Floyd broke the glass.
    b   The glass was broken by Floyd.

Harris wanted the transformation to help tame syntactic diversity. It was a tool, plain and simple, with no particular implications for the general Bloomfieldian program beyond a welcome increase in scope, bringing order to the third, and highest, level of linguistic analysis. What Chomsky wanted the transformation for isn't at all clear in his earliest writings, but his brilliant first book, *Syntactic Structures*, leaves the impression that he is carrying out his mentor's Bloomfieldian intentions. Right at the outset, Chomsky says

> During the entire period of this research I have had the benefit of very frequent and lengthy conversations with Zellig S. Harris. So many of his ideas and suggestions are incorporated into the text below and in the research on which it is based that I will make no attempt to indicate them by special reference. (1957a:6)

Chomsky followed this acknowledgment with the observation that "Harris' work on transformational structure . . . proceeds from a somewhat different point of view" than his own, and there are some hints in the book that the differences might run fairly deep. But "somewhat different point of view" is mild, and most linguists took *Syntactic Structures*—a lucid, engaging, persuasive argument that transformations are the most promising syntactic tool—largely as a popularization of Harris's theories. (Harris can be a very forbidding author.)

Chomsky's contributions were recognized as far-ranging, even renovating, for Bloomfieldian linguistics—one review called the goals of *Syntactic Structures* "Copernican" in scope (Voegelin, 1958:229), and Bloch was "convinced that transformational theory (or whatever you want to call it) is a tremendously important advance in grammatical thinking"[25]—but not threatening. Yet, in a few short years, *linguistics* effectively wore a new name-cum-adjective, *Chomskyan,* and the dispossessed Bloomfieldians were vehemently denouncing this "perversion of science," this "Anglicizing straitjacket," this "theory spawned by a nest of vipers." "Let's stop being polite," one Bloomfieldian implored the flagging troops, "reject any concession to the heresy, and get back to linguistics as a science, an anthropological science."[26]

They did stop being polite (in fact, somewhat before the exhortation), but it was too late. The heresy had become orthodoxy.

But—sponsored by Harris, nurtured by Bloch, adopted by Householder—how did it become heresy in the first place? The simple answer, and the fullest one, is that Chomsky changed his rhetorical stance to the holy Bloomfieldian church from extension to rejection and replacement. He was no longer just taking American linguistics boldly into syntax, where it had once feared to tread, or had trod very timidly; he was systematically dismantling the Bloomfieldian program and erecting his own in its place. One by one, he attacked the foundations of recent American linguistics: behaviorism, positivism, and the descriptive mandate. One by one, he attacked the cornerstones of its specific theories: not just its Immediate Constituent syntax, but also its phonology, its morphology, and its very conception of language. Some of his opposition to the architecture of Bloomfieldianism was implicit in *Syntactic Structures*—a generative theory, after all, is a theory of knowledge; a theory of mind; a theory of mental structure—but many developed, or were revealed, after the first flushes of success that followed its publication.

Those flushes, as flushes are wont to do, came among the young, "especially," in Bloch's estimation, with "the most brilliant among them" (Murray, 1980:79). They found Chomsky electromagnetic and the older linguists lost their heirs to the heresy. Younger linguists deserted the Bloomfieldian mainstays—sounds, words, and the description of indigenous languages—to work on syntax, even on semantics, and to plunge deeply into one language, rather than looking broadly at many. The classic example is Paul Postal, who did his doctoral research on Mohawk in the Bloomfieldian citadel of Yale, but did it with a growing unease:

> In the back of my mind had been this sense of strangeness. I guess I was pretty unhappy about what passed for linguistics, but I had no sense of why, or what was wrong, or what counted as an alternative. It seemed to me that it was probably based on some wrong assumptions, but I couldn't put my finger on them.

When Chomsky put *his* finger on certain issues, though, "it seemed exactly right."

The general views [Bloomfieldians held] were very primitive and silly—ideas involving rigorous collection of data and cataloging, and all that kind of stuff. They took this to be science, very genuinely and kind of sadly. And, as far as one could see, this had no connection with what modern science was like.

Chomsky's conception of science, which was closely in tune with contemporary philosophy of science (now invoking positivism almost exclusively as a whipping boy), was one of the most attractive features of his program. The promise of getting at meaning was another. The promise of getting at mind was another. The brew was intoxicating, and customers were lining the bar.

The Chomskyan Revolution

In the late forties . . . it seemed to many that the conquest of syntax finally lay open before the profession.

At the beginning of the fifties confidence was running high. Many linguists felt that a new synthesis of the discipline was needed and that a suitable time was rapidly approaching. This would continue the Bloomfield tradition taking into account the results achieved in two decades. Indeed, some spoke of the need for a "revision of Bloomfield," not a replacement but an updating. No one, however, felt able to undertake the task.

H. Allan Gleason

We shall have to carry the theory of syntactic structure a good deal beyond its familiar limits.

Noam Chomsky

Looking for Mr. Goodstructure

In one reading of linguistic history, the Bloomfieldians of the 1950s were biding their time for some convincingly complete model to displace their picture of language. The "fullest flowering" of Bloomfieldian grammar-construction was Trager and Smith's *Outline of English Structure* (Stark, 1972:414). It was at once a reworking and a practical application of Trager's earlier classic (with Bloch), *Outline of Linguistic Analysis*. Since the application was to English, it had enormous educational advantages, and it was a self-conscious exemplar of the program, illustrating by example "a methodology of analysis and presentation that we believe to be representative of the scientific method as applied to a social science—linguistics" (Trager and Smith, 1957 [1951]:7). It was brief. It promised significant inroads into syntax; even—though this was relegated to an area of concern labeled *metalinguistics*—into meaning. Reflecting the growing confidence of the field, it was also a deliberate attempt to put the best Bloomfieldian foot forward into the scholarly world at large: "Educators, diplomats, anthropologists, and others were presented with a promise of a linguistics that was rigorous, central, expanding, and useful" (Hymes and Fought, 1981 [1974]:138). But, as everyone could see, it leaked.

Some of its weaknesses came up for discussion at the first Texas Conference on Problems of Linguistic Analysis in English, when the participants fell into a discussion about scientific progress. Paradigm shifts cropped up, and Archibald Hill told his colleagues how science works in such circumstances:

> When things don't fit, scientists labor to patch up the system by adding things, taking them away, or rearranging. It is only when a complete new system is presented, a system more complete, more consistent, and simpler in its totality than the old system, that any real change is made. (1962a [1956]:17)

The obvious analogy surfaced—the Ptolemaic-to-Copernican cosmological shift—with Trager and Smith in Ptolemy's hot seat. James Sledd, one of the discussants and an outspoken critic of Trager and Smith's *Outline*, could only manage a backhanded compliment for their model:

> The great strength of the Trager-Smith system is that it has pretensions to completeness, and Mr. Hill is right that if we want to overthrow the Trager-Smith system we can do it only with a system which has more justifiable pretensions to completeness. (Hill 1962a [1956]:17)

This exchange looks for all the world like a symptom of the historical stage in the growth of a science that Thomas Kuhn calls *a crisis,* when the science goes through "a period of pronounced professional insecurity," the prelude to a revolution (1970:67f). A science in crisis, says Kuhn, is a science looking to shuck whatever program gave rise to its insecurity, looking for a new, more complete, more consistent, more simple system than the old one, to give it back some confidence, looking, in many ways, for a messiah.

And perhaps linguistics, as an abstract and collective entity, *was* looking for a savior. Subsequent events suggest as much—in particular, they suggest there was some generational discontent, with younger members impatient to get at the good stuff that had been kept at bay for twenty-five years, meaning and mind. But there is little indication in the literature of the period that there was a crisis on any front, and this exchange in Texas is certainly not a symptom of messianic longings. Aside from Sledd (who was putting words in Hill's mouth about wanting to overthrow Trager and Smith), there was no serious talk of doing away with Trager and Smith's model at the conference—the comments are more on the order of patching it up—and not the slightest hint of frustration at the Bloomfieldian program underwriting their model. Indeed, Robert Stockwell, who had just been talking with Trager, passed on the good news that the *Outline* was, even as the Texas discussants spoke, being overhauled in a direction which promised to satisfy some of the system's pretensions. Trager and Smith were aiming for a good deal more completeness. Word-formation processes (morphology) were to get increased attention, and "the syntax, further, will be completely redone and much expanded." This syntax, phonological syntax, played very well at conferences in the early and mid-fifties, attracting a good many adherents, especially among younger linguists eager to get at new material. As the name implies, it built systematically on the very attractive base of Bloomfieldian phonology, representing the natural and desired expansion of the field: incremental science at its best.

Linguistics was changing and expanding in the fifties, showing sporadic dissent over the central tenets, increased tolerance for other approaches, and some dalliance in the banned domain of psychology. But measured dissent, pluralism, and exploration, at least in this case, represent the exact opposite of Kuhn's definition of crisis. They were symptoms of a pronounced sense of professional security. The earlier hostility toward Europe, and meaning, and mind, and the undue reverence for method, and the chest-thumping war cries of "I'm a scientist and you're not": *these* were the signs of insecurity. By the fifties, paranoid aggressiveness had given way to a quiet satisfaction and optimism (in some quarters, as we have seen, to an almost gloomy optimism that all the real problems had been solved). The other major Bloomfieldian codification published in the fifties, off the presses almost in a dead heat with Trager and Smith, was Zellig Harris's (1951 [1947]) *Methods in Structural Linguistics* and it was hailed as "epoch-marking in a double sense: first in that it marks the culmination of a development of linguistic methodology away from a stage of intuitionism, frequently culture-bound; and second in that it marks the beginning of a new period, in which the new methods will be applied ever more rigorously to ever widening areas" (McQuown, 1952:495). A glorious period of advancement may have been over, but a new and more glorious one, building on those advances, was just beginning.

Into this atmosphere came *Syntactic Structures,* published the year after the First Texas Conference. It couldn't have fit the mood better. It appeals calmly and insistently to a new conception of science. It promises the transformational taming of syntax. And it elegantly walks the tightrope of the signified—supporting Bloomfield's argument that they couldn't be allowed to taint the analysis of signifiers, but offering persuasive suggestions that linguists could get at meaning anyway.

Chomsky's book was welcomed. But it was not—and this point is often missed in histories of the period—taken to herald the arrival of a complete new system, more consistent and simpler, that would revolutionize linguistics. Chomsky was not hailed as the messiah, not immediately. For one thing, *Syntactic Structures* had virtually nothing to say about the old system's strongholds, sounds and words. But more importantly, its implications for the Bloomfieldian superstructure were almost entirely submerged. Chomsky's program looked much more like the projected steady expansion of Bloomfieldianism, ever more rigorous, to ever-widening areas; all the more so as Chomsky was the favored son of Harris, author of the double-epoch-marking *Methods.*[1]

Soon there was talk from Chomsky and his associates about plumbing mental structure; then there was a new phonology; then there was an explicitly new set of goals for the field, cut off now completely from its anthropological roots and hitched to a new brand of psychology. By this point, in the early sixties, it was clear that the old would have to be scrapped for the new. These last developments—accompanied for the most part with concerted beatings of one or more of the Bloomfieldians' sacred cows—caught most of the old-line linguists somewhat unawares. They reacted with confusion, bitterness, and ineffective rage. Rapidly, the whole kit and kaboodle of Chomsky's ideas swept the field. The entrenched Bloomfieldians were not looking for a messiah, but, apparently, many of their students were. There was a revolution.

Syntactic Structures

> Chomsky's *Syntactic Structures* is a striking and original book, which forced its
> readers to look at familiar things from a fresh angle. But in taking this view, he
> did not destroy his predecessors' basic concept of the structure of language.
> Rather he gave new life to it.
>
> Carl F. Voegelin and Florence M. Voegelin

For the 1950s, Chomsky was, in the terms of one lapsed Bloomfieldian, "a very
aberrant young linguist" (Gleason, 1988:59). He was something of an outsider,
always an advantage for seeing the limitations and weaknesses of an established
program. His exposure to the field came almost entirely through Harris, and Harris
was a card-carrying Bloomfieldian, but *in extremis,* representing, in many ways, the
best and the worst of the program. He had a fixation on esoteric, if not peripheral,
issues, and a preoccupation with methodology which far outstripped even that of
his contemporaries. He, too, had a somewhat unusual background for a Bloomfieldian—coming not from the rolled-up-sleeves-and-loosened-collar world of
anthropology, but the bookish, intensely logical world of Semitic philology—and,
except for Hockett, he was the only linguist of the period pursuing the natural, but
largely ignored, ramifications of the Saussurean conception of *langue* as a "rigid
system," the only linguist of the period seriously exploring the mathematics of language. Chomsky's education reflected Harris's interests closely. It involved work in
philosophy, logic, and mathematics well beyond the normal training for a linguist.
He read more deeply in epistemology, an area where speculation about the great
Bloomfieldian taboo, mental structure, is not only legitimate, but inescapable. His
honors and master's theses were clever, idiosyncratic grammars of Hebrew, and—
at a time when a Ph.D. thesis in linguistics was almost by definition a grammar of
some indigenous language, fieldwork virtually an initiation rite into the community
of linguists—his doctorate was granted on the basis of a highly abstract discussion
of transformational grammar, with data drawn exclusively from English. When his
thesis made the rounds at the Linguistic Institute in the summer of 1955, it looked
completely alien, "far more mathematical in its reasoning than anyone there had
ever seen labeled as 'linguistics'," and, predictably, it fell utterly flat:

> A few linguists found it very difficult; most found it quite impossible. A few thought
> some of the points were possibly interesting; most simply had no idea as to how it might
> relate to what they knew as linguistics. (Gleason, 1988:59, 60)

That was, of course, the rub, the dragging friction on any acceptance of his ideas:
how to make his work palatable to linguists. His thesis—"Transformational Analysis"—was not only forbiddingly technical, but completely unrelated to the daily
activities of Bloomfieldian linguists. And it was only one chapter of a massive
manuscript—*The Logical Structure of Linguistic Theory*—he had feverishly
worked up while on a fellowship to Harvard in the early fifties. A few copies of *Logical Structure* were available here and there in mimeograph, but it was known
mostly by rumor, and had the whiff of Spinoza or Pierce or Wittgenstein about it,

or some other fecund, mathematical, relentlessly rational, but cloistered mind. Chomsky must have been considered, when considered at all, somewhat the way Crick recalls the feeling about his collaborator on the structure of DNA: "Watson was regarded, in most circles, as too bright to be really sound" (1988:76).

With this particular background, Chomsky was not, despite acknowledged brilliance, the ideal candidate for a job in an American linguistics department, and found himself in the Research Laboratory of Electronics of the Massachusetts Institute of Technology. His research was open-ended, allowing him to continue his abstract modeling of language, but the appointment was only partial and he had to teach to round out his income: German, French, philosophy, logic. And linguistics. Since there was no one there to tell him otherwise (MIT had no linguistics department), he taught *his* linguistics, and the lecture notes for this course became the answer to the rhetorical gulf between the audience for *Logical Structure* (written for Chomskyan linguists when there was only one, Chomsky) and everyone else in the field.

These notes, revised and published as *Syntactic Structures,* constitute one of the masterpieces of linguistics. Lucid, convincing, syntactically daring, the calm voice of reason calling from the edge of a semantic jungle Bloomfield had shooed his followers from, it spoke directly to the imagination and ambition of the entire field. The most ambitious, if not always the most imaginative—the young—responded most fully, but the invitation was open to all and the Bloomfieldians found many aspects of it very appealing.

Science and Generative Grammar

> By a generative grammar I mean simply a system of rules that in some explicit and well-defined way assigns structural descriptions to sentences. . . . Perhaps the issue can be clarified by an analogy to a part of chemical theory concerned with the structurally possible compounds. This theory might be said to generate all physically possible compounds just as a grammar generates all gramatically 'possible' utterances.
>
> Noam Chomsky

Especially attractive to the Bloomfieldians was the conception of science Chomsky offered in *Syntactic Structures.* The first few sentences of the book advance and defend the conception of linguistics as an activity which builds "precisely constructed models" (1957a:5), and building precisely constructed models was the mainstay of Bloomfieldian linguistics (though they were happier with the word *description* than with *model*). But Chomsky also made the motives behind such construction much more explicit than they previously had been. There are two, he says. One motive is negative: giving "obscure and intuition-bound notions" a strict formulation can quickly ferret out latent difficulties. The other is positive: "a formalized theory may automatically provide solutions for many problems other than those for which it was explicitly designed" (1957a:5). In short, the clear and precise formulation of a grammar has the two most important attributes that recommend one scientific theory over another, greater fragility and increased scope. If you can break a scientific theory, it's a good one, since that means it has clear and testable connections to some body of data; if you can break it in principle but not in practice, so much the better, since not only can it be tested against data, the testing proves it compatible with that data. The law of gravity you can test by dropping a pen and measuring its descent; if it floats upwards, or zips sideways, or falls slowly to the ground, then the law is in trouble. But the pen never does (unless you're someplace weird, like a space capsule or a centrifugal chamber, when the bets have to change), so gravity is fragile in principle, resilient in practice. And the more coverage a theory has, the more efficient it is. The law of gravity is (more or less) equally applicable to falling pens and orbiting planets. Two laws for those phenomena, rather than one, mess things up, and scientists like to be tidy whenever they can.

Two definitions are crucial for Chomsky to achieve these scientific virtues: a *language* is "a set (finite or infinite) of sentences" and a *grammar* is "a device that generates all of the grammatical sequences of [that language] and none of the ungrammatical ones" (1957a:13): a grammar is a formal model that predicts which strings of words belong in the set of sentences constituting a language and which strings do not belong.[2] An adequate grammar of English, then, would generate sequence 1, but not sequence 2 (which is therefore stigmatized with a preceding asterisk, following the now-standard linguistic practice).

1   Kenny is one cool guy.

2   *guy cool one is Kenny

Now, a grammar which aspires to generate all and only the set of sentences possible in a language—a generative grammar—by Chomsky's definition, is a scientific grammar:

> A [generative] grammar of the Language L is essentially a theory of L. Any scientific theory is based on a finite number of observations, and it seeks to relate the observed phenomena by constructing general laws in terms of hypothetical constructs such as (in physics, for example) "mass" and "electron." Similarly, a grammar of English is based on a finite corpus of utterances (observations), and it will contain certain grammatical rules (laws) stated in terms of the particular phonemes, phrases, etc., of English (hypothetical constructs). These rules express structural relations among the sentences of our corpus and the indefinite number of sentences generated by the grammar beyond the corpus (predictions). (1957a:49)[3]

Beyond this very attractive identification of grammar and theory, Chomsky also offered a new philosophy of science. By 1957 philosophy of science had shifted considerably, and Bloomfield-endorsed positivism had sunk from almost complete dominance to an approach that dared not speak its name—the 1957 presidential address to the American Philosophical Association was "Vindication of L*G*C*L P*S*T*V*SM" (Rynin, 1957).

Methodological fretting had fallen into disrepute and all that now counted was the results, however obtained. Linguistics should proceed, went Chomsky's articulation of this new methodological indifference, by way of "intuition, guess-work, all sorts of partial methodological hints, reliance on past experience, etc."

(1957a:56). The crucial interests of linguists *qua* science should be those revolving around whether the grammar stands up once it has been formulated. Does it generate the sentences of L? Does it preclude non-sentences of L? Does it fit established scientific constraints like fragility, elegance, and generality?

There is a measure of antagonism in this move for those Bloomfieldians who cared about such things (Trager and Hall, for instance), and some no doubt found Chomsky's methodological nonchalance distasteful—even, in the familiar curse-word, *unscientific.* But most linguists weren't very troubled by foundational issues of this sort. More importantly, *Syntactic Structures* doesn't frame its philosophy of science in antagonistic terms. It comes, in fact, in a frame that couldn't help but appeal to the Bloomfieldians' scientific fondness—defining their principal concern, grammars, as on a par with physical or chemical theories. Chomsky was, from a Bloomfieldian perspective, confirming and elaborating their notions of what makes for good science.

What most of them didn't notice (though their students did) is that Chomsky changed the focus of linguistics radically—from discovering good grammars to justifying and evaluating them. Linguistics was slipping from a primarily descriptive enterprise into a theoretical enterprise directed toward exploring the general principals underlying descriptions.[4]

Syntax and Transformational Grammar

> I find myself differing with Harris and Chomsky chiefly on points that I regard as minor: I am glad to see syntax done well in a new format.
>
> Ralph B. Long

By far the most attractive aspect of *Syntactic Structures* for the Bloomfieldians was its titular promise to advance the structuralist program into syntax. Chomsky's first step was to translate Immediate Constituent analysis into a more testable format. Immediate Constituent analysis was a body of "heterogeneous and incomplete methods" (Wells, 1947b:81), which had begun hardening into a more systematic theory of syntactic structure—most attractively in the phonological syntax of Trager and Smith—but was still a long way from the rigid formalism called for by Chomsky's notion of generative grammar. Out of the relatively loose group of Immediate Constituent procedures, Chomsky extracted a notation based on variables and arrows such that a simple rule like $X \rightarrow Y + Z$ defined the relations among the variables in an easily diagrammable way; that is, in the way illustrated by figure 3.1.

From this notation, Chomsky built a rule system for English of the following sort.[5]

3 a $S \rightarrow NP + VP$
   b $NP \rightarrow Det + N$
   c $VP \rightarrow V + NP$
   d $Det \rightarrow the$
   e $N \rightarrow \{dog, duckling, sandwich, farmer, affix \dots\}$
   f $V \rightarrow \{bite, chase, hop, kill, passivize \dots\}$

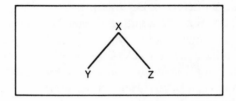

Figure 3.1. A diagram of the abstract, hierarchical relationship generated by the formal rule, $x \rightarrow y + z$.

The symbols in 3 are all mnemonic: *S* stands for *sentence*, *NP* for *noun phrase*, *VP* for *verb phrase*, and so on. (The only one that may not be immediately apparent from a grade-school knowledge of language, *Det*, stands for *determiner*, and isn't especially important for our purposes; its main function here is to help identify one of the members of noun phrases—namely, *the*, as specified by 3d.) The rules, then, are descriptions of how sentences, noun phrases, and verb phrases hang together. They express such notions about the syntax of English as "sentences have noun phrases and verb phrases, in that order" (in more traditional terms, sentences have subjects and predicates), and "verbs are such things as *bite* and *chase*."

The rules of 3—*phrase structure rules*—cover only the tiniest portion of English syntax, of course, but they illustrate conveniently the type of expressions that Immediate Constituent Analysis (or, in Chomsky's rechristening, phrase structure grammar) handles most efficiently. Consider how they work. Each rule is an instruction to rewrite the symbol to the left of the arrow as the symbol(s) to the right of the arrow, yielding a derivation of a sentence in exactly the sense that word has in calculus. Representing this derivation graphically, we get a tree diagram (or *phrase marker*) of the sort that has become ubiquitous in modern linguistics, illustrated by PM-1 (where S dominates NP and VP, as called for by rule 3a; NP dominates Det and N, as called for by 3b; N dominates *duckling* in one instance, *farmer* in another, as allowed by 3e; and so on).

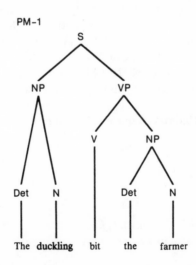

Once Chomsky has this machinery in place, he briefly demonstrates that phrase structure grammars are superior to the only legitimate candidates proposed as formal models of syntax—Shannon and Weaver's (1949) nascent finite state grammars—which, not coincidentally, had been endorsed by the Bloomfieldian boy-wonder, Charles Hockett (1955:2). Any satisfaction Immediate Constituent fans might have felt from this triumph, however, was very short lived. Applying to grammar the same principles necessary for a good scientific theory, Chomsky demonstrates that whatever phrase structure grammar's representational virtues, its treatment of some fundamental phenomena is surpassingly ugly: "extremely complex, *ad hoc,* and 'unrevealing'" (1957a:34). That is, they may be adequate as flat descriptions of the data, in the way that randomly ordered lists adequately describe all the elements of a compound, but they lack the simplicity and concision found in a chemical formula.

Lo, in the east, a transformation.

Several transformations, in fact; a small flock; and Chomsky shows how they can, rather effortlessly, clean up after phrase structure analyses. Two of these transformational analyses, centering on rules which became known as *Affix-hopping* and *Passive* (or *Passivization*), rapidly achieved the status of exemplars in the next few years, as transformational grammar solidified into a paradigm.

Affix-hopping depends on too much detail about the English auxiliary system to treat very adequately here, but it was extremely persuasive. Bloomfieldian linguistics was fundamentally a distributional pursuit, fundamentally about accounting for the distribution of sounds and words—what comes before what—and getting the distributions right for English auxiliary verbs is a very complicated matter when it is left in the hands of phrase structure rules alone. As one sliver of the problem consider the progressive aspect (4b, in contrast to the simple present, 4a):

4   a   Andrew skateboards.
     b   Andrew is skateboarding.

The tricky part about 4b is that progressive aspect is clearly coded by two chunks separated by another one: *is* and *-ing* are both necessary, but that darn *skateboard* gets between them. Chomsky made a number of innovations to the phrase structure rules in order to describe the discontinuous distribution, *is . . . -ing,* but the really ground-shifting move was his proposal of this elegant little transformation:

5   Af V $\Rightarrow$ V Af

(The structure to the left of the double arrow "becomes" the structure to the right.)

Rule 5 simply attaches the affix preceding a verb to its backend, making sure that the suffix *(-ing)* in fact shows up where it's supposed to show up, abutted to the verb's hindquarters. The modifications Chomsky made to the phrase structure rules ensured that they produced something like 6:

6   Andrew is *-ing* skateboard

Then Affix-hopping would kick in, leapingfrogging, the *-ing* over the *skateboard,* and the (distributionally correct) 4b was the result. What's the big deal? Well, the phrase structure rules generate *is* and *-ing* side-by-side, capturing the fact that they serve as a unit to signal progressive aspect, and Affix-hopping redistributes that unit, capturing the fact that they don't in fact occur side-by-side in people's speech.

If Affix-hopping isn't very convincing about the merits of Chomsky's system, consider how badly a phrase structure account of sentences like those in 7 does. It leaves completely unexpressed the important fact that actives and passives have very clear syntactic and semantic parallels.

7   a   The duckling bit the farmer.
    b   The farmer was bitten by the duckling.

A grammar that handles syntax exclusively with phrase structure rules would generate the sentences in 7 independently, with two sentence rules like these ones:

8   a $S \rightarrow NP + V + NP$
    b $S \rightarrow NP + be + V + by + NP$

Two rules for two phenonena (8a for 7a, 8b for 7b), necessarily implies that any connection between active sentences and passive sentences is wholly accidental; an active is an active, a passive is a passive, and the only loosely connecting point about them is that they are both sentences. However, as every native speaker of English knows, there is an obvious pattern to these correspondences. For instance, 9a and 9b are clearly legitimate, sensible, English sentences; 10a and 10b are clearly illegitimate and nonsensical (or, legitimate only under a bizarre construal of *sandwich*):

9   a   The farmer bit the sandwich.
    b   The sandwich was bitten by the farmer.
10   a   *The sandwich bit the farmer.
     b   *The mailman was bitten by the farmer.

An adequate grammar of English—that is, one which meets Chomsky's criterion of enumerating all and only the legitimate sentences of English—must therefore generate the first pair of sentences and preclude the second; the phrase structure account can only do so at the expense of unintuitive redundancy. For instance, it must stipulate independently what subjects and objects the verb *bite* can take in an active sentence and what it can take in a passive sentence, although the two sets are strictly inverse (the subject must be able to bite and the object must be biteable in actives; the subject must be biteable and the indirect object must be able to bite in passives). But supplementing the phrase structure rules of 3 with the following transformation (rather than with the rules of 8) gives a much more satisfactory account of the obvious correspondences between 7a and 7b, between 9a and 9b, and even between the anomalous 10a and 10b.

11   $NP_1\ V\ NP_2 \Rightarrow NP_2\ be\ \text{-}en\ V\ by\ NP_1$

(The subscripts simply mark NPs which are identical on either side of the double arrow.)

The Chomskyan Revolution 45

In a grammar organized along these lines—a transformational grammar—the phrase structure rules generate 7a (or 9a), which can then become the input for the transformation, Passive (rule 11), with 7b (or 9b) as the output, or it can "surface" without engaging 11 at all.

In short, a transformational grammar explains the systematic correspondences between actives and passives by deriving them from the same source.[6]

But 11 can't do the job on its own: since the rule introduces *be* and *-en* side-by-side and since passive sentences have a verb in between them, Affix-hopping also needs to apply in the derivation, gluing the *-en* onto the butt of the main verb. Chomsky's transformations occur in tandem. They are ordered; in this case, Passive applying before Affix-hopping. Notice, then, that we have another—and in terms of the subsequent history of the field, a much more important—application of the notion, "derive," to consider. We spoke earlier of a tree (or phrase marker) as derived from phrase structure rules. Now we are talking about the derivation of a sentence, the transformational derivation of a sentence. In fact, with 9a and 9b we are talking about two transformational derivations, one in which Passive applies, one in which it doesn't. For 9a, only one rule applies, Affix-hopping, so its derivation is relatively simple—though quite abstract, since Affix-hopping moves the tense marker, PAST, over *bite* and the final result doesn't really have an affix at all. For 9b, two rules apply, Passive and Affix-hopping, in that order, making for a slightly more complicated derivation.

Moving up a level of abstraction to phrase markers, consider this process graphically, as shown in PM–2 through PM–5.

In the first derivation (PM–2 ⟹ PM–3), only Affix-hopping applies; in the second derivation (PM–2 ⟹ PM–4 ⟹ PM–5), two rules apply, Passive and Affix-hopping. The job isn't complete here—later sound-based rules have to apply in order to get *bit* out of *bite* + PAST, to get *was* out of *be* + PAST, and to get *bitten* out of *bite* + *-en*—but these were all quite straightforward in the Bloomfieldian sound-and-word scheme of things.

In both cases, the rules ensure the quintessential Bloomfieldian goal of getting the distributions right, but the most important feature of these two derivations for

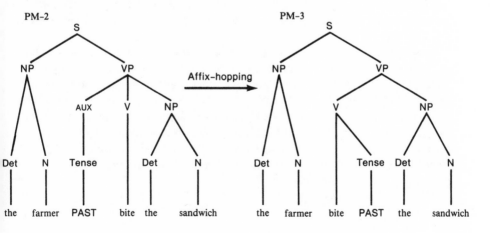

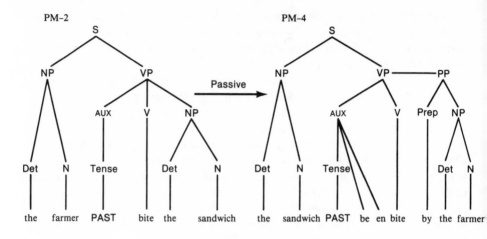

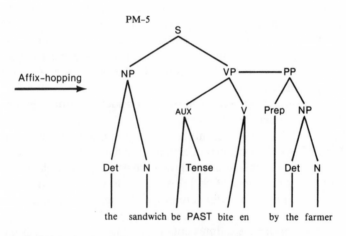

many people is that they start from the same place (PM–2): two derivations from a common source, yielding two distinct but clearly related sentences corrects "a serious inadequacy of modern linguistic theory, namely, its inability to account for such systematic relations between sentences as the active-passive relation" (Chomsky, 1962a [1958]:124).

In sum, phrase structure rules establish basic patterns and introduce words; they say such things as "a determiner followed by a noun is a legitimate noun phrase" (rule 3b, NP → Det + N)," and "*the duckling* is a legitimate example of that pattern" (rules 3d, Det → *the,* and 3e, N → { . . . *duckling,* . . . }). Transformations alter those basic patterns to account for a wider range of sentences and phrase types; they say such things as "if *the farmer killed the duckling* is a legitimate English sentence, then so is *the duckling was killed by the farmer*" (rule 11, Passivization, NP₁ V NP₂ ⟹ NP₂ *be -en* V *by* NP₁, along with rule 5, Affix-hopping, Af V ⟹ V Af, which helps get the affix-and-verb order right).

The grammar that emerges from Chomsky's discussion is extremely rudimentary, accounting for only the tiniest fragment of English. Chomsky sketches a number of other transformational solutions to syntactic problems, and outlines a division of labor into singularly and generalized transformations; the former for such phenomena as affix-placement and active-passive relations, the latter for such phenomena as relative clauses and conjoined clauses, capturing the intuition that sentences 12a (with a relative clause) and 12b (two conjoined clauses) are "made up of" 13a and 13b.

12 a Logendra abused the duck which had buzzed him.
     b The duck buzzed Logendra and he abused it.
13 a Logendra abused the duck.
     b The duck buzzed Logendra.

But even after Chomsky has laid out a nice sample of equally appealing solutions, the case for transformational grammar in *Syntactic Structures* is grossly underdetermined; the book is in many ways, remember, a summary of his massive *Logical Structure*. Still, by the time Chomsky is through: (1) the only other explicitly proposed generative grammar (the Hockett-endorsed finite state grammar) is disconfirmed; (2) the case for phrase structure rules working on their own (therefore, Immediate Constituent analysis) is eviscerated; and (3) the outline of a very powerful, novel approach to syntax is served up in a few, short, compelling strokes. This approach (schematized in Figure 3.2[7]) does the main Bloomfieldian work better than any previous syntactic model and does a few additional jobs to boot.

A set of phrase structure rules generates a core of underlying phrase markers, which feed into a set of transformations that massage them into their final, observable shapes, the ones we talk and write with (with all the affixes in place, for instance): the system purring harmoniously to generate all and only the grammatical sentences of a specific language.

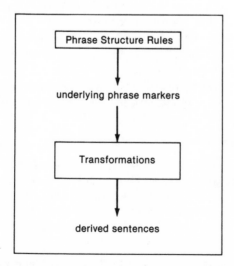

Figure 3.2. The transformational grammar sketched in *Syntactic Structures*.

There is still something missing from this picture, however: a privileged notion that Chomsky inherited from Harris and subtly altered, the kernel sentence. For Chomsky, the kernel sentence hinges on the fact that transformations come in two flavors, obligatory (like Affix-hopping) and optional (like Passive). Obligatory transformations go to work whenever their structural requirements are met (that is, whenever the conditions on the left of the arrow occur; for Affix hopping, whenever the sequence "Af + V" shows up in a derivation). Optional transformations only go to work sometimes, without any real guiding mechanism (so, Passive would apply some of the times that its structural requirements are met, some of the times that the phrase structure rules generated the sequence "NP V NP").

All generalized transformations are optional.

The optional/obligatory distinction may look peculiarly unnecessary, but consider the alternative. If Passive and Affix-hopping, for instance, weren't different in this regard, the model would be in all kinds of trouble—generating some sequences that aren't English, and failing to generate some that are. If Affix-hopping were optional, then the grammar would produce gibberish like "Andrew is -ing skateboard," since the affix would fail to be moved. If Passive were obligatory, then the grammar would fail to produce sequences like "The dog bit the mailman," since every time the phrase structure rules generated such a sequence, Passive would turn it into a passive.

If the generalized transformations (the ones which made complex sentences out of simple ones) were obligatory, then the grammer would again fail to produce some sentences (namely, all simple ones, since the relevant transformations would necessarily combine them all).[8]

The distinction was crucial, which is where the kernel comes in.

Kernel sentences in the *Syntactic Structures* model are those derived sentences which had only undergone obligatory transformations. More than just kernels, they were also said to be the kernels *of* other sentences—parallel ones which had undergone optional transformations. A derived active sentence, then, was the kernel sentence of a derived passive (7a for 7b, for instance, and 9a for 9b). Two or more derived simple sentences were the kernels of a derived complex sentence (13a and 13b for 12a, for instance, and also for 12b). All of this probably sounds unduly complicated; the important point is simply that the grammar generated two classes of sentences, kernels and everything else, and that kernels had more cachet.

The kernel was the seed of meaning in transformational grammar.

The Appeal of Meaning

> We should like the syntactic framework of the language that is isolated and exhibited by the grammar to be able to support semantic description, and we shall rate more highly a theory of formal structure that leads to grammars that meet this requirement.
>
> Noam Chomsky

Chomsky's distributional interests—virtually inevitable under the tutelage of Harris—were not the only elements of his Bloomfieldian heritage. He also had a deep methodological aversion to meaning, and his work reinforced one of the key elements of the Bloomfieldian policy toward meaning: it had to be avoided in formal analysis.

But *Syntactic Structures* was instrumental in reversing a far more problematic trend in Bloomfieldian linguistics: that meaning was unavailable for study. To some extent, Chomsky was catching a wave. Just as syntax saw increased action in the fifties, meaning was making a tentative comeback from the periphery. The anthropological linguist, Floyd Lounsbury, was beginning his soon-to-be influential work on componential analysis (1956, 1964 [1962]). The missionary linguist, Eugene Nida, had published his "System for the Description of Semantic Elements" (1951), in the European emigré journal, *Word.* Dwight Bolinger had even argued (also in *Word*) that, as defensible or desirable as meaning-aversion might be in phonology, it was a handicap for higher levels of analysis. "Meaning is the criterion of the morpheme," he said, and, therefore linguists have a duty to "develop a theory of meaning and apply it consistently" (1950:120). Martin Joos had even hailed Harris's transformational analysis as "a beginning . . . on a structural semantics," calling it "the most exciting thing that has happened in linguistics for quite a few years" (1957:356).

Joos's characterization is off the mark for Harris, but Chomsky's extension of Harris *can* be viewed as such a beginning. Chomsky says his work was, from the outset, "an effort to construct a semantically responsible theory of language," and the way to tackle meaning for him is through structure:

> The focus in both *LSLT* [*Logical Structure of Linguistic Theory*] and *Syntactic Structures* is on trying to figure out what formal structures languages must have in order to account for the way we understand sentences. What's the point of trying to figure out what the structures must be, except to see how they mean? The evidence is all semantic evidence. The facts are: Look, we understand the sentences this way, that way, the other way. Now how must language be working so as to yield those results?[9]

The structure of utterances—syntax—has long looked like the way to study meaning. That was the route taken by the Modistae, for instance, and by most philosophers of language in this century. For good reason: whatever sounds and words do, however they function in language, it takes syntax to make assertions and claims about the world, to really mean something. *Apple* is an orthographic symbol which stands in for a certain class of fruit, but it doesn't get seriously involved in meaning until it participates in a structure like "John ate an apple" or "Did John eat an apple?" or "Who ate an apple?"—to borrow some of Chomsky's examples in *Syntactic Structures* (1957a:71). In other terms, turning the chair briefly over to one of the most accomplished syntacticians ever, Otto Jespersen, the Bloomfieldian strongholds of phonology and morphology look at language from the outside; not syntax. Syntax "looks at grammatical facts from within, that is to say from the side of their meaning or signification" (Jespersen, 1954.2:1).

Back to Chomsky: "This purely formal investigation [in *Syntactic Structures*] of the structure of language has certain interesting implications for semantic studies" (1957a:12).[10] And, after he has established their syntactic worth, Chomsky proceeds to argue for transformations in explicitly semantic terms. For instance, he asked his readers to consider the phrases in 14.

14   a   the shooting of the hunters
     b   the flying of the planes

Both of these phrases are ambiguous between readings where the nouns are objects of the verbs (the hunters are being shot, the planes are being flown) and where they are subjects (the hunters are shooting, the planes are flying). Again, we are faced with a problem about which the Bloomfieldian program has nothing to say, but which reflects clear facts of English, and again, transformational grammar has an answer. The best the Immediate Constituent approach can do with phrases like this is—using the phrase structure apparatus Chomsky supplies—to treat them as members of the same class, with the structure given in 15.

15   *the* V *-ing of* NP

But transformational grammar can easily formulate rules of the following sort:

16   a   NP V ⟹ *the* V *-ing of* NP
     b   NP₁ V NP₂ ⟹ *the* V *-ing* of the NP₂

Transformation 16a changes an NP like "the hunters" followed by a V like "shoot" into structures like 14a; transformation 16b changes structures like "someone shoots the hunters" into the same structure. That is, the two senses of 14a each have a distinct transformational history—the same post-transformational structures, but two different pre-transformational structures—offering an explanation for the ambiguity.

   Chomsky's goal is to chart a small part of the huge and daunting semantic rainforest, to construct a "theory of linguistic structure [which] may be able to incorporate a part of the vast and tangled jungle that is the problem of meaning" (1957b:290). The ambition is a guarded one to be sure, but far more enterprising than Bloomfield's attempt to turn his back on meaning altogether, shucking it off on other disciplines. *Syntactic Structures* offers an impressive general outline of how linguists could begin to talk meaningfully about meaning, and it is clear in retrospect that many linguists found this outline to be the single most compelling feature of Chomsky's program. Three of his most prominent recruits, in particular—Paul Postal, Jerrold Katz, and Jerry Fodor—soon set to work on an explicit incorporation of semantics into the *Syntactic Structures* model, and it was this work which inspired the more thorough incorporation of meaning that defined the appearance of generative semantics.

   The Bloomfieldians were ready for Chomsky. They were ready for his notions of science—explicitly defining a grammar of a language as a theory of that language, subject to the criteria for any theory: simplicity, generality, testability. In fact, Hockett had said pretty much the same thing a few years earlier (1954:232–3). They were ready for his advances in syntax. No area of linguistics was more ripe—indeed, overripe—for investigation, and everyone knew it. They were even ready, despite the injunctions of their great, defining, scientific benefactor, Leonard Bloomfield, to follow Chomsky's (or, in their minds, Harris's) transformations into the uncharted jungle of meaning—well, into the edges of that jungle. Hill says that most of the leading linguists of the period, while all followers of Bloomfield, were nevertheless all "eager to break into semantics when they felt it possible" (1991:79),[11] and one of Bloch's students in the fifties recalls that even Bloch, an old wouldn't-touch-meaning-with-a-ten-foot-pole hardliner if there ever was one, "was poised to accept semantics," at least in the tightly manageable, formal methods of symbolic logic. It's just that he, along with perhaps most of the defining Bloomfieldian theorists, "didn't feel up to doing it himself. He said he would wave encouragement as the logicians took off." Certainly, he waved encouragement as Chomsky took off.

The Bloomfieldians were ready for some elaboration of their program, some revisions and extensions. They were ready for *Syntactic Structures*. They weren't ready for a replacement. They weren't ready for what followed *Syntactic Structures*.

Chomsky *Agonistes*

> I was told that my work would arouse much less antagonism if I didn't always couple my presentation of transformational grammar with a sweeping attack on empiricists and behaviorists and on other linguists. A lot of kind older people who were well disposed toward me told me I should stick to my own work and leave other people alone. But that struck me as an anti-intellectual counsel.
>
> Noam Chomsky

There are myths aplenty in linguistics these days surrounding Chomsky's spectacular rise, celebrating his brilliance and prescience, his predecessors' obtuseness and dogmatism. We have already seen the finished-field myth, which, if we take Harris to fill Planck's shoes, puts Chomsky in Einstein's. There is also that recurrent feature of scientific breakthrough stories, the Eureka Moment, Chomsky's moment putting a nice twist on the archetypical Archimedes in his more literal tub:

> I remember exactly the moment when I felt convinced. On board ship in mid-Atlantic, aided by a bout of seasickness, on a rickety tub that was listing noticeably—it had been sunk by the Germans and was now making its first voyage after having been salvaged. (Chomsky, 1979 [1976]:131)

Less dramatically—with neither nausea nor Nazis—but still good copy, Chomsky's introduction to linguistics is said to have come by way of reading the proofs to Harris's dense, highly technical *Methods in Structural Linguistics* (1951 [1947]), which is roughly akin to an introduction to mathematics by way of Russell and Whitehead's *Principia*. After this abrupt immersion, the stories go, he toiled in virtual obscurity, turning out masterpieces, first for the uncaring eye of Zellig Harris, then for the indifferent publishing world, convinced all the while that his work would never amount to more than a private hobby. Fortunately for science, however, clear-eyed and forceful supporters persuaded him that he owed his work to the world, as Copernicus's supporters had persuaded him, and Darwin's supporters, and, in the most extreme case of reluctance overcome, Saussure's exhuming supporters. When he followed this advice, he was confronted by phalanxes of blindly opposed Bloomfieldians, whom he demolished effortlessly, enfeebling the arguments and dumbfounding the arguers. He is said, in short, to have rescued linguistics from a long dark night of confusion, to have pulled back the curtain Bloomfield mistakenly drew over the mind; to have finally—and we could see this one coming for some time—made linguistics a science.

Like all good myths, these ones are true, and, of course, false.

To fan away the unpleasant smell that usually attends such comments: the falseness of these origin myths doesn't involve specifics. There is no implication here, in the word *myth* or even the word *false,* that anyone is a liar. Harris may or may not have had the feeling that linguistics was so successful it was about to go out of business as a science (Harris was an inscrutable character), but Chomsky certainly developed that impression himself, working under Harris. Nor was the impression exclusive to him and (possibly) Harris; the finality of John Carroll's early fifties overview of linguistics, for instance, recalls Lord Kelvin's remarks that physics had little more to look forward to than increasingly precise measurements—"Since the publication of Bloomfield's work in 1933, theoretical discussions among linguistics have been largely on matters of refinement" (1953:30). And if the proofs of Harris's *Methods* did not constitute Chomsky's first exposure to linguistics (his father, William, was a respected Semitic philologist; little Noam was reading historical linguistics by the age of ten and studying Arabic in his teens), they were certainly his first serious exposure to the themes, techniques, and motive forces of Bloomfieldianism; he had not, for instance, taken so much as a first-year college course in structural linguistics.

And his transformational-generative research was carried out in relative obscurity: Holding a prestigious fellowship at Harvard, he was a lively, precocious, influential member of an early fifties intellectual scene in Cambridge that included philosophers like W.V.O. Quine, Nelson Goodman, and J. L. Austin, psychologists like George Miller, Jerome Bruner, and John Carroll, and itinerant intellectuals like Benoit Mandelbrot and Marvin Minsky; but his ties with linguists were limited and unorthodox. He was at least as isolated from the Bloomfieldian community as, say, Saussure in Geneva was from the neogrammarians, or Sapir in Ottawa was from the Boasian community. And he certainly produced masterworks in this obscurity—most notably, the massive *Logical Structure of Linguistic Theory.* (It isn't clear how indifferent either Harris or the publishing world was to his efforts, but *Logical Structure* wasn't published for another twenty years; see Chomsky, 1975a[1973]:1ff, 1988b:78n2; Newmeyer, 1980a:33–35, 1986a:28–31; Murray, 1980.)

And a small group of supporters (most notably, Morris Halle and philosopher Yehoshua Bar-Hillel) undoubtedly convinced him that his ideas were valuable, not just as his own cerebral toy, but for the entire field. And the generative light bulb surely clicked on for him exactly where he remembers it clicking on, above a sea green face, reeling and listing in mid-Atlantic. And, along with the accommodation of Bloch and others, Chomsky also encountered resistance, increasingly vociferous resistance as he developed and spelled out the implications of his thought for the Bloomfieldian infrastructure. And Chomsky dealt with the resisters very effectively, if not to the satisfaction of all his opponents, certainly to the satisfaction of a far more crucial element in the debate (in any debate), its audience; Chomsky is one of the hardest arguers in modern thought. The supporters and resisters and supporters-cum-resisters among the old guard were swept aside indiscriminately, if not by Chomsky, certainly by the quickly growing cadre of transformationalists in the audience. And, while linguistics was a science before he came along—as it was before Jones, and Saussure, and Bloomfield came along—it was, also as with those men, a much different science once his ideas took root.

No, the falseness is not in details. It is in the routinely extreme interpretations put on these details by the great majority of post-revolutionary linguists: that the study of language begins in real terms with Chomsky; that all linguists before him "were hopelessly misguided bumblers, from whose clutches Chomsky has hero-ically rescued the field of linguistics" (Lamb, 1967:414). Listen to Hockett's bitter lament:

> I . . . view as genuinely tragic the success of the "eclipsing stance" of [Chomskyan lin-guistics.] We have currently in our ranks a large number of young people, many of them very bright, from beginning students up to and including a few full professors, who know nothing of what happened in linguistics before 1957, and who actually believe (some of them) that nothing *did* happen. (Hockett, 1980:105)[12]

Hockett has reason to complain—not least because he was the Bloomfieldian-most-likely, the late master's favored son, and he was, along with Nida, Householder, Hill (even, aside from a sort of John-the-Baptist role in linguistic folklore, Harris)— pretty much swept aside in the prime of his career. None of this is new, of course, nor peculiar to science. "The first eruption," Priscilla Robertson says, in her nice refraction of Tocqueville's volcano image for the 1848 French revolution, blew off "not only the King but also, indifferently, the top layer of men who had hoped to reform the monarchy and who had by their criticism helped prepare for the revo-lution" (1952:14), an observation which generalizes to almost every abrupt social or scientific shift. Among the more spectacular political examples this century has provided, two from Russia spring most readily to mind, Kerensky and Gorbachev.

If many linguists' view of history is not exactly tragic, then, a word more appro-priate for the daily curses of lives much harsher than the ones lived out in academic hallways—in revolutionary France, for instance, and in the turbulence and oppres-sion surrounding the various revolutions in Eastern Europe, and in South-Central Los Angeles—it is certainly wrong. The falseness of the Chomskyan myths, again, resides in the general mood enveloping their ritual retellings that all was for naught between the 1933 publication of Bloomfield's *Language* and the 1957 publication of Chomsky's *Syntactic Structures*.

But part of their truth resides here as well. Bloomfield and his progeny had not ushered in a linguistic night of the living dead, grammar-zombies lurching from longhouse to longhouse, stiffly cataloging the phomemes, morphemes, and rudi-mentary syntactic patterns of language after language after language after language. But things *were* getting a little mechanical. And, more crucially, the most compel-ling aspects of language had not only been relegated to the bottom of a very long-term agenda, they had been given over to other disciplines altogether. Meaning and mind could be treated only in the distant future, and only by sociologists, psychol-ogists, ethnologists; seemingly, everyone but linguists were given Bloomfield's license to hunt down meaning in the deer park of the mind. Linguists had to stick to their sounds and words.

Chomsky—and here another aspect of the myths' truth resides—almost single-handedly shook linguistics free of its description-sodden stupor, and gave linguists leave to talk about meaning, and to talk about mind; indeed, the force of Chomsky's argument on the latter point was such that linguists were virtually the only ones with leave to talk about mind. *Almost* single-handedly. He was not without cowork-ers and proselytes—most notably, Morris Halle and Robert Lees—who fed his theories, and milked them, and brought his wares to market. Nor would it do to forget that there *was* a market; that Saussure and Bloomfield and Harris had made the mathematicization of linguistics possible; that Harris and Wells and Trager and Smith were making some headway with syntax; that Nida and Lounsbury and Bol-inger were clamoring about meaning. It certainly makes some sense to talk of language studies BC, Before Chomsky, but the linguistic calendar, even for generative and transformational and semantic notions, does not begin in 1957.

Calendars aside, Chomsky is the hero of the story. He is a hero of Homeric pro-portions, belonging solidly in the pantheon of our century's finest minds, with all the powers and qualities thereof. First, foremost, and initially, he is staggeringly smart. The speed, scope, and synthetic abilities of his intellect are legendary. "Most of us guys who in any other environment would be absolutely brilliant," one col-league says, "are put to shame by Noam." He is dedicated to his cause, working long, full hours; in fact, he is dedicated to a constellation of causes, linguistic, psy-chological, and philosophical (and social; like Russell and Einstein, Chomsky has deep political convictions, for which he also labors tirelessly). He is, too, a born leader, able to marshal support, fierce, uncompromising support, for positions he develops or adopts. (Inversely, he is many linguists' Great Satan, certain to marshal fierce, uncompromising opposition to almost anything he says or does.) Often, it seems, he shapes linguistics by sheer force of will. And—the quintessential heroic trait—he is fearless in battle.

Peeling Off the Mentalist Blinders

> HILL: If I took some of your statements literally, I would say that you are not studying language at all, but some form of psychology, the intuitions of native speakers. ,
> CHOMSKY: That is studying language.
>
> Exchange at the Third Texas Conference

The first unmistakable battleground of the Bloomfield-to-Chomsky changing of the guard was mentalism, though it is unmistakable only in retrospect. The generative challenge to mentalism looms so large in the rearview mirror that it is difficult to see how the old guard missed it. But they did.

Despite a general expansion of Bloomfieldian interests, mentalism was still taboo. Morris Swadesh, for instance, published a stinging attack on "the fetish that anything related to the mind must be ruled out of science" (1948; cited in Hymes and Fought, 1981:159). Swadesh was one of Sapir's most respected students. He had a formidable reputation in fieldwork and several influential papers, including one of the earliest distributional discussions of the phoneme (1934). Yet his critique couldn't even make it into a linguistics journal (it was published way out of the mainstream, in *Science and Society*), and had absolutely no impact on the field. Even the increased linguistic interest in psychology that marked the early-to-mid-fifties, spawning the term *psycholinguistics,* was distinctly behaviorist, psychology without the mind.

Chomsky came to see any study of language that didn't attend to its mental tentacles as completely sterile, and began promoting linguistics as a fundamentally psychological enterprise, coupling this promotion with a crushing attack on behaviorism.[13] The triumph on both fronts was staggering. Within a few years, behaviorism, Bloomfield's inspiration for a new and improved science of language, was virtually extinguished as a force in linguistics, invoked only in derision. It was also in rapid retreat at home, where psychologists hailed Chomsky as a champion in the promising emergent program, cognitive psychology (the term is too complex for proper treatment here, but, very roughly, cognitive psychology is oriented around the systems of knowledge behind human behavior; in principle, it is completely the inverse of behaviorism).

There were some murmurs of dissent toward behaviorism in mid-fifties psychology, especially in Cambridge, out of which the new approach was emerging, an approach whose birthday, according to George Miller, is 11 September 1956, the second day of a symposium at Harvard which ended with Chomsky outlining the arguments behind *Syntactic Structures.* We can't be sure what Chomsky said in that lecture, but his attitude to behaviorism at the time is apparent in *Syntactic Structures'* unambiguous rejection of "the identification of 'meaning' [that Bloomfield effects in his foundational tome—1933:22–32] with 'response to language'" (1957a:100). Chomsky was in fact extremely important to the emergence of cognitive psychology. In particular, his arguments against behaviorism (published a few years later in a review of Skinner's *Verbal Behavior*) were considered absolutely devastating.[14] Like most of Chomsky's finest arguments, his case against Skinner is as effective emotionally as it is intellectually. The reaction of Jerome Bruner, one of the founding voices of the cognitive psychology movement, is representative. He recalls the review in very charged terms: "Electric: Noam at his best, mercilessly out for the kill, daring, brilliant, on the side of the angels . . . in the same category as St. George slaying the dragon" (1983:159–60).

Dragon does not overstate the case. Behaviorism was tied up with some ethical perspectives that many intellectuals in the fifties were beginning to see as irredeemably vicious. There was, in the wake of the bloodshed and madness early in this century, a great deal of interest in the human sciences about the control of individuals and groups. Some of this interest was manifestly evil, where *control* meant *building better soldiers* or *making citizens more docile,* but much of it was very well intentioned, with the goal of happier, less aggressive, more fulfilled people, individually and collectively: in both cases, evil and good, behaviorist psychology, stimulus-response psychology, was the shining light of these interests. It held out the mechanical promise that getting people to behave would just be a matter of finding out which buttons to push, and pushing them. If you wanted a certain response, behaviorists would find the right stimulus for you. And linguists, since language is the cheapest, most omnipresent stimulus, were very concerned observers of this project. Bloomfield, for instance, heartened the troops in his 1929 LSA address with this prediction:

> I believe that in the near future—in the next few generations, let us say—linguistics will be one of the main sectors of scientific advance, and that in this sector science will win through to the understanding and control of human conduct. (1970 [1929]:227)

With the stunningly bad behavior of the Second World War—millions dead in Europe, apocalyptic explosions over Asia—segueing into the worldwide existential trembling of the cold war, and with the ever-growing reverence for science that accompanied these events, some linguists' faith in the powerful future of their field increased until they found themselves "at a time when our national existence—and possibly the existence of the human race—may depend on the development of linguistics and its application to human problems" (McDavid, 1954:32).[15]

Nowadays, there is a disturbingly Orwellian ring to such talk, even in its best-intentioned varieties. Understanding human conduct is fine, desirable in fact, but *control* had begun to stir a chilling breeze in the fifties (cued, in part, by the publication of *Nineteen Eighty-Four*). *Control* and its various synonyms *(manipulate, cause)* therefore play a large role in Chomsky's review, as does Skinner's principal source of authority, his bar-pressing rodent experiments. The first mention of Skinner, stuck awkwardly (therefore, prominently) into a more general discussion, is this sentence: "Skinner is noted for his contributions to the study of animal behavior" (1959 [1957]:26). Animals, especially rats, recur incessantly thereafter, Chomsky repeatedly stressing the vastness of the gulf between a rat navigating a maze for a food pellet and even the most elementary verbal acts.

Even Bloomfield, in the heady early days of behaviorism, realized the distance between a stimulus and a response in linguistic terms was formidably wide; that was the chief reason he outlawed meaning (considered, essentially, as the response to some stimulus) and mind (the mediative organ between stimulus and response). But Chomsky tattoos home the point that this gulf renders a stimulus-response, billiard-ball model of language completely vacuous:

> A typical example of a *stimulus control* for Skinner would be the response to a piece of music with the utterance *Mozart* or to a painting with the response *Dutch.* These responses are asserted to be "under the control of extremely subtle properties" of the physical object or event [Skinner, 1957:108]. Suppose instead of saying *Dutch* we had said *Clashes with the wallpaper, I thought you liked abstract work, Never saw it before, Tilted, Hanging too low, Beautiful, Hideous, Remember our camping trip last summer?,* or whatever else might come into our mind when looking at a picture. . . . Skinner could only say that each of these responses is under the control of some other stimulus property of the physical object. (1959 [1957]:31; Chomsky's italics; the interpolation replaces his footnote)

Chomsky is on the side of the angels here, all right, St. George to Skinner's dragon, but he is also on the side of free, dignified, creative individuals, people who belong to a tradition that includes Mozart and Rembrandt, people who cannot be controlled: his audience.[16]

The intellectual aspects of Chomsky's case, complementing the emotional aspects, are wide-ranging and damning. The long review has a steady commentary bulging from its footnotes, some of it bitingly glib ("Similarly, 'the universality of a literary work refers to the number of potential readers inclined to say the same thing' . . . i.e., the most 'universal' work is a dictionary of clichés and greetings."— 52n42), but most of it detailing the counter-evidence, qualifications, and questionable claims Chomsky has gleaned from the vast literature of learning theory; Chomsky, the reader can never forget, has done his homework. The most effective part of Chomsky's attack for almost every reader, however, is not the extent of the counter-evidence he marshals, but the two brief and devastating arguments he levels at behaviorism. One argument is based on the notion of creativity. The other goes by the name (presumably inspired by the dragon's own terminology), the *poverty of stimulus argument.*

Chomsky is a steadfast champion of creativity in the review, a notion broad enough to evoke Mozart and the Dutch Masters in its own right, but which has a very specific, narrow, and technical meaning in his work, coupled intimately with generative grammar. With a moment's reflection (as the conventional argument in an introductory course in Chomskyan linguistics now runs), it is clear that there are innumerable grammatical pieces of potential verbal behavior which have never been performed before, innumerable grammatical pieces of language which have never been uttered, never been a stimulus, never been a response; for instance,

17   Nanook put a pinch of yellow snow between his cheek and gums.

A simple behaviorist model has huge difficulties accounting for such facts. The sentence is not just unpredictable, in the sense of "Remember our camping trip last summer?" It is *unique.* Yet speakers of English have no trouble recognizing 17 as a legitimate, if unsavory, sentence of their language. They understand it immediately, and they would have no trouble, in the unlikely event that the circumstances become appropriate, producing it themselves. In a word, sentence 17 illustrates that human grammars are *creative:* they produce output which is not part of their input.

Output and input are important for Chomsky because he came to see the single most important factor about human language to be the ability children have to move rapidly from the input data of language they hear to a full command of that language, to a controlled and grammatical output. From this review on (anticipated by Lees, 1957, and to some degree by Hockett, 1948), language acquisition becomes an essential component of Chomsky's argumentation: the central problem for linguistics to solve, Chomsky insists, is how this creative ability establishes itself so quickly in the brain of a child. This problem is the one with which he most successfully flays not only Skinner but all things Skinnerian.

Behaviorist learning theory, Chomsky says, is based on a "careful arrangement of contingencies of reinforcement" and on the "meticulous training" behaviorists regard as "necessary for a child to learn the meanings of words and syntactic patterns" (1959 [1957]:39). This position, he hastens to add, "is based not on actual observation, but on analogies to lower organisms," so we are behooved to see if in fact these ingredients are necessary. They aren't. As the poverty of stimulus argument goes, one of the most remarkable facts about human languages—which are highly abstract, very complex, infinite phenomena—is that children acquire them in an astonishingly short period of time, despite haphazard and degenerate data (the "stimulus"). Children hear relatively few examples of most sentence types, they get little or no correction beyond pronunciation (often not even that), and they are exposed to a bewildering array of false starts, unlabeled mistakes, half sentences, and the like. Sounds and words, the principal Bloomfieldian units of focus, are amenable to stimulus-response acquisition; the child says "ice cream" and gets some ice cream. Syntax, Chomsky's natural medium, is not. Sentences are too variable, too dynamic, too creative, to have any significant correspondences to a rat and its bar.

Neither psychologists nor philosophers (to whom the review is also pointedly addressed—1959 [1957]:28) would have had any difficulty seeing the significance of Chomsky's critique for contemporary views of the mind. With linguists, the matter isn't so clear. For one thing, Bloomfieldians had a poacher-shooting tradition, and many were probably happy to cheer an up-and-comer's participation in the sport, thrashing a big-shot psychologist with the audacity to hunt in the preserve of linguists; Bloch, who published the review, "delighted [in] this superb job of constructive destruction" (Murray, 1980:80). More importantly, psychology was largely peripheral for most Bloomfieldians. It is noteworthy, for instance, that no other linguist reviewed *Verbal Behavior,* which was published two years before Chomsky's bludgeoning. And the constructive part of Chomsky's assault, the part that really threatened Bloomfieldian assumptions, was still somewhat amorphous in 1959. Poverty of stimulus has long been a well-known fact of language (Whitney had observed that children generally "get but the coarsest and most meagre of instruction"—1910 [1867]:12), but building a positive program around that observation was something new.

Chomsky started slowly. He ended his tanning of Skinner with the poverty of stimulus argument, but his clues for a replacement to behaviorist learning theory are suggestive at best:

> The fact that all normal children acquire essentially comparable grammars of great complexity with remarkable rapidity suggests that human beings are somehow specially designed to do this, with data-handling or "hypothesis-formulating" ability of unknown character and complexity. (1959 [1957]:57)

Chomsky adds the invitation that

> The study of linguistic structure may ultimately lead to some significant insights into this matter.

And he thereby—with his *somehow* and his *unknown character*—makes it clear that the door is now open, for anyone bold enough to follow him through, on the exploration of mental structure. The door is open for a younger generation, but it is not yet closed on the older one. Chomsky's review rehabilitates mentalism in the clearest terms since Bloomfield eclipsed Sapir on language and the mind, but it does not spell out in any detail the essential differences between Chomsky's view of mental structure and Skinner's view. These differences, when he does spell them out over the next few years, cut to the very bone of the Bloomfieldians' picture of science; therefore, of themselves as scientists.

Meanwhile, the Bloomfieldians had more to worry about than Chomsky's skinning of the behaviorist dragon in 1959, the year Morris Halle published his *Sound Pattern of Russian.*

Morris Halle and the Phoneme

> I could stay with the transformationalists pretty well until they attacked my darling, the phoneme.
>
> Archibald Hill

Chomsky met Morris Halle in 1951. They "became close friends, and had endless conversations" over the next several—extremely formative—years (Chomsky, 1979 [1976]:131). Like Chomsky, Halle was something of an outsider. Although he came to the U.S. as a teenager and later earned his doctorate from Harvard, his intellectual heritage—especially what it meant to be a "structuralist"—was much more European than American. Certainly he never swam, or even waded, in the Bloomfieldian mainstream. His doctorate was under the great Prague School structuralist, Roman Jakobson, from whom he inherited both mentalism and a certain friendliness to meaning (Halle's influence on Chomsky in both these areas was very likely much more substantial than has generally been appreciated, though Chomsky also had a great deal of direct contact with Jakobson). His thesis was on the sound system of a venerable European language, Russian; there was no Amerindian imperative, no description-for-the-sake-of-description compulsion, and it was published (1959a [1958]) under a title that paid deliberate homage to Bloomfield's partial rival, Sapir. Halle had also studied engineering for a while before entering linguistics, so there were mathematico-logical interests in his background, as in Chomsky's, beyond those of most American linguists.

He helped Chomsky get his position at MIT. He was also instrumental in establishing first a transformational research group there, then a doctoral program in linguistics (under the auspices of Electrical Engineering), and finally an independent linguistics department, of which he became the first chair. And he joined Chomsky in his first clear challenge to the orthodoxy—a paper on English stress phenomena which challenged a critical Bloomfieldian assumption about the independence of phonology from other grammatical processes (Chomsky, Halle, and Lukoff, 1956).[17]

Most importantly, at least in the short run of the late 1950s, when Chomskyan linguistics was gaining its polemical stride, Halle had an argument.

The argument is highly corrosive to a cornerstone of the Bloomfieldian program, the phoneme, and many linguists, then as now, regarded it as absolutely devastating. For the emerging Chomskyans of the early sixties, the argument—or, as Sadock later called it (1976), the *Hallean syllogism*—was totemic, a clear and present sign that even the most respected and impressive, the most beloved, of Bloomfieldian results, was made of unfired clay. For the fading Bloomfieldians of the early sixties, the argument was exactly the inverse, a sign of absolute and unwarranted hostility to an object of scientific beauty, and it earned the new movement their undying enmity. The Bloomfieldian resistance movement begins here.

The Bloomfieldians could not claim sole proprietorship over the phoneme. It

crystallized in Europe, in Kazan and Prague, about the same time it was crystallizing in America, and the lines of influence are quite complicated. But it was their
darling. Chomsky and Halle went after it like pit bulls (as did Lees, their student,
who gave the first presentation of Halle's anti-phoneme argument at the 1957 LSA
Annual Meeting).

Halle's argument is an impressive, persuasive, dismissive assault on a cornerstone of Bloomfieldian phonology, but it was neither powerful enough on its own
to cheese off the guardians of linguistic orthodoxy nor compelling enough on its
own to win over a band of revolutionaries. It was not, however, on its own. It came
with an elegant new phonology, whose virtues Halle demonstrated in a winning
treatment of the "highly complex patterns of phonological relationships in Russian" (Anderson, 1985:321). Negative arguments have a very short shelf-life, and,
regardless of conviction and oratorical prowess, if they don't come with a positive
program, there is little hope for widespread assent. Indeed, only a very weak form
of assent is called for by an exclusively negative rhetoric—a consensus of dissent, a
communal agreement that something is wrong, without a clear idea of how to put
it right. Einstein and Schrödinger, as passionate, eloquent, and sharply reasoned as
their attacks on probabilistic models of subatomic behavior were, had no remotely
comparable program to offer if Bohr's work had been overturned. Their arguments
failed. Scientists need something to do. Halle gave the new generation something
to do.

Moreover, this new phonology, Chomsky and Halle both insisted, was part of a
package. If you liked the syntax, and many people loved it, you had to take the
phonology.

At this point, it was teeth-rattlingly clear to the old guard that they were, in fact,
the old guard, that Chomsky, Halle, Lees, and the other MITniks (as the genera-
tionally charged term of derision tagged them) meant to shove them aside. Trager
and Smith's codification of Bloomfieldian phonology (actually, Bloomfieldian *pho-
nemics;* even the label has changed since the fifties) had a few loose belts, perhaps
some squeaky pulleys, but it was the foundation upon which they thought syntax
would have to be built. Even Sledd, who was fairly harsh about that phonology,
spliced it to Fries's syntax for his textbook (1959), and Stockwell had proposed
hitching it to Chomsky's syntax in 1958 (Hill, 1962c:122). Halle's *Sound Pattern of
Russian,* and Chomsky's presentation of Halle's work in 1959—again at Texas—
ruled this splicing out completely. It was all or nothing at all.

The Bloomfieldians, of course, were unmoved. The whole anti-phoneme argument rests on only a very few scraps of data—four words, both in Halle's original
presentation (1959a [1958]:22-3), and in Chomsky's more famous representations
(1964d [1963]:88–90; 1966b [1964]:78–82)—which hardly seems warrant enough
to throw out twenty years of effort, and the data was known to be problematic
before Halle worked it into his assault. Hallean phonology, cried Hockett, was
"completely bankrupt" (1968 [1966]:3; 1965 [1964]:201–2). "Worse than 'bankrupt'!" Trager chimed in: "a product of a fantastic never-never land" (1968:80).
They felt that the phoneme bought them more expressiveness than it cost, and were
unprepared to discard it on the basis of a minor anomaly. Less rationally, Hallean
phonology also borrowed rather heavily from Jakobson's work, and the Bloom-

fieldians had a history of antagonism toward the Prague Circle. But the argument was considered absolutely crushing by the Chomskyans—primarily because it was embedded in a carefully developed and comprehensive phonological theory which fit more closely with their syntactic work (it was called generative phonology and had very close parallels to syntactic transformations).

The first concerted counterattack came from Fred Householder, one of the earliest supporters of transformational syntax, teaching it at Indiana and implementing a number of early innovations. But he drew the line at this new phonology, launching an urbane and nasty assault in the inaugural number of the new generative-flavored *Journal of Linguistics* (1965).

The response from Cambridge was immediate (the lead article in the very next number of the very same journal), extensive, and brutal (Chomsky and Halle, 1965).[18] It is almost twice as long as Householder's original critique, and brimming with thinly veiled *ad hominems*. Actually, it would be more accurate to call them *ad homineses*—attacks to the *men*—since Householder is recurrently taken to represent overall Bloomfieldian blockheadedness (pp. 103, 105, 106, 107n4, 109n6, . . .). Chomsky and Halle suggest that Householder and his ilk don't understand the nature of problems confronting the linguist, "or, for that matter, the physical scientist" (104). They turn his mock-humility (Householder regularly expresses puzzlement over Chomsky and Halle's arguments) back against him, implying incompetence (119, 127, 129n26). They hector him like a schoolboy ("To repeat the obvious once again . . ."—127n24; also 103, 133n27, 136). He is inattentive (126, 127, 128n25). He is confused *(passim)*. He doesn't even understand Sapir's classic paper on the "Psychological Reality of Phonemes" (136; Sapir, 1949b [1933]:46–60). He trucks with inconsistencies, and "a linguist, who, like Householder, is willing to accept inconsistent accounts—in fact, claims that such inconsistency is ineliminable—has . . . simply given up the attempt to find out the facts about particular languages or language in general" (106): he isn't even doing linguistics. It is numerology (108).

Householder answered right away, but briefly and anemically, giving only a two-page policy statement reiterating some earlier points and wholly ignoring Chomsky and Halle's arguments. Hockett (1968 [1966]:4n3), for one, thought the reply sufficient, and Trager quotes Hockett approvingly, with a slight reproof to Householder for taking Halle's work seriously enough to dignify it with comment in the first place (1968:79, 80). But Chomskyans, and most non-Bloomfieldian observers, considered the matter closed: Chomsky and Halle had been challenged, they answered the challenge, and completely dumbfounded the opposition.[19] The new phonology was here to stay and one of the Bloomfieldians' most sacred possessions, the phoneme, was tagged as a worthless trinket.

There was more.

Enlisting the Grandfathers

It seems to me that the traditional analysis is clearly correct, and that the serious problem for linguistics is not to invent some novel and unmotivated alternative, but to provide a principled basis to account for the correct traditional analysis.

Noam Chomsky

Syntactic Structures was no threat to the Bloomfieldian program, so it must have been something of a surprise at the 1958 Texas conference—a deliberately staged contest of several emerging syntactic programs—when Chomsky came out battling. He was very active in all of the post-paper discussion periods, particularly so (and at his sharpest) following Henry Lee Smith's presentation of the only real competitor to transformational syntax in terms of rigor or prestige, phonological syntax. His own presentation essentially condensed *Syntactic Structures,* but put more of an edge on its notions. The paper argues that transformations are an important advance over Immediate Constituent analysis, and that generative grammar is an important advance for the field as a science, and that transformational-generative grammar can make important semantic inroads—all the carrots come out.

But Chomsky also wove in his mentalist concerns (his review of Skinner was written in this period, but still to be published), introduced some noxious data for certain Bloomfieldian principles, and sketched Halle's argument against the darling phoneme. He also said that Harris's work on transformations brought to light "a serious inadequacy of modern linguistic theory"—the inability to explain structural relatives, like active and passive versions of the same proposition—and that this inadequacy was the result of ignoring a major "chapter of traditional grammar" (1962a [1958]:124). These two elements, explanation and traditional grammar, became the primary themes of his anti-Bloomfieldian rhetoric over the next few years.

The following year he came back to Texas with an exclusively phonological paper ("The discussions were animated and sharp."—Chomsky, 1979 [1976]:133), establishing unequivocally that his program was a replacement of Bloomfieldian linguistics, not an extension.

The most pivotal event in the campaign against Bloomfieldian linguistics, however, was another conference, the 1962 International Congress of Linguists, where Chomsky was the invited speaker at the final plenary session. The four other plenary speakers that year were august Europeans (Nikolaj Andreyev, Emile Beneviste, Jerzy Kuryłowicz, and André Martinet), which gave young Chomsky "the appearance of being THE spokesperson for linguistics in the United States" (Newmeyer, 1980a:51; Newmeyer's emphasis). He used the moment brilliantly, putting his work, on the one hand, into very sharp relief against the Bloomfieldian program, and, on the other, aligning it closely with traditional grammar, the amorphous prestructuralist program which Bloomfieldians delighted in "grandly berating" (Sledd, 1955:399), but which was still favored in many parts of Europe. Better yet, the whole Bloomfieldian program, which left many Europeans sour, was subjected to a withering attack.

Chomsky's paper, in these and many other ways, also makes inescapably clear that his work isn't just a new way to do syntax. The bulk of the paper, in fact, is devoted to phonological issues, to showing how thoroughly the Bloomfieldians had mismanaged an area everyone regarded as their strongest, and how, therefore, "the fundamental insights of the pioneers of modern phonology have largely been lost" (1964b [1962]:973).[20] His arguments are wide-ranging, compelling, and extremely well focused. The number of themes Chomsky smoothly sustains, and the wealth of detail he invokes, are remarkable, but the paper effectively comes down to:

• traditional grammar was on the right track, especially with regard to uncovering the universal features shared by all languages;

• Bloomfieldian work, despite some gains, is on completely the wrong track—in fact, has perverted the course of science—especially in its disregard of psychology and its emphasis on the diversity among languages;

• the only real trouble with traditional grammar is its lack of precision;

• fortunately, in the last few decades, the technical tools have become available, through work in logic and the foundations of mathematics;

• transformational-generative grammar, which incorporates these tools, is therefore exactly what the field has been waiting for, the ideal marriage of modern mathematics and the old mentalist and universal goals that American structuralists had discarded.

The emblem of traditional grammar in Chomsky's 1962 address was one of the pre-structuralist Wills, Wilhelm von Humboldt, whom he quotes early and at length on the enterprise of linguistics generally. "We must look upon *language,* not as a dead *product,*" he quotes Humboldt, "but far more as a *producing.*" And "the *speech-learning* of children is not an assignment of words, to be deposited in memory and rebabbled by rote through the lips, but a growth in linguistic capacity with age and practice." And "the constant and uniform element in this mental labour . . . constitutes the *form* of language."[21] Coseriu (1970:215) says that the person speaking in these quotations is not Wilhelm, but Noam, von Humboldt, and Chomsky later admits to a certain "interpretive license" (1991a [1989]:7). The quotations are unquestionably selective; as the title of Humboldt's essay suggests, *On the Diversity of Human Language-structure (Über die Verschiedenheit des Menschlichen Sprachbaues),* he was at least as caught in the tension between uniformity and uniqueness, between inner form and outer realization, as Sapir. But these are still the words of Humboldt and they reflect important concerns—creativity, language learning, and linguistic universals—that the Bloomfieldians had largely disregarded, and that Chomsky was resurrecting. The linchpin in Chomsky's case is in the first quotation from Humboldt, through a slight but natural refraction of *producing (Erzeugung)* to *creating*—that is, exactly the feature of language Chomsky used so effectively in hiding Skinner.

In other published versions of his International Congress paper (there were at least four—1962c, 1964b [1962], 1964c [1963], 1964d [1963]), Chomsky heralded two seventeenth-century texts as even better representatives of the traditional grammar Bloomfield had banished from linguistics, both from the Port-Royal-des-Champs abbey outside of Paris, the *Art of Thinking* and the *General and Rational Grammar.* These books (now more commonly known as the Port-Royal *Logic* (Arnauld and Nicole, 1963 [1662]) and the Port-Royal *Grammar* (Arnauld and Lancelot, 1975 [1660]) epitomize the "general grammar idea" that Bloomfield (1933:6) saw as wielding a long and pernicious influence over linguistics. Bloomfield had reason to complain. The Port-Royal linguistic work implied that the common mental structure underlying all language was that bane of American descriptivism, Latin. But Chomsky saw something very attractive in the general grammar idea which Bloomfield had ignored and disparaged: that there *is* a common mental structure underlying all languages.

Moreover, beyond the clear mentalism that Port-Royal linguistics shared with Humboldt, it exhibits a far more transformational style of reasoning, particularly as a manifestation of creativity. One example that Chomsky got a good deal of mileage from illustrates the point very well. Consider sentences 18a–18d.

18   a   Invisible God created the visible world.
     b   God is invisible.
     c   God created the world.
     d   The world is visible.

The Port-Royal *Grammar* says that 18a is a proposition which includes the other three propositions, 18b–18d, and that 18b is the main proposition, in which 18c and 18d are embedded (Arnauld and Lancelot, 1975 [1660]:99). That is, the *Grammar* here is talking, in a very natural interpretation, about kernel sentences, and its rather vague idea of "inclusion" looks like the Harris-cum-Chomsky notion of generalized transformation (which splices one kernel sentence into another). In short, Chomsky has little trouble supporting his position that the *Syntactic Structures* model "expresses a view of the structure of language which is not at all new" (1964b [1962]:15); in fact, that it is "a formalization of features implicit in traditional grammars," or, conversely, that traditional grammars are "inexplicit transformational grammars" (1964b [1962]:16).

Bloomfieldian linguistics (or, as Chomsky took to calling it in the 1962 ICL address, the *taxonomic model*), it seems, had sinned in two interrelated and horrid ways when it left the garden of general grammar. It neglected universals, and it avoided explanations. The master, of course, has the definitive words here:

> Features which we think ought to be universal may be absent from the very next language that becomes accessible. Some features, such as, for instance, the distinction of verb-like and noun-like words as separate parts of speech, are common to many languages but lacking in others. The fact that some features are, at any rate, widespread, is worthy of notice and calls for an explanation; when we have adequate data about many languages, we shall have to return to the problem of general grammar and to explain these similarities and divergences, but this study, when it comes, will not be speculative [as with the Modistae and the Port-Royalists] but inductive. (Bloomfield, 1933:20)

Now, the Bloomfieldians were certainly interested in general, even universal features of language. It is telling that not only Sapir, but Bloomfield and the LSA embraced the title *Language*. They didn't choose *Languages* or *Tongues*, or *A Bunch of Unrelated Facts about the Noises We Make When We Want Someone to Pass Us the Salt*. But the master's pervading cautiousness, always looking over his shoulder for another language that could sink his inductive generalizations, had led the Bloomfieldians to avoid all talk of universals. Taking the descriptive mandate to its logical extreme, in fact, means that there are *no* universals: "languages could differ from each other without limit and in unpredictable ways" (Joos, 1957:96).[22] So much for the first sin, ignoring universals.

Chomsky cites Joos's without-limit expression of sin in his 1962 ICL paper to illustrate Bloomfieldian misguidedness on universals; a few years earlier, he had paraphrased another Joos extremity, expressing the other primary Bloomfieldian sin, "that the search for explanations is a kind of infantile aberration that may affect philosophers and mystics, but not sober scientists whose only interest is in 'pure description' . . . [a position] which can find little support in well-developed sciences" (Chomsky, 1962a [1958]:153n25).[23] Returning to this theme with a vengeance in 1962, Chomsky says that there is only one real virtue to a theory of language, it explains the structure of specific languages, and the Bloomfieldian aversion to universals made explanation completely unattainable.

Jakobson's work, as the best illustration of this goal, involved a theory of phonetic universals: a finite inventory of features that characterizes all the possible phonemic differences in human languages, just as a finite inventory of atoms characterizes all possible chemicals. The existence of a chemical is explained by combinatory possibilities of atoms. Now, Jakobson's inventory (adopted in principle by Halle's *Sound Pattern*) included articulatory and acoustic features that, for the most part, the Bloomfieldians subscribed to as well. But the extreme descriptivism of the languages-can-differ-from-each-other-without-limit-and-in-unpredictable-ways position is completely antithetical to an inventory that could be considered universal in any meaningful way. If the differences between any two languages are unpredictable, they are likewise unexplainable.

Or, so went Chomsky's argument at the International Congress, and, with that argument, almost all the essential pieces were in place for unseating Bloomfieldian linguistics: it ignored the mind; it failed to recognize language acquisition and creativity as the fundamental problems of linguistics; its phonology was off base; it perverted linguistics from the search for universals; it was concerned with taxonomy when it should be concerned with explanation. But there was one more problem with Bloomfieldian linguistics. It was irredeemably empiricist.

The Rational Chomsky

Empiricism insists that the mind is a tabula rasa, empty, unstructured, uniform at least as far as cognitive structure is concerned. I don't see any reason to believe that; I don't see any reason to believe that the little finger is a more complex organ than those parts of the human brain involved in the higher mental faculties; on the contrary, it is not unlikely that these are among the most complex structures in the universe.

Noam Chomsky

Chomsky took something else from his Port-Royal grandfathers, their epistemology, and among his main projects in the few years after his International Congress presentation was championing their views of knowledge and the mind. Those views, usually bundled up in the word *rationalism,* had long been in a serious state of disrepair. Their patron saint is Descartes, and Whitehead had defined the general disregard for rationalism by saying "We no more retain the physics of the seventeenth century than we do the Cartesian philosophy of [that] century" (1929:14). It was *passé* philosophy. Its perennial opponent in the epistemic sweepstakes was, largely due to the work stemming out of the Vienna Circle, on top. Empiricism was *au courant.*

To rehearse these terms:

> **Empiricism:** all knowledge is acquired through the senses.
> **Rationalism:** no knowledge is acquired through the senses.

Nobody in the history of epistemology, naturally, has bought (or tried to sell) either position; the only function they have served is as straw men in various polemics. The members of the loose philosophical school known as British Empiricism—a school with a varying roll, but which usually includes Locke, Hume, Berkeley, and Mill—held positions that fall more fully within the first definition than within the second, along with several other eminent minds, such as Epicurus, Aquinas, and Ayer. The opposing tradition is ably represented by Plato, Descartes, Spinoza, and Leibniz. But even the most casual reading of any these thinkers makes it clear that the only useful definitions here are fuzzy rather than discrete, and that the quantifiers should be tempered to reflect genuinely held beliefs:

> **Empiricism:** most knowledge is acquired through the senses.
> **Rationalism:** most knowledge is not acquired through the senses.

Even with this tempering, however, we have to keep in mind that *knowledge* refers to domains like mathematics, language, and hitting an inside fastball, not to the name of your sixth-grade teacher or where you left the car keys. But the definitions are workable.

Getting back to Chomsky, his attraction to rationalism goes hand-in-glove with his involvement in the late fifties emergence of cognitive psychology. Behaviorism was undergoing reconsideration in the early sixties, in part because of Chomsky's recent excoriation of Skinner, and behaviorism rests heavily on empiricism. The big problem with empiricism for cognitive psychology is that the more sophisticated mental functions don't look like they could arise from a blank slate. The nascent cognitivists believed it to be "a hopelessly wrong epistemological base from which to view the higher functions of the mind" (Bruner, 1988:91). Besides, Bruner says, pointing out that cognitivists could take courage from the growing rationalism in related fields, "There were, so to speak, such nearby figures as Von Neumann, Shannon, Nelson Goodman, Norbert Wiener, and the vigorous young Noam Chomsky who were making such claims loudly and convincingly." The vigorous young Chomsky, in fact, not only made his rationalism explicit and backed it up with bold arguments in mid-sixties books like *Aspects of the Theory of Syntax* and *Language and Mind,* he entitled another book adjectivally after Saint René, *Cartesian Linguistics,* to make sure the point couldn't be missed.

And Chomsky's rationalism is radical. Rationalism, stripped of its straw-man status, makes the unobjectionable claim that some mental capacities come as part of the start-up kit of the mind. One of the best formulations of rationalism is by Leibniz, who compares the mind to "a block of marble which has veins," and who says that learning is essentially a "labor to discover these veins, to clear them by polishing, and by cutting away what prevents them from appearing" (1949 [1705]:45–46). For Chomsky, in his starkest formulation of rationalism, one of these genetic veins in the marble of our minds enables us to grow a language. That's right: *grow* a language, just as we grow an arm or a leg or a kidney.

A prominent subcomponent of this claim is that such growth could take place only in human brains; it is not that we have a quantitatively more sophisticated command of symbols than other species, the way we have, say, a more sophisticated thumb than apes, or better vocal control, or more acute phonological discrimination, but that we have a qualitatively different "mental organ." To many Bloomfieldians, rationalism was bad enough, but topping it off with species specificity made it look as if Chomsky was placing man outside the natural world. It was claims of this order that finally convinced them that his grammatical elevator didn't go all the way to the top floor.

The grow-a-language position is actually quite compelling, absurd as it looks at first pass, and follows rather naturally from the poverty of stimulus argument. It might be, as Chomsky suggests in his review of *Verbal Behavior,* that the relevant innate endowment of humans is no more (but certainly no less) specific than general-purpose data-handling or hypothesis-formulating abilities, that the same cognitive properties which guide the growth of vision also guide the growth of language: for the visual cortex, they handle data like "horizontal" and "vertical" and "in-front-of"; for the language faculty, they handle the data like "noun" and "verb" and "sentence." Or it might be, as Chomsky began forcefully articulating in the sixties, that the language faculty is itself a highly specific mental organ with its own special and independent character, that such things as noun and verb and sentence are not just in the data, but genetically prewired into the brain. But, in either case, rationalism is a necessary part of the explanation and a strictly interpreted (strawman) empiricist philosophy of mind must be discarded.

Rationalism and empiricism are very important for a later part of our story, when epistemological foundations came back under scrutiny in the generative-interpretive brouhaha, but, for the moment, the central point is that they illustrate just how deep the Bloomfieldian-Chomskyan division rapidly became. What looked to most of the old guard like a new way to do syntax mushroomed in less than a decade into a new way to do linguistics, a new way to look at human beings, and a new way of doing science; new, and completely inverse. They were baffled and enraged.

Many Bloomfieldian camels had collapsed by the time Chomsky's rationalism became explicit, but that was the last straw for Hockett. In 1964, giving his presidential address to the LSA, Hockett was hailing *Syntactic Structures* as one of "only four major breakthroughs" in the field, placing it in the company of Jones's Asiatic Society address and Saussure's *Course,* and as late as 1966 he was working in generative grammar (1965 [1964]:185). But after Chomsky's rationalism had become inescapably clear, Hockett began fulminating about "the speculations of the neomedieval philosopher Noam Chomsky" (1967:142–44). Hall, playing on Hockett's theme (but with fancier spelling), joined in to rail about Chomsky "threatening to negate all the progress achieved over four centuries . . . [and] dragging our understanding of language back down to a state of mediaeval ignorance and obscurantism" (1968:128–29). Trager, keying on the mysticism most Bloomfieldians equated with rationalism, condemned Chomsky as "the leader of the cult [that has] interfered with and interrupted the growth of linguistics as one of the anthropological sciences for over a decade, with evil side-effects on several other fields of anthropology" (1968:78). The sky was falling. The sky was falling.

Burying the Bloomfieldians

> Is it really true that young linguists use my name to frighten their children?
>
> Fred Householder

In and among these early polemics about behaviorism, the phoneme, and rationalism, Chomsky and Halle attracted some of the best young minds in the field to the Research Laboratory of Electronics, the eclectic and very well funded branch of MIT which was the incubator of Chomskyan linguistics. The group—including Lees, Postal, Katz, Fodor, Edward Klima, and Jay Keyser—quickly formed very close intellectual ties and began hammering out the details of transformational grammar. As Fodor recalls,

> It's not much of a hyperbole to say that *all* of the people who were interested in this kind of linguistics were at MIT. That's not quite true. There were others scattered around. But for a while, we were pretty nearly all there was. So communication was very lively, and I guess we shared a general picture of the methodology for doing, not just linguistics, but behavioral science research. We were all more or less nativist, and all more or less mentalist. There was a lot of methodological conversation that one didn't need to have. One could get right to the substantive issues. So, from that point of view, it was extremely exciting.

It was also very successful. The group made rapid headway on a number of very thorny issues, particularly in the Bloomfieldians' weakest areas, syntax and semantics. Success, we all know, is heady, and the group's most definitive character trait was cockiness: they were young, they were bright, and they were working on a novel and immensely promising theory in collaboration with one of the finest intellects of the century. "In a situation like that," Katz notes, "it's quite natural for everyone to think they have God's Truth, and to be sure that what they're doing will revolutionize the world, and we all thought that."

Developments spread rapidly. Everyone spoke in the hallways, attended the same colloquia, and saw each other's papers long before they reached publication. They also saw many papers that never reached publication at all, the notorious *samizdat* literature that still characterizes work at MIT: arguments and analyses circulated in a mimeograph (now electronic) underground, never making their way to the formal light of day but showing up in the notes of important works that did. This situation, quite naturally, infuriated (and infuriates) anybody trying to follow the theory but failing to hook into the right distributional network.[24]

The most famous of these quasi-publications was naturally Chomsky's massive *Logical Structure of Linguistic Theory* (1975a [1955]), which is cited a dozen times in *Syntactic Structures* despite extremely modest and dog-eared circulation.

Though still programmatic, it is far more detailed, far more closely argued, far more mathematically dense than Chomsky's published arguments, and it gave the impression that the foundations of his model were firmly in place. It looked to be the iceberg of which *Syntactic Structures* formed the tip (see, in fact, Halliday's remarks in Lunt, 1964 [1962]:988). Chomsky's *A Fragment of English Grammar*, the mimeographed notes for his Third Texas Conference paper, was also cited widely, and Halle's suitably evangelical *Seven Sermons on Sounds in Speech*, was available through IBM. Mostly, though, the citations were to little more than memoranda floating around Cambridge.

In the publications that did issue formally, the program took clearer and clearer shape. The most important early publication, next to *Syntactic Structures*, was Lees's review of it. Chomsky overstates the case wildly when he says that Lees "was basically their [the Bloomfieldians'] hit man. He was the guy they sent around to denounce this, that, and the other thing. They heard about this heresy brewing at MIT, and he came down to take care of it for them." But Lees came to Cambridge (to work on a machine language project) with firm structuralist convictions, with a good standing in the Bloomfieldian community, and with a confrontational personal style. He found Chomsky's work arresting and effectively became his first doctoral student.[25] Lees was in part an expositor, and his review provided a rather careful account of Chomsky's key principles and solutions, but it was also the first resounding shot in the campaign against the Bloomfieldians. Using the familiar we're-doing-science-and-you're-not war cries, the review put Chomsky's work in very sharp relief against the rest of the field: transformational-generative grammar was chemistry, everything else in linguistics was alchemy. Lees's dissertation was also a major contribution to the emerging Chomskyan paradigm. It came out in 1960 as *The Grammar of English Nominalizations*, and was, as Benfey said of Bopp's *Conjugationssystem*, "the first work to be totally imbued with the spirit of the new linguistics" (Hoenigswald, 1986:177). Almost instantly, it became an exemplar for the program—a template for how to do transformational syntactic analysis, the perfect complement to Halle's *Sound Pattern*, a template for the new phonology.

Katz was also very influential. He teamed up with Fodor to contribute an extremely important paper to the Chomskyan enterprise, "The Structure of a Semantic Theory" (Katz and Fodor, 1964b [1963]), an article which made the first explicit proposals on how transformational grammar could accommodate semantics, and then he teamed up with Postal (Katz and Postal, 1964) on a book which brought those proposals closer to the heart of transformational grammar and precipitated the next major technical advance in the theory, the notion of deep structure.

But the publications streaming from Cambridge were not restricted to positive proposals. Many were attacks, following the lead of Chomsky's keel-hauling of Skinner, and his obstreperous performance at the 1958 Texas conference, and his International Congress attack on the theoretical underpinnings of Bloomfieldian descriptivism, and Halle's attack on the phoneme, and Chomsky and Halle's joint pummeling of Householder. But the disciples outdid their masters. The most famous polemic is Postal's *Constituent Structure* (1964), something of a negative exemplar, or an exemplar of negativity—a template not for working in the new program, but for eviscerating the opposition. It is a methodical, closely-reasoned, and withering argument to the effect that all varieties of structuralist syntax collapse into Chomsky's phrase structure notation, and consequently are decidedly inferior to transformational analyses. The book's reputation for brutality is so firm that one of Postal's colleagues describes it as

> a character assassination of all the major players in syntax: Bloch [under whom Postal had studied], and Hockett, and Sid Lamb, and Ken Pike. Immediate Constituent analysis, he said, was all hopelessly inferior and inadequate. So, his personality in the early days was . . . well, he was just a mad dog.

The mad-dog assessment is a little harsh, perhaps reflecting Postal's conference performances, or his later *Aspects of Phonological Theory* (1968 [1965]), but it does capture the unstoppable, unalterable tone of absolute certainty that pervades the book, and virtually everything else Postal wrote on transformational-generative grammar; one gets the sense that there is just no point trying to reason with Postal. He'll just come up with another argument. If that doesn't work, he'll find another, and another. This attitude suffused MIT, and gave rise in many Chomskyans to the "pretentious and cavalier" style that Bar-Hillel (1967:542) deplored in Katz—they had all the answers and most everyone else was hopelessly misguided. The attitude bewildered and aggravated even the most sympathetic, smooth-tempered linguists. Einar Haugen, for instance, as catholic and openminded a linguist as there was in the Bloomfieldian period, called Chomsky's program "a great advance," but lamented that

> once one begins to have discussions with the people who advocate this new approach, one discovers a certain dogmatism . . . and I wish that somehow the people who are so enthusiastically pursuing this new form, would understand some of the problems in presenting their ideas to other people, so that those others could accept them willingly. (Dallaire and others, 1962:41)

The result, for many, was the one reached in "On Arguing with Mr. Katz" by Uriel Weinreich (another broad and generous *indépendiste* from the Bloomfieldian period), that, since his opponent has completely abandoned "the ordinary conditions of scholarly fair play," the argument simply has to be abandoned (1968:287).

But the antagonism that surfaced in print was only a dull echo of the clamoring at conferences, the tone being set by Chomsky's featured appearance, the year after the publication of *Syntactic Structures,* at the Third Texas Conference on Problems of Linguistic Analysis in English—an event, in retrospect, almost significant enough to warrant a title so cumbersome. Both the motive behind this invitation and its results in the Bloomfieldian community are subject to some dispute. Some analysts suggest that the conference organizers invited Chomsky to give him a deserved comeuppance (Newmeyer, 1980a:46; Anderson, 1985:314); others find the organizers more benign (Murray, 1983:184).[26] Some Bloomfieldians apparently came away persuaded that the brash young Chomsky had been put in his place; others left the conference openly sympathetic to the new program, or at least its syntax. But the importance of the conference was not in its impact on the members of the entrenched paradigm (though it clearly helped to enlist at least one Bloomfieldian, Robert Stockwell, an erstwhile fan of phonological syntax). Rather, it played very well to the youth of the field, Chomsky's performance at the conference occupying a substantial role in the mythology formed among the growing cadre of young transformationalists, particularly once the proceedings reached publication (Hill, 1962c [1958]):

> Here we see linguistic history documented as nowhere else: Chomsky, the *enfant terrible*, taking on some of the giants of the field and making them look like confused students in a beginning linguistics course. (Newmeyer, 1980a:35, 1986a:31)[27]

The Bloomfieldians were not entirely outraged by the terrible infant, though, and invited Chomsky back the following year, when he gave a paper on the application of generative principles to phonological analysis. This second appearance was a more decisive, and divisive, sociological event than the 1958 conference, since Chomsky attacked the Bloomfieldians on their theoretical home court, phonology, armed with Halle's work on Russian.[28] Chomsky's performance at the 1962 International Congress served a similar role; again the proceedings document contention, and again Chomsky appears to take most of the points soundly. The conference galvanized the transformationalists (who were, of course, present *en bloc*), and the various published versions sparked a good deal of interest outside Cambridge.

But Chomsky has always been very careful about how and where his public disputations occur, and he has never been a very avid conference-goer. Most of the frontline proselytizing fell to other partisans, particularly students, who took up the cause with "missionary zeal" (Newmeyer, 1980a:50, 1986a:42), a phenomenon for which Holton offers a very useful illustration:

> It was not Cortéz but the men he had left in charge of Mexico who, as soon as his back was turned, tried to press the victory too fast to a conclusion and began to slaughter the Aztecs. (1988:35)

While it is not exactly Holton's point, his analogy suggests that there is frequently an aspect of intellectual genocide to the onset of a new scientific program, and the emergence of Chomskyan linguistics is a textbook example, though it would be a considerable stretch to talk about Chomsky's back being turned while the slaughter went on. The level of the attacks was often so excessive that it is difficult to believe they were uniformly condoned, but he and Halle strongly encouraged their students to enter the fray. Too, they had coupled their work inseparably with a rejection of all things Bloomfieldian. A big part of guiding their students toward the light was steering them away from the darkness. One of the most efficient ways to define an approach is in opposition to something, or someone, else—what those guys are/ were doing is hopelessly misguided, and we're not going to commit the same errors. Ostoff and Bruggmann beat up on the comparativists. Boas and Sapir beat up on the Latinizing missionaries. The Bloomfieldians took their habit of grandly berating traditional grammar so far as to personify it into a crusty old cipher, one Miss Fiddich, a symbolic schoolmarm whom they regularly cited with contemptuous bemusement as the source of some grammatical observation that they wanted to dismiss as trivial or of an attitude that they wanted to ridicule.

Both Chomsky and Halle deny any excesses in their presentation of previous work, but their students of the period recall classes on the Bloomfieldians that hall-way banter labeled "Bad Guys Courses," and it is noteworthy that contributors to transformational grammar from outside MIT—Charles Fillmore, for instance, and Emmon Bach, and Carlota Smith—were far less polemical than Lees, or Postal, or Katz, or Bever, or Chomsky and Halle. Inside the citadel, the mood was us-against-them. Infidels were rushed to the stake. This recollection is from Robin Lakoff, a Harvard linguistics student in the early-to-mid-sixties (and later an important gen-erative semanticist) who was a frequent and enthused spectator to the carnage:

> I remember well the times that non-transformationalists would speak at MIT, in those early years when the field still saw itself as fighting for survival in a hostile world. Rather than attempting to charm, conciliate, find points of connection, the circle at MIT reg-ularly went for blood. Points were made by obvious public demolition; the question or counterexample that brought the offender to his knees [was] repeated for weeks or months afterwards with relish. (R. Lakoff, 1989:967–68)

On the other coast, where an early convert, Robert Stockwell, had set up shop, Vic-toria Fromkin remembers that "the weekly seminars at the Rand Corporation in Santa Monica more resembled the storming of the Winter Palace than scholarly discussions" (1991 [1989]:79).

The two most fervent revolutionaries were Lees and Postal. Lees was the earliest, and the most flamboyant. A very direct man, he employed a style calculated to shock and enrage which he now describes (with characteristic bluntness) as "getting up at meetings and calling people *stupid.*" These tactics made him a legend among the transformationalists, but they did not endear him to the other side; Householder cautiously begins a review of Lees's *Grammar of English Nominalizations* with the remark that Lees "is noted as a redoubtable scholarly feuder and cutter-down-to-size" (1962:326), probably the mildest terms used by his opponents.

Postal was even less loved by the Bloomfieldians. Like Lees, he is warm and genial in personal settings, and quite tolerant of opposing viewpoints. But his rep-utation for intellectual savagery is well-deserved, rooted firmly in his public demeanor at conferences, especially in the early years. The stories are legion, most of which follow the same scenario. Postal sits through some anonymous, relatively innocuous, descriptive paper cataloguing the phonemic system of a little-known language. He stands up, begins with a blast like "this paper has absolutely nothing to do with the study of human languages," and proceeds to offer a barrage of argu-ments detailing its worthlessness—often making upwards of a dozen distinct counter-arguments against both the specific data used and the framework it is couched in. The performances were renowned for both intellectual precision and rhetorical viciousness. One tirade against Joos was so ruthless that it was stricken from the record of a Linguistic Society meeting (Hill, 1991:74), and some sense of his style is apparent in the casualness with which he categorizes his opponents' posi-tions as "empirically and logically contentless remarks" (of Hockett) and "substan-tively empty assertions" (of Gleason) and "tortured with a kind of intellectual schizophrenia" (of the whole Bloomfieldian program) in his published counterat-tacks (respectively, 1968 [1965]:4, 5, 6). And this (of the descriptive mandate):

> One cannot argue with someone who wishes only to classify utterances. People have a right to do what they want. We can ask, however, whether this has the right to be called

'linguistics'; whether it has the right to claim to be a significant field of inquiry. (Dallaire and others, 1962:10)

Complete and utter dismissiveness is not unusual in these circumstances. Of a similar contemptuousness and smugness among Oxford philosophers in the thirties, Isaiah Berlin says, "This was vain and foolish and, I have no doubt, irritating to others." But, he adds, "I suspect that those who have never been under the spell of this kind of illusion, even for a short while, have not known true intellectual happiness" (1980:115). Arnold Zwicky, an MIT graduate student at the time, recalls the mood in exactly these terms. The viciousness, he says, was propelled by an intense conviction that Chomsky's program was closing rapidly in on the Truth:

> there was a kind of holy war aspect to some of this, a feeling that some people had that they had to turn people's minds around, and that it was *important,* and that any device that did this, including ridicule, was legitimate.

Frederick Newmeyer, who entered the field just at the tail end of these events, finds the overall effect of the Chomskyans' confrontational tactics to be salutary, because the encounters showed an entire generation of linguists that language and science are important enough to arouse the passions, and because they showed clearly that the Bloomfieldian program was on the defensive; indeed, on the retreat (1980a;50f; 1986a:42). Still, there is a somewhat apologetic tone in his observation that "even undergraduate advocates of the theory embarrassed their teachers by ruthlessly lighting into linguists old enough to be their grandparents" (1986a:40). Postal, too, shows some empathy for their position:

> It was really a psychologically painful situation, because [Bloomfieldian linguistics] was itself a revolutionary linguistics that had gained its ascendancy by proclaiming that it was the scientific way to study language, and that traditional linguistics was unscientific. They had, themselves, trampled on people rather forcefully, made a lot of enemies, did a lot of unpleasant things. Now, bang, not very long after they were really in place, *they* were suddenly being attacked, and in a way that was incomprehensible to them. They were being told that *they* weren't being scientific. That just had to be a nightmare for them.

It was. They reacted with horror and lasting bitterness. But the sky had fallen. As early as 1963, the more dispassionate Bloomfieldians were beginning to admit defeat (Wells, 1963:48). By the middle of the decade it was clear to everyone, friend and foe alike, that "neither linguists nor psychologists [were] doing to language what they did as recently as five years ago" (Saporta, 1965:100); just ten years after the publication of *Syntactic Structures,* "the great majority of the papers" at the 1967 LSA summer meeting "were now firmly in the Transformational-Generative area" (Hill, 1991:89). And the Bloomfieldians had become, quite literally, jokes to the new generation. A parody of a table of contents page from the journal *Language* was compiled at the 1964 Linguistic Institute, including, among other burlesques and cruelties, an entry for a review by Henry Lee Smith of a book attributed to George Trager, *How to Publish and Perish.*

The Beauty of Deep Structure

The hidden harmony is better than the obvious.
Heraclitus

In general, as syntactic description becomes deeper, what appear to be semantic questions fall increasingly within its scope
Noam Chomsky

Spreading the Word

Chomsky's work very quickly rippled into neighboring academic ponds. One of the first, naturally, was psychology. Behaviorists and Bloomfieldians had been getting together since the brink of the fifties, the latter finally deciding to disregard the warnings of the master. There were hybrid conferences and seminars at linguistic strongholds like Cornell and Indiana, a growing stream of publications, and a shiny new word, *psycholinguistics.* Almost all of this work shared a positivist philosophy of science and a commitment to stimuli and responses as the principal components for modeling language behavior, *parole.* Chomsky, first very directly, through conversations, colloquia, and collaborations with the emergent cognitive community in Cambridge, then somewhat indirectly, through his deadly assault on *Verbal Behavior,* had a huge impact in clearing away the positivism and the stimulus-response models. Within a year of Chomsky's review, *Plans and the Structure of Behavior* came out as the flagship text of the new psychology (Miller, Galanter, and Pribram, 1960). It was heavily influenced by Chomsky, not just in its chapter on language, which rehearsed and propounded the major arguments of *Syntactic Structures,* but in its general attack on behaviorism, enthusiastically citing Chomsky's skinning of the dragon.

The following year, the first book of readings in the crossbred discipline of psycholinguistics came out, and it included four pieces by Chomsky, all extracted from *Syntactic Structures* (not, on the surface, a book much concerned with psycholinguistics, but by the sixties being read through the lens of Chomsky's later explicit mentalism), as well as papers by Halle, George Miller, and other Chomsky comrades from Cambridge. It also began, despite including some of the behaviorist work of the preceding decade, with the distinctly Chomskyan condemnation of behaviorist theories for their endemic failure "to account for some of the most obvious facts of language" (Saporta, 1961:v). Psycholinguistics, despite its behaviorist origins, became entirely cognitive in orientation, deeply influenced by Chomsky, especially in his collaborations with Miller on the cornerstone chapters of the *Handbook of Mathematical Psychology* (Luce and others, 1963)—"holy writ," Eric Wanner calls them. "Even now," he says, "these chapters seem to me the clearest foundational statement of the field" (1988:143). And psychology in general had not only Chomsky's corrosive attack on behaviorism to steer them from the shoals of empiricism but also the positive beacon of his linguistics to guide them into new and richer waters; his "linguistic arguments had shown that an activity could be rule governed and yet infinitely free and creative" (Neisser, 1988:86). Chomsky shows up, deservedly so, as a full-chested revolutionary hero in histories of cognitive science (Gardner, 1985; Baars, 1986; Hirst, 1988).

Chomsky's effect on English studies was more diffuse, but equally rapid. American structuralists had little to offer people studying literature and composition; little, that is, except scorn. Bloomfieldian work was heavily sound-based, rarely extended in any systematic way to units as big as sentences, and was congenitally nervous about meaning. In themselves, these characteristics were enough to discourage composition folk, rhetoricians, and literary critics—people who spend a great deal of time with written texts, who are very interested in sentences, paragraphs, discourses, and who fret passionately about meaning. At best, Bloomfieldian work left the English people cold and empty-handed. But its presentation also left them hostile. Bloomfieldian work was regularly coupled with strident attacks on traditional grammar, the only place where these people could find text-and-sentence analyses of any depth. Alienation was almost complete; noisy, often nasty skirmishes broke out regularly over such issues as prescription and description, the English department types wanting to pursue norms, or even ideals, that they could use to make little Johnny speak and write *correctly,* while the linguists insisted that Johnny already spoke his language fine and that the only role for a scientist was to describe the way language came out of his mouth or (in principle) off his pen. The Bloomfieldians looked like Philistines to the English gurus. "With them," sniffed A.S.P. Woodhouse in a typical, and typically snobby, complaint, "whatever is is right" (1952). Linguists responded like chimpanzees waving their scientific genitalia from the other side of a watering hole, as in the exclamatory polemics of Hall's *Leave Your Language Alone!* (1950).

The atmosphere was not good. But the English folk very much wanted to get something they could use from the scientists. So Chomsky was a dream come true for them. He starts with the sentence, his work devolved from Harris's discourse studies; he promises to help crack meaning; and he embraces the traditional grammars on the English professors' shelves. He found a ready market. There was a little marketing, a few well-placed papers and conference appearances by Stockwell, Lees, and Postal contrasting Bloomfieldian vices with transformational and generative virtues. But it was barely necessary: Chomsky and Stockwell met one English professor, Paul Roberts, at the 1958 Texas conference, and he practically leapt into their arms, publishing several books and articles over the next few years, often with Chomsky's close cooperation (Roberts, 1962:viii; 1964:vii). Roberts had some allegiance to Bloomfieldian methods, so he didn't engage in the negative polemics of Lees or Postal, but he sang the virtues of Chomsky's program far and wide in English circles—"this grammar is traditional grammar made explicit and rigorous" (1963:334)—and, very shortly, there was an English choir raising the roofbeams with claims that it was impossible to confront grammar "without using the brilliant work of Noam Chomsky" (Catwell, 1966:xix).[1]

Chomsky's impact on philosophy was different again, but the most successful of all. There were important contacts from the very beginning of Chomsky's career, preceding even his contacts with psychology. At the University of Pennsylvania Chomsky took several courses with Nelson Goodman, who strongly recommended him for the fellowship which he took up with the Harvard Society of Fellows and which in turn brought him into contact with, among others, Quine, Austin, and Bar-Hillel—contacts which were a fertile mixture of agreement and disagreement. For instance, Bar-Hillel was an enthusiastic supporter of the formal tack on syntax pioneered in linguistics by Harris and championed by Chomsky. His support was instrumental in Chomsky's pursuit of generative theory. There were certainly some differences (a few of them visible in an early Chomsky paper that takes issue with Bar-Hillel's criticisms of Harris—Chomsky, 1955b; Bar-Hillel, 1954), but they were reflected against a background of shared assumptions about language and its models. Austin, on the other hand, disagreed with much of that background, but he was nevertheless quite impressed with *Syntactic Structures,* incorporating it into his final lectures. And Chomsky found Austin's general, ordinary-language approach to meaning very congenial. Quine's famous criticism of logical positivism stimulated Chomsky, as did his complementary endorsement of the role simplicity played in science, though Chomsky disagreed rather violently with Quine's behaviorist semantic notions.

But, unlike the impact on psychology and on English studies, there was no immediate storm of interest in philosophy. Chomsky was the featured speaker of a section on "Explanatory Models in Linguistics" at the important 1960 International Conference on Logic, Methodology and Philosophy of Science, and his work began playing a role in serious philosophy at least as early as Putnam (1975 [1961]:85–106). But interest grew gradually, and it only reached significant levels after his frankly rationalist publications of the mid-sixties. By the early seventies, Gil Harman could say "nothing has had a greater impact on contemporary philosophy than Chomsky's theory of language" (1974:vii), in large measure because of its epistemological implications.

Chomsky has become well known in philosophy, then, not just as the leading figure in a related discipline, but as one of their own, the most forceful advocate of an erstwhile-discredited theory of mind, and his arguments have made rationalism respectable again.

Meanwhile, back in linguistics, Chomskyan theories were undergoing very substantial changes. The post-*Syntactic Structures* years, particularly with the increase of researchers, saw a great deal of activity—Lees's work on nominalizations, Halle's work in phonology, Katz and Fodor and Postal's work splicing semantics into the shifting transformational model. At the heart of all this activity, prompting and proposing and promoting, was Noam Chomsky. While not every early transformational grammarian would agree with Lees' self-deprecating assessment that "we all rode on Chomsky's coat tails then," none would deny that his work was driving everything and everyone else.[2]

Into the Great unNoam

> Noam is not a human being. He's an angel.

> [Chomsky is] uncommonly dishonest . . . a crank and an embarrassment.
> Martin Peretz

Now that we have seen his phenomenal rise but before we look into the war which grew out of the crowning document of that rise, *Aspects of the Theory of Syntax,* perhaps this is the place for a brief excursus on Chomsky's personality. Perhaps not: we are in dangerous and enigmatic territory; opinions really do range from Keyser's angel to Peretz's cranky devil, with significant clusters at both ends, and there really is motivation for both extremes, and I certainly have no special competence for the job of disentangling the elements of his character that have sponsored, in almost equal proportions, devotion and demonization. The one consolation here is that very few people do have special competence on Chomsky's personality. Háj Ross, who was his student, his colleague, and his opponent, who worked under him and with him for more than two decades in the MIT linguistics department, says he is almost completely in the dark about the man's personality:

> Chomsky is a real mystery man. What he's like as a person, I don't know. . . .

> There was never any personal contact outside of the university. When people came, they would meet Noam in his office, never even go to lunch or anything. You never go to Chomsky's house for a party or something, even among staff members, faculty members. . . .

> I know very little about Chomsky, where Chomsky's heart is.

Even Keyser's assessment has to do more with the seemingly extra-anthropic level of Chomsky's intelligence than with his personality; Peretz's, more with political differences than with personality. Of the people involved in our story, probably Halle is the only one who knows where Chomsky's heart is. But there are some rather extreme and inescapable characteristics which contribute to virtually every aspect of our tale and which call for some notice. We have already seen several—in particular, his immense intellectual and eristic gifts—and these qualities alone are enough to trigger worship from some quarters, bitter jealousy from others. At least three more call for some specific attention.

First, his graciousness: Chomsky, a man who lectures tirelessly, who works long, hard hours, who publishes several books a year, will spend hours with absolutely anyone who is interested in talking with him. Sometimes the wait is a long one—his calendar is booked months in advance—but he will patiently, kindly, and in exactly the right level of detail, explain his ideas to linguists, to lay people, to undergraduates, to other people's students, to journalists and directors and candlestick makers, and carefully explore their ideas with them. People come to MIT in droves to see him. His linguistics classes, which usually have quite small enrollments, are held in a lecture theater to accommodate the scores of visitors every week, and the chairs outside his office are always warm. To graduate students, he is ceaselessly helpful. Virtually all the theses he has supervised (well over sixty), and a large number he has had no official role in, at a wide range of universities, include heartfelt acknowledgments of the time and thought he has given the work.[3] He answers letters, from anyone and everyone, about anything, in great detail. He gives away ideas for free. He has a deeply admirable commitment to scholarly exploration, at all levels.

Second, the curious counterpole to graciousness, his spleen: Chomsky can, and often does, sling mud with abandon. His published comments can be extremely dismissive, his private comments can reach startling levels of contempt. He has many specific objects of scorn, but his most general one is every linguist—often wrapped together in a collective noun such as "the field"—who holds opinions about language and linguistics which depart significantly from his. He often speaks, for instance, about the immaturity of linguistics, contrasting it with the "more mature" sciences, and the context of these remarks invariably betrays the forked intentions behind the term. He does not mean simply that linguistics is in an earlier developmental stage than physics or chemistry, that linguists have only to solve methodological problems and reach empirical consensuses that workers in those other sciences have already achieved. He also means that linguists are whiny, irrational, petulant adolescents. There are, he allows, "a few quite serious people" (or, in one of his peculiar locutions, "a tiny majority of the field")—who happen coincidentally to be sympathetic with Chomsky's goals—but virtually everyone else in linguistics is intellectually and emotionally and even morally callow.[4] In part, this seems to reflect the need to work in an us-and-them, or even a me-and-them, intellectual climate. In part, it is delusion. His program is hugely successful, yet he always seems to feel embattled. In part, it is just flat arrogance.

Third, returning to the positive, extremely positive, side of Chomsky, his compassion: Chomsky is clearly very moved by, and very moving on, the downtrodden, especially when his own government or its agents have done the treading. The quotation from Ross above, in fact, was cut off a little too abruptly. He says he doesn't know where Chomsky's heart is, "except that his heart is for people who are oppressed by nasty politics. He works very hard for people like that." He works *very* hard for people like that. Beginning in the mid-sixties, just as generative semantics was starting to take wing, he opposed the war in Vietnam, and the leaders responsible for it, and the intellectuals supporting them, with every ounce of attention and strength he could wrest from his linguistics. What this meant, in the climate of the time, was

> speaking several nights a week at a church to an audience of half a dozen people, mostly bored or hostile, or at someone's home where a few people might be gathered, or at a meeting at a college that included the topics of Vietnam, Iran, Central America, and nuclear arms, in the hope that maybe the participants would outnumber the organizers. (Chomsky, 1987:54–55)

He had the courage to match his compassion, and as the anti-war movement gathered some steam, he moved to the forefront of the protests, leading marches, engaging in civil disobedience, getting arrested, and inevitably, ending up on Nixon's enemy list ("Has any other linguist received [that] accolade?" asked Bolinger admiringly—1991 [1974]:28).[5] His political interests have not subsided, nor has the energy he pours into them, which eats deeply into his linguistics. The time he allocates to political activism has forced him almost completely to abandon his work in phonology, in mathematical modeling, and in the history of linguistics "up to the point of barely reading about them" (Chomsky, 1982a:57). His political speaking engagements match his linguistic ones, and the huge bibliography of his political writings is almost as long as the huge bibliography of his linguistic and philosophical writings, despite the fifteen-year head start of the latter (Koerner and Tajima, 1986).

He acts graciously, he spouts invective, he displays great compassion and courage. He works hard. He's smart. He argues compellingly. That's about as much as we can say; who can tell where his, or anyone's, heart is?

But these characteristics are clear, and they are dramatically magnified through the tremendous stature Chomsky has achieved. As early as 1970, not even fifteen years after the publication of *Syntactic Structures*, he was given his own monograph in the Cambridge University Press Modern Masters series, alongside Einstein, and Freud, and Marx (Lyons, 1970a). That book, stripping its title down from *Noam Chomsky* to just *Chomsky*, has now gone into its third edition, the only one in the series to have done so (Lyons, 1991), and it has been joined by dozens of other celebrations—several more with only his name as title, along with such entries as *On Noam Chomsky, Reflections on Chomsky, Chomsky's System of Ideas, Challenging Chomsky, The Chomsky Update, The Chomskyan Turn*, even *The Noam Chomsky Lectures*, a play—not counting the hundreds of books which assume or teach or attack or mangle his notions in more general terms.[6] Chomsky, in fact, is one of the all-time citations kings—again in the company of Freud and Marx, and a long way beyond Einstein—with thousands upon thousands of references to his work. According to the Institute for Scientific Information, the Top Ten list for U.S. academic journals over the past seven years looks like this (Kesterton, 1993):

1. Marx
2. Lenin
3. Shakespeare
4. Aristotle
5. The Bible
6. Plato
7. Freud
8. Chomsky
9. Hegel
10. Cicero

Chomsky is the only one with a pulse, let alone an active role on the intellectual stage.

This stature means that virtually every twitch of his eyebrow sends off waves of influence in linguistics. His political concerns, as the mildest example, have had a small influence, attracting socially conscious students to linguistics, and inspiring others through his ceaseless example. But his graciousness and his spleen have had far more immediate consequences. He has friends, vehement ones. He has enemies, vehement ones. His withering polemical style has led emulators to embarrassing extremes, and he has many emulators on this score, both defending him and attacking him. Recently, James McCloskey wondered why it is that "phonologists, morphologists, and semanticists can all survive and cooperate in courteous disagreement, but syntacticians seem to thrive on a more robust diet of anger, polemic, and personal abuse" (1988:18). Certainly there is nothing inherently factionalizing about syntax, and there may be several answers, but one of them, and unquestionably one of the most significant, is Noam Chomsky. His demeanor has defined the field, not just because he hurt some feelings by displacing the Bloomfieldians and quashing the generative semanticists and contemptuously dismissing stray attackers, but because of the style of argument he uses to those ends, and because of the people he has attracted and trained and provoked with that style of argument, because of the gunslinger mentality that has suffused the field since the late fifties.

   The first ripples of his tremendous influence on the sociology of the field, once the Bloomfieldians were dispatched, began with his followers' exegesis of *Aspects*—a process epitomized by the shift in perspective the generative semanticists had toward their former leader. At the outset, they "all felt they owed an allegiance deeper than professional commitment to Chomsky—it verged on worship," Robin Lakoff remembers. But "once Chomsky was seen not to be an idol," once his erstwhile devotees fell on the receiving ended of his terrible rhetoric, "he was recast as satanic, the Enemy" (1989:963, 970).

   This process began, innocently enough, with the exploration of *Aspects'* central notion, deep structure.

Stalking the Hidden Harmonies

Deep structure goes back at least to the Port-Royal-des-Champs abbey. The Port-Royalists (chiefly, Antoine Arnauld) were at the tail end of the Modistic mission, looking for something underneath language, something deeper, more profound, that determined language. But their deeper, more profound somethings looked an awful lot like Latin, with a side order of logic, a combination plate which got little more than a disdainful chuckle from the Bloomfieldians. Latin had been such a dismal failure as a descriptive tool for Amerindian languages that its presence alone was enough to earn the Bloomfieldians' scorn, but logic, something artificial, was even worse. It was synonymous with consistency and prescription for Bloomfield, an odious part of the hated "school tradition, which seeks to apply logical standards to language" (1933:6), and Harris warned of the dangers of obscuring the differences among languages by forcing them all into "a single Procrustean bed, and . . . imposing on all of them alike a single set of logical categories" (1951 [1947]:2). Throw in mentalism (logic being the "rules for all activities of the mind"—Arnauld and Nicole, 1963 [1662]:20), and the whole project looked piteously misguided; the search for universals was therefore yet another project in Bloomfield's program for the very distant future (1933:20). Even Sapir had said wistfully that spading up "the great underlying ground-plans" of language was a task for a long-off "some day" (1949a [1921]:144).

But one of the principal weapons in Chomsky's rout of the Bloomfieldians was exploiting their failure to look underneath language, and, aside from a few mopping-up operations (Chomsky's counterattack on Reichling and Uhlenbeck; Lakoff's counterattack on Hockett; McCawley's attack on Hockett; Postal's attack on everyone who endorsed constituent structure; Postal's attack on Bloomfieldian phonology; Postal's attack on Martinet; Postal's counterattack on Dixon . . .),[7] the rout was all but complete by 1965. Looking underneath language was back, with a vengeance. The sign, seal, and delivery of this return was *Aspects of the Theory of Syntax,* Chomsky's magnificent summary of the developments introduced or incited by *Syntactic Structures.* In the argot of the day, *Syntactic Structures* was the Old Testament; *Aspects,* the New.

The designations are apt in many ways, especially in how the semantic speculations of *Syntactic Structures* prefigure technical developments in *Aspects.* The clearest case of this typology is the early model's kernel sentence—a notion with an indeterminate role but clearly relevant to meaning—which *Aspects* reincarnates as the semantic portal, deep structure. Several other elements of the new model also put flesh to the spirit of *Syntactic Structures,* but deep structure was, unquestionably, the crux of the new model. For any given derivation, it was the point at which the words showed up, at which restrictions between words were enforced, at which grammatical relations like subject and object were specified, and at which certain abstract morphemes were incorporated to trigger the later application of the model's exemplary transformations. All of these characteristics have very clear semantic links. Words, for instance, are units of meaning, and rules which ensure that a color-term such as *green* modifies concrete words (like *eggs*) but not abstract ones (like *ideas*) is a way of preventing the grammar from generating nonsense. Subject and object are, whatever else they may be, clear signals of propositional meaning (*Fideau bit the cat* means something very different from *The cat bit Fideau*), and the abstract morphemes distinguish among sentence types, like question and command, as well as between negative and positive assertions. *Aspects* fulfilled the promise of *Syntactic Structures* for a responsible and revealing semantic program.

The shape of the semantic solution in *Aspects* is simple, intuitive, and therefore, on both counts, extremely attractive. The integrated model it served up was beautiful. At its heart the generative engine of *Syntactic Structures* hummed, producing a deep structure. The deep structure fed the semantic component, accounting for the meaning. It also fed the transformations, which turned it into a surface structure. The surface structure, in turn, fed the phonological component, accounting for the sound. The derivation linked them.

There were epicycles, lots of them, and there were large gaps, but the general outline was compelling enough to warrant the optimism, if not the arrogance, of the early Chomskyans. They had, in a few short and feverish years, hammered out an elegant framework which accomplished the ultimate goal of all linguistic work from at least the time of the Stoics. They had formally linked sound and meaning.

But the Old Testament/New Testament cryptonyms for *Syntactic Structures* and *Aspects* are exactly wrong in a few important respects. Chomsky's first book has the clarity, univocality, even, in a sense, the narrative drive, of the New Testament. *Aspects* is far less direct.

Syntactic Structures is a one-man show. By 1965, transformational grammar had been thoroughly transformed—added to, subtracted from, permuted. While the theory was still very much Chomsky's, it showed signs of refraction through a committee. *Aspects* is more comprehensive, deeper in its implications, and much more detailed technically than *Syntactic Structures*. But it is also blurrier around the edges, and more cagey in its assertions. This caginess is nowhere more apparent than with the central hypothesis of the *Aspects* model, the Katz-Postal principle, which says that transformations have no semantic impact. *Aspects'* endorsement of that principle is forceful in the main text (if anything, there are indications it might not go far enough—1965 [1964]:158–59), but the endorsement is accompanied by a discursive note hinting that, on the other hand, it could be "somewhat too strong" (1965 [1964]:224n5). *Aspects'* hermeneutical potential is much closer to the prophetic books of the Old Testament than the New, and subsequent generations of linguists have found support in it for an amazing range of positions.[8]

The first two positions to grow out of *Aspects*—generative semantics and interpretive semantics—can in fact be traced to exactly that tension between the main text's claims about the Katz-Postal hypothesis and its but-then-again note: Chomsky says it is perhaps too weak, and then again, well, perhaps it is too strong. Generative semantics went with the too-weak position; interpretive semantics went with the too-strong position. Both sides took *Aspects* as their defining document; Chomsky, as their spiritual leader.

The Inevitability of Deep Structure

> It would be absurd to develop a general syntactic theory without assigning an absolutely crucial role to semantic considerations.
>
> Noam Chomsky

There are several ways to read the early history of transformational grammar, but one of the most revealing ways is by the light of two adjectival lamps that John Goldsmith and Geoffrey Huck propose, *distributional* and *mediational*. A distributional linguistics program, Goldsmith and Huck (1991) say, is one that aims primarily at getting the sequences of signifiers—the sounds and the words—right. A mediational program aims primarily at charting the relations between sound and meaning.

These are orientations, not absolutes, and they can come in any proportions. But one often provides the driving genius in programs where the other is badly slighted. The Bloomfieldians, we have seen, were relentlessly preoccupied with distributional concerns, and so wary of mediational ones that they ruled them out of the field. The Modistae went so far in the other direction that they followed the reverse policy, ruling all interest in sound as unscientific and sublinguistic.

Read by the light of these terms, the history of transformational grammar from *Syntactic Structures* to *Aspects of the Theory of Syntax,* is a dramatic move from distributional goals to mediational goals. The hero of the story, the grand mediator, is deep structure.

A great deal of the transformational-generative energy expended in and around MIT was used to dismantle the obstacles blocking the formulation of deep structure. The earliest work on the transformation arose, remember, in Harris's investigations of discourse patterns, and in a context that Chomsky viewed with a peculiarly fatalistic sanguinity:

> I remember as a student being intrigued by [linguistics]—the problems were fun and everything—but we were all wondering what we were going to do in ten years. Suppose you've done a phonemic analysis of every language. Suppose you've done an IC [Immediate Constituent] analysis of every language. It's fun to do. It's like a cross-word puzzle. It's challenging and hard. But it's going to be over in ten years. That's the way the field looked. It looked as if it were essentially over.
>
> Well, at that point, Harris had this idea of trying to do something new by looking at the structure of discourse. He tried to use the features of linguistic analysis for discourse analysis.

Applying standard Bloomfieldian procedures to discourse was no easy task, however, since they depend crucially on frequent recurrence throughout a corpus, on lots of distributional data. For instance, the argument that [p⁻] and [pʰ] are both members of the same English phoneme, /p/, depends on finding many occurrences of each, in a number of environments (like *pit* and *tip* and *stipple*). The argument that {-d}, {-t}, and {-əd} are all instances of the same English morpheme, {-PAST}, depends on a corpus which includes a lot of past tense verbs (in particular, verbs like *barred, rushed,* and *dated*). But the units of discourse, sentences, are a motlier crew than phonemes or words. The same sentence type recurs rarely, if at all, even in very long corpora.

So Harris needed a tool to normalize the texts he was analyzing, to reduce them to a core of elementary sentence types. He needed transformations.

Once he had his transformations, despite very deep-set semantic qualms, he saw clearly that they could serve a mediational role in grammar:

> Meaning is a matter of evaluation, and cannot be fitted directly into the type of science that is developed in structural linguistics or in transformational theory. Still, for various purposes it may be possible to set up some practical evaluation of meaning; and with respect to most of those evaluations, transformations will have a special status. That many sentences which are transforms of each other have more or less the same meaning . . . is an immediate impression. . . . To what extent, and in what sense, transformations hold meaning constant is a matter for investigation; but enough is known to make transformations a possible tool for reducing the complexity of sentences under semantically controlled conditions. (Harris, 1970 [1957]:449–50)

This impulse to operate under semantically controlled conditions even suffuses the terminology. The distillate of a text, after the surface variations are boiled away, was what Harris called *kernels,* and the notion of a kernel is incoherent without some recourse to meaning. The kernel sentence could not contain the syntactic essence, since transformations, by definition, alter syntax. Nor could it be the morphological essence, since transformations add and delete morphemes; nor phonological, since transformations alter sound patterns. Harris's kernels are the semantic, propositional essence of the discourse.[9]

Chomsky's early system grew directly out of Harris's work. The most obvious debt is terminological, but there are subtler obligations: in particular, the notion of kernel and transformation as somehow semantically privileged. But it is the differences between Chomsky and his teacher that mark the important steps toward deep structure. The most crucial difference is that Chomsky's grammar is not an analytic device for texts. Harris wanted his transformations for such practical purposes as machine translation and automated information retrieval (1970 [1956]:388; [1959]:458ff). Chomsky wanted them to model the linguistic knowledge in a native speaker's head, and he is very explicit that a central feature of his approach is "to provide a satisfactory analysis of the notion of 'understanding'" (1957a:87), to get directly at meaning, and the kernel was vital. His early grammar treats meaning as a triplet, distributed over three levels of analysis—phrase structural, kernel, and transformational—but the kernel was the fulcrum:

> In order to understand a sentence it is necessary to know the kernel sentences from which it originates (more precisely, the terminal strings underlying these kernel sentences) and the phrase structure of each of these elementary components, as well as the transformational [or derivational] history of development of the given sentences. The general problem of analyzing the process of "understanding" is thus reduced, in a sense, to the problem of explaining how kernel sentences are understood, these being considered the basic "content elements" from which the usual, more complex sentences of real life are formed by transformational development. (1957a:92)

Chomsky and his early collaborators had a strong intuition that all the real semantic action took place in and around the kernel.[10] Lees, the polemical recruiter, expressed the intuition as plainly as possible, proposing an "effective research program" within Chomskyan linguistics to reduce the problem of sentence-meaning to a problem of kernel-meaning, and to provide "a syntactic analysis for certain apparently semantic notions, wherever possible" (1962 [1960]:7).

There were two big complications to this project, though, which required a major remodeling of the *Syntactic Structures* grammar: (1) kernels were not rich enough semantically to represent fully the meaning of sentences, and (2) transformations changed meaning. Both complications are easy to see, easy enough that one example illustrates both, our Cormorant-Island sentence pair:

1   a   Everyone on Cormorant Island speaks two languages.
     b   Two languages are spoken by everyone on Cormorant Island.

Sentence 1a is the kernel of sentence 1b, but it doesn't fully represent the meaning of 1b; it can be true when 1b is false. The transformation relating 1a and 1b (Passive) changes the meaning. So, Lees's proposed research program had two complementary jobs, to beef up the kernel and to slim down the transformation.

The first stage came with Katz and Fodor's "The Structure of a Semantic Theory" (1964b [1963]), the earliest attempt to construct an explicit semantic theory that dovetailed with Chomsky's syntactic framework. The paper, along with sundry companion pieces by Katz, was a self-conscious attempt to do for semantics what Chomsky had done for syntax, exploring such issues as anomaly, ambiguity, and redundancy. The first problem was just to provide an explicit way of representing meaning. Katz and Fodor did it with essentially two components—a dictionary and a set of semantic interpretation rules. The dictionary contained lexical entries specified as to syntactic behavior and semantic content. The meaning of *bachelor,* for instance (more specifically, one meaning of *bachelor*), was represented as [+noun, +male, +human, −married].[11] The syntactic feature, [+noun], specified that *bachelor* could participate in a noun phrase (follow a determiner, take an adjective, accept a relative clause, and so forth). The other features ensured its semantic behavior: [+male], for instance, is what makes 2a nonsense, [−married] makes 2b redundant, and so on.

2   a   Logendra is a buxom bachelor.
    b   Logendra is a bachelor, and he's not married either.[12]

The use of these specifiers provides a notation for meaning, and something like the use of phrase structure rules provides a notation for syntax, paving the way for semantic interpretation rules that would, for instance, identify a [−male] adjective like *buxom* modifying a [+male] noun like *bachelor* as a *non sequitur.* Katz and Fodor, that is, provided a new wing for the grammatical house blueprinted in *Syntactic Structures.* But insofar as the program went of reducing sentence-meaning to kernel-meaning, the primary plans were still flawed. Transformations still changed meaning.

"It would be theoretically most satisfying," Katz and Fodor said (1964b [1963]:515), "if we could take the position that transformations never changed meaning." One of the main reasons that they couldn't take that position, they further noted, was the way transformations were formulated. Katz wasted no time getting to this job, reformulating the troublesome transformations. He enlisted Postal, the premier semanticist of the period teaming up with the premier syntactician. Their book, *An Integrated Theory of Linguistic Descriptions,* adopted and defended the very strong position that (most) transformations held meaning constant. This position, central to the story of generative semantics, was the one that led to some of Chomsky's more famous two-stepping in *Aspects,* Katz-Postal hypothesis.

The clearest example of a transformation which altered meaning in the early theory is *Syntactic Structures'* use of T_{not} which turned sentences like 3a into their own negation (3b).

3   a   Nirm kissed David.
    b   Nirm did not kiss David.

Unquestionably, sentences 3a and 3b mean different things; therefore, the operation of T_{not} alters meaning; therefore, the kernel sentence of 3b cannot represent its meaning. Fortunately, though, both Lees (1968 [1960]:19) and Klima (1964 [1959]) had happened on a device that would make Katz and Postal's job much easier. They had suggested—for independent, syntactically motivated reasons— that the phrase structure rules should generate an abstract marker, NEG, a marker which then triggered the transformation. The reasons for this innovation aren't important for our purposes, but Katz and Postal seized on its potential effects, eagerly pointing out that it changes the semantic picture substantially (1964:73f).

The strings underlying sentences 3a and 3b, they noticed, are no longer identical. The NEG-proposal does not affect 3a, but 3b now derives from a string like 3c.

3   c   NEG Nirm kissed David

The kernel was richer (since it contained NEG), the transformation poorer (since its range of application was restricted to only those sentences with the marker), and the change-of-meaning problem went politely away (since 3b and 3c mean the same thing).[13]

Lees and Klima, in fact, provided an entire class of arguments for neutralizing the semantic effect of transformations. Chomsky (1957a:90f), for instance, had noticed that transformations supplied a convenient syntactico-semantic typology for sentences like 4a–4c.

4   a   Nirm could have kissed David.
     b   Could Nirm have kissed David?
     c   Who could Nirm have kissed?

Sentence 4a is the kernel; 4b comes from applying T_q (a transformation which moves the auxiliary verb to the front of the sentence) to 4a; 4c comes from applying to 4b a transformation (T_w) which inserts a sentence-initial wh-word, moves the *could,* and deletes the object. Notice that Chomsky's T_q and T_w not only alter meaning in the narrow sense (for instance, T_w loses information; namely, the identity of the kissee), but they change the function of the sentence. Sentence 4a asserts, 4b requests confirmation (or disconfirmation), and 4c requests specific information.

The light bulb went on. Katz and Postal (1964:79–117) argue very forcefully in *Integrated Theory* for a few parallel changes to the phrase structure rules such that the kernels of 4a–c are no longer the same.[14] Sentence 4a has the same kernel it always had, but 4b now comes from 4d; 4c, from 4e.

4   d   Q Nirm could have kissed David
     e   Nirm could have kissed wh + some + one

The effect once again is to rob from the transformation and give to the kernel. The transformations lose their range of application (T_q being triggered now by the presence of Q; T_w, by the presence of wh), along with their power to change meaning, and the kernels become syntactically and semantically more explicit. Katz and Postal work similar magic with imperative sentences, supplying an IMP marker which triggers the relevant transformation. Where earlier theory would derive both 5a (an assertion) and 5b (a command) from the same underlying source, Katz and Postal propose a source for 5b which looks like 5c.

5   a   You eat chicken.
     b   Eat chicken!
     c   IMP you eat chicken.[15]

Once again, semantic duty is transferred from the transformation to the kernel.

There was a larger obstacle in the path of kernel enrichment, however. Transformations with an explicit semantic reflex, like T_{not} and T_q, are clear exceptions to the policy that transformations don't preserve meaning: every time they operate, they affect the meaning of the structure they operate on, and they affect the meaning in exactly the same way. As such, they might have been handled by a variety of adjustments other than underlying triggers. For instance, the Katz-Postal proposal excepted one entire class of transformations (generalized transformations; discussed below) by allowing the operation of a different class of semantic rules for them; similar maneuvers could have accommodated the small class of rules which clearly altered meaning or function.

Rules which are generally neutral to meaning and yet have semantic repercussions in a few scattered cases are a much bigger class of headache. The real terror in this regard was the passive transformation. It was a rhetorical mainstay of the early Chomskyan program—trotted out regularly to exemplify both the distributional and mediational virtues of transformational grammar—and it usually left the kernel of meaning intact. But not always.

The headache was well known from the beginning. The only discussion in *Syntactic Structures* on the inability of transformations to hold meaning constant centers on precisely this situation—the interaction of the passive transformation with quantifiers, returning us once again to our Cormorant-Island sentences (repeated again for convenience).

1 a Everyone on Cormorant Island speaks two languages.
   b Two languages are spoken by everyone on Cormorant Island.

Under normal interpretation, of course, 1a describes a situation where everyone on the island knows two languages, but makes no claim about what these languages are; each person could know a completely distinct pair. Fred might know Basque and Korean while Wilma knows Kikuyu and Dyirbal. Call this the *different-two* interpretation. Sentence 1b (again under normal interpretation) claims that there are two languages such that everyone on the island knows them (say, to be completely arbitrary, English and Kwakwala); each person on the island knows at least that set of two languages. Fred might still speak Basque and Korean, but he also has to speak English and Kwakwala. Ditto for Wilma and her languages. Call this the *same-two* interpretation.

Integrated Theory handles these sorts of sentences—examples of intriguing phenomena revolving around the scope of quantifiers (like *every* and *all*)—with smoothness to spare. Katz and Postal's argument involves refining the semantic duties of a grammar: it needn't account for the "normal" reading of a sentence, they say, just for its basic content, and the basic content of both 1a and 1b, they further say, is the same; both sentences are ambiguous in exactly the same way. Both describe the same-two situation, and both describe the different-two situation, and therefore the transformation has only a low-level stylistic effect, not a genuine semantic impact.[16] This sort of solution is extremely common in Chomskyan work, and we will see much more of it. One aspect of the data is spirited away (the normal reading is essentially declared irrelevant), and the model efficiently handles what is left.

Even with all the synthesizing and innovating of Katz and Fodor and Postal, there remained one very substantial obstacle to the formation of deep structure, generalized transformations. Chomsky took on this one himself.

One of the principal functions of transformations, dating back to Harris, was to deal with compound sentences. For Harris, transformations disentangled independent clauses (to extract the kernels); for Chomsky, they spliced together independent clauses (putting kernels together). For both jobs, transformations worked beautifully, so well that the very first argument Chomsky offers in *Syntactic Structures* for the existence of transformations is the ease with which they handle compound sentences. For instance, sentences 6a and 6b could be neatly spliced with T_{rel}, forming either 6c or 6d.

6 a The cat chased the dog.
   b The cat ate the Kibbles.
   c The cat that ate the Kibbles chased the dog.
   d The cat that chased the dog ate the Kibbles.

The problems for deep structure are obvious: no amount of finagling and enriching of the underlying phrase markers can predict the semantic consequences of incorporating another sentence.

Chomsky argues in *Aspects* that generalized transformations are a sufficiently different kettle of fish from singularly transformations (all the rest) that they should be eliminated altogether. As we have seen from Katz and Postal's labors, changing something at one point has ripples in a transformational grammar. Adding IMP and Q and their ilk to the phrase structure rules calls for a change to the transformations (which now apply only when they are specifically triggered). Chomsky's flushing of a whole class of transformations requires some adjustments elsewhere. Like Katz and Postal, Chomsky turns to the phrase structure rules. He changes the noun phrase rule, for instance, thusly.

7 NP $\rightarrow$ Det + N + (S)

This one simple change has very significant repercussions. It adds a property to the base that linguists call *recursion,* because it allows the phrase structure rules to apply recurrently (the S rule introduces an NP, which can now introduce an S, which introduces an NP, which . . .). But the most important effect of the change for our purposes is that it gets rid of generalized transformations altogether. Look at how smoothly a rule like 7 works and how smoothly it interacts with transformations. Rule 7 generates strings like 6e (containing the noun phrase "the cat the cat chased the dog," which includes the sentence, "the cat chased the dog") and 6f.

6 e the cat the cat ate the Kibbles chased the dog
   f the cat the cat chased the dog ate the Kibbles

These two strings are perhaps a little more bizarre than the ones resulting from Katz and Postal's work, but they involve no bigger step in abstraction than adding an IMP or a Q, and Chomsky's proposal gets rid of an entire class of transformations. Rather than a generalized transformation to splice 6a and 6b together, Chomsky only needs a run-of-the-mill singularly transformation to introduce a pronoun: 6e underlies 6c, and 6f underlies 6d, and the relativization transformation just substitutes *that* for the second occurrences of *the cat.*

Base-recursion was a grand and brilliant and typically Chomskyan stroke. Not only were generalized transformations gone, leaving behind a tidier box of transformations, all the discarded transformations were troublesome meaning changers. Chomsky's move also, for good measure, eliminated an entire class of semantic interpretation rules that Katz and Fodor had matched to generalized transformations (a proposal that made some embarrassment—1964b [1963]:514–15), leaving behind a tidier box of semantic rules. Things were getting better all the time, and the improvements warranted a terminological change. *Kernel* was out. *Deep structure* was in.

"This," Chomsky said of deep structure, "is the basic idea that has motivated the theory of transformational grammar since its inception" (1965 [1964]:136). And the *Aspects* model finally got it right. Where Harris's kernel had neither the structure nor the transformational resilience to serve as the semantic control he envisioned Chomsky's (Katz-and-Fodor-and-Postal-abetted), deep structure had both. It brought a great deal of information with it from the phrase structure rules, and it determined all of the potentially meaning-altering transformations.[17]

There were flies stirring in the semantic ointment, and they would soon come out to buzz. Briefly, though, deep structure was the triumphant product of a highly accelerated evolution, and *Aspects of the Theory of Syntax* was the brilliant synthesis of a hugely productive period in linguistic history.

The Model

> Thus the syntactic component consists of a base that generates deep structures and a transformational part that maps them into surface structures. The deep structure of a sentence is submitted to the semantic component for semantic interpretation, and its surface structure enters the phonological component and undergoes phonetic interpretation. The final effect of a grammar, then, is to relate a semantic interpretation to a phonetic representation—that is, to state how a sentence is interpreted.
>
> Noam Chomsky

The *Aspects* model of transformational grammar was a hit, both in the burgeoning Chomskyan community and in the academic world at large. We have already discussed most of its general features, or at least broached them, but two more technical innovations require a little space: Δ-nodes ("delta-nodes") and the separate lexicon. The first development is relatively small potatoes in terms of the *Aspects* model itself, but it has implications for later work by Chomsky that figures heavily in our story. The second development figures even more heavily, since the generative-interpretive brouhaha was in great measure about the complexion of the lexicon; it also entails the discussion of three further modifications, feature notation, lexical insertion rules, and complex symbols.

Chomsky proposes (adapting a suggestion by Katz and Postal) the Δ as a "dummy symbol" (1965 [1964]:122), essentially a lexical place-holder inserted by the phrase structure rules. The critical feature of Δ-nodes for later discussion is that they are completely empty—phonologically and semantically null. The strings in 8 illustrate their principal function in the *Aspects* model: where earlier theory would derive 8c from 8a, *Aspects* derived it from 8b.

8 a   someone flipped the flapjack
   b   Δ flipped the flapjack
   c   The flapjack was flipped.

In 8b, Δ holds the subject slot open so that *the flapjack* can be moved in and substituted for it. This may look arbitrary, as many innovations do in isolation, and it was fairly minor in the *Aspects* scheme of things, but it had important consequences. One of them, though it seems not to have been discussed at the time, is semantic (McCawley, 1988.1:81–82): using Δ for the subject rather than *someone* eliminates the unpleasant semantic side effects of the latter. Sentences 9a and 9b, for instance, were transformationally related in the pre-*Aspects* world, but they clearly mean something different (9a implying that *someone* ≠ *Chomsky*).

9 a   Someone wrote Chomsky's *Aspects* in 1965.
   b   Chomsky's *Aspects* was written in 1965.

With a semantically null Δ, that problem is eliminated. More importantly, Δ-nodes were the thin edge of the wedge for a lot of grammaticizing over the next several decades. In concert with trigger morphemes and other abstractions, Robin Lakoff says, Δ-nodes "functioned as an Open Sesame" to generative semantics (1989:948), and they opened a completely different magical cave for Chomsky in the late seventies, who introduced a whole bestiary of similarly empty categories.

In the immediate grammatical remodeling of *Aspects,* though, a far more substantial and far-reaching innovation was the creation of an independent lexicon, a dictionary listing all the words and their attributes. This project involved a great deal of very detailed work, but only three developments call for attention here: feature notation, complex symbols, and lexical insertion rules.

The *Syntactic Structures* model introduces words into the derivation by way of phrase structure rules. One rule determines that a single noun can constitute a noun phrase NP → N), another rule determines that *flapjack* is a noun (N → {*flapjack, . . .* }); there is no formal distinction between a syntactic category and a lexical item. But this procedure entailed an unwieldy proliferation of phrase structure rules. For instance, *flapjack* is not the same type of noun as *Indira.* In traditional terms, the difference is between a count noun (something discrete, of which you can have more than one) and a proper noun (which is in a sense a property of the thing it names), but these differences have distributional reflexes; *flapjack* can occur with determiners, adjectives, and prepositional phrases *(the fat flapjack on the griddle),* for instance, which *Indira,* except in markedly odd circumstances, cannot (**the grinning Indira with the spatula*). And *milk* is another noun type yet (traditionally, a mass noun), one which identifies nondiscrete entities; in distributional terms, like *Indira,* unlike *flapjack,* it cannot be pluralized (**milks*), but like *flapjack,* unlike *Indira,* it can occur with determiners and adjectives *(the cold milk).* All of this means that a *Syntactic Structures* grammar is forced to subcategorize nouns along the lines of 10.

10   a $N_c \rightarrow$ {*flapjack, griddle*, . . . }
     b $N_p \rightarrow$ {*Indira, Noam*, . . . }
     c $N_m \rightarrow$ {*snow, milk*, . . . }
(where c = *count;* p = *proper;* m = *mass.*)

It gets more complicated. Some nouns are categorically unstable; *steak,* for example, can be a count noun *(Tom is eating two steaks),* but it can also be a mass noun *(Tom likes steak).* So, the grammar must register such nouns multiply, so that *steak* occurs in both rule 10a and 10c.

A separate lexicon, though, gets rid of this redundancy (more properly, controls it). Rather than a separate rule listing all the various flavors of nouns (and verbs, and adjectives, and adverbs, . . .), *flapjack* is listed in the lexicon, by way of feature notation, as [+count], *Indira* as [+proper], *milk* as [+mass], *steak* as [+count, +mass], and so on.

Bringing in feature notation also allowed for some improvements to the lexical insertion rules. In *Syntactic Structures,* they are just phrase structure rules. In *Aspects,* they are transformations. They are also far more expressive, coming in two general types: strict subcategorization rules and selectional rules. The first set of rules ensured the appropriate distribution of lexical types; for instance, that a count noun followed a determiner, but a proper noun did not (11a but not 11b); or that an intransitive verb occurred on its own, but not with an object (12a but not 12b).

11   a   The cook flipped the flapjack.
     b   *The Indira flipped the flapjack.
12   a   Noam disappeared.
     b   *Noam disappeared the kernel.

The second set of rules ensured that more constraints with a closer connection to meaning were not violated; for instance, that verbs of cognition appeared only with things that could be cognized (13a rather than 13b), or that color adjectives co-occurred only with concrete nouns (14a rather than 14b).

13   a   Avashinee believed her mother.
     b   *Avashinee believed her cantaloupe.
14   a   Naveen's eyes are brown.
     b   *Naveen's theories are brown.

Both rule types were regarded as syntactic, even though selectional restrictions reflected the content of the words a little more directly (in traditional terms, 11b and 12b are ungrammatical, a syntactic notion, 13b and 14b are anomalous, a semantic notion). Both rule types were enforced primarily through something Chomsky called *complex symbols* (and the vast majority of the lexical innovations were Chomsky's). The phrase structure rules (following a few modifications) now generated trees like PM–1; in which the branches terminate in clusters of features. These clusters are complex symbols.

PM–1

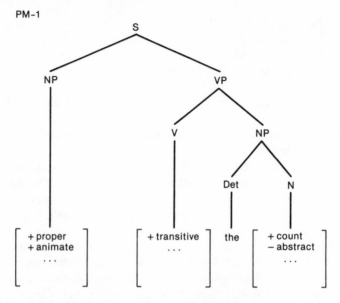

The lexical insertion rules (that is, both the strict subcategorization and selectional rules) then replace these clusters with any entries from the lexicon compatible with the features they contain. *Indira,* which is [+proper, +animate, . . .], can substitute for the complex symbol under the first NP; *flip,* for the verb complex symbol; and *flapjack,* for the count noun. Or, respectively, *Floyd, break,* and *glass* can fill the same slots. After these rules have fired, and the complex symbols have been replaced, the result is a tree like PM–2, the deep structure.

PM–2

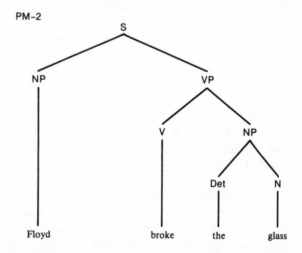

The deep structure, that is, was the product of phrase structure rules and lexical insertion rules. It preceded all transformations. And it represented the semantic core of the sentence. Deep structure, in sum, was the glowing star of a model of language (or, more properly, a model of linguistic knowledge), which had mutated quite radically between *Syntactic Structures* and *Aspects*.

The earlier model's two rule types, phrase structure rules and transformations, had swapped some of their duties, the former getting richer and more powerful, the latter getting poorer and more restricted. Phrase structure rules still defined the basic phrase types, but now they also helped define some additional sentence types, by way of trigger morphemes. This innovation included the new job of policing the transformations. Question formation could now only apply if the S rule supplied a Q to trigger it; Imperative-formation needed an IMP, Negative-formation a NEG. More radically, phrase structure rules were overhauled to apply recursively, making one whole class of transformations (generalized transformations) obsolete.

New components and new rule types also showed up, some taking over jobs previously done by either phrase structure rules or transformations, some taking up completely new tasks. There was a lexicon, which catalogued the words. The lexicon included some new devices, like subcategorization rules and the building blocks for complex symbols. Lexical insertion rules showed up, to take over a job (in a transformational fashion) that was previously handled by phrase structure rules. And there was a semantic component, along with its semantic interpretation rules.

All of this remodeling—remaking a distributional grammar into a mediational grammar—produced the conception of linguistic knowledge hinging on deep structure, represented in figure 4.1.[18]

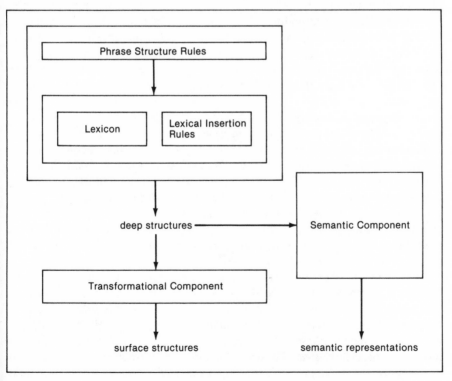

Figure 4.1. The *Aspects* model.

This grammar is certainly more intimidating than the simple—in retrospect, almost quaint—model of *Syntactic Structures* (Figure 3.2), but it is also more complete. The phrase structure rules generate trees which terminate in complex symbols. The lexicon and the lexical insertion transformations work in concert to plug a legitimate set of words into the tree (substituting for the complex symbols). These three components (constituting the base) generate a deep structure, which feeds into the semantic component to produce a representation of its meaning, and into the transformational component to produce a representation of its surface syntactic structure.

And one more, absolutely crucial, point about the model: it rested very heavily on the Katz-Postal principle that transformations don't change meaning. If transformations could change meaning—if the deep structure and the surface structure could mean different things—then the *Aspects* model is seriously undermined.

If some elements of the diagram are still a little vague, don't worry. They were vague at the time, too. Some of the arrows and boxes in Figure 4.1 were little more than that, arrows and boxes, with no clear specification of what they meant. No one bothered to specify what a semantic interpretation rule was, for instance, and no one said what a semantic representation looked like. The whole right side of the diagram was shrouded in obscurity, but it was promising obscurity, and the left side was much better explored. Phrase structure rules were pretty well investigated by this point, and so were transformations. The lexicon and the lexical insertion rules were immediately subjects of serious attention. Deep and surface structures bloomed profusely in all the papers of the period. Productive, innovative, challenging work was afoot. And no one was too concerned about the lack of detail concerning the semantic component. It would come, and, in any case, Katz and Fodor had defined semantics as an essentially residual matter. Their equation was "linguistic description minus phonology and syntax equals semantics" (1964b [1963]:482), what was left over when the sounds and the syntax were taken care of, and syntax wasn't finished yet (nor was phonology, but that's another story).[19] Syntax, in fact, was expanding, or, in the terms of the period, it was getting deeper: becoming increasingly abstract; welcoming arcane theory-internal devices like triggers and empty categories and complex symbols; and embracing the content of sentences much more directly, with feature notation and selectional restrictions and, of course, deep structure. There was a lot of semantic action in the *Aspects* model, that is, but most of it was happening on the syntactic side of the diagram.

Much of it was happening around the grand mediator, deep structure. To get a closer look at how deep structure worked, we will have to stop simplifying the trees a bit, though we can stick to very straightforward examples. Look over the two phrase structure trees, PM-3 and PM-4.[20]

The two phrase markers are, respectively, the deep structure and the surface of the sentence represented orthographically as *The man has been fired.* The surface structure captures all the important distributional facts of the syntax—the sequence of the words and affixes (it is one of the peculiarities of English that the past participle of *fire,* represented in PM-4 by the sequence *fire* + EN, is written and pronounced just like the past tense, *fired*). The deep structure captures more abstract facts about the sentence—that *the man* isn't the "logical subject" of the sentence, since he is on the receiving end of the action of the verb, getting fired; that someone else, some-

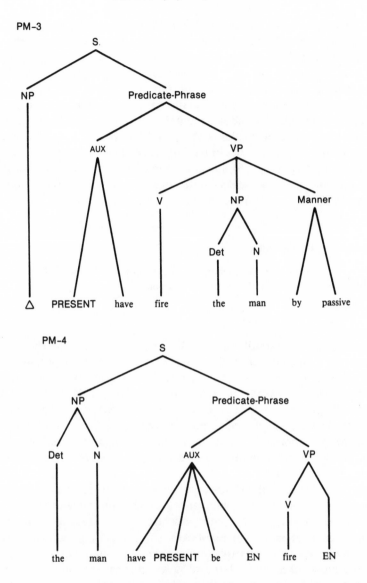

PM-3

PM-4

one unspecified, is doing the firing; that the sentence is a passive; that the verb is really transitive, even though it isn't followed by an object noun phrase at the surface.

Deep structure received by far the most attention of the two structures at the time, inside and outside of linguistics, but it is impossible without surface structure: it is the pairing that made the *Aspects* model so appealing. Eric Wanner, for instance, recalls that "in the mid-1960s, deep structure was an exciting idea," but when he explains that excitement, it hinges on both structures. "[Deep structure] provided a formal justification for the intuitively appealing notion that sentences have an underlying logical form that differs from the surface arrangement of words and phrases" (1988:147).

It was easy to lose sight, and many people did, of the fact that neither structure has any privileged position in the theory. The charismatic power of the term, *deep structure;* and the fact that deep syntactic relations were virtually indistinguishable from semantic relations; and the constant longing in and around linguistics for a direct principled connection between meaning and sound; and the vagueness engulfing the only potential competitor in the *Aspects* model for questions of meaning, semantic representation; and statements from Chomsky like "one major function of the transformational rules is to convert an abstract deep structure that expresses the content of a sentence into a fairly concrete surface structure that indicates its form" (1965 [1964]:136); and much more, led most people to think of deep structure on a much grander scale—as the long-awaited grammarian's stone that linked the drossy material of speech to the golden realm of thought.

Competence and Performance

We cannot but conclude that linguistic form may and should be studied as types of patterning, apart from the associated functions.

Edward Sapir

The *Aspects* model is a beautiful model. But what is it a model *of?* The answer is "competence," which takes us back to one of Saussure's most important distinctions in parceling out the jungle of language into idealized, and therefore more manageable, plots—between *langue* and *parole*. We roughly identified these notions two chapters back with, respectively, language and speech, or grammar and behavior. *Langue* is the stable and abstract pattern of a language, *parole* is its manifestation. More importantly for doing the job of linguistics, the difference is between "what is essential [and] what is accessory and more or less accidental" (Saussure, 1966 [1916]:14).

For the Bloomfieldians, this distinction corresponded more or less to the difference between a corpus (a body of observations, a collection of linguistic specimens) and its grammar (a body of rules and representations describing those specimens, an abstract account of patterns manifest in the corpus). The corpus was not *parole,* of course, but it was a representative sample of *parole;* the grammar was not *langue,* but it was a model of *langue.* (Harris, always a little extreme, defined *langue* as "merely the scientific arrangement of [*parole*]"—1941:345.)[21] Heavily influenced by positivism, they stuck close very close to the data, and venerated corpora. If they wanted to investigate English, they read English texts and they listened to English speakers. Fries's two major books on English, for instance, are based on a collection of letters to the government (1940) and a collection of recorded telephone conversations (1952).

Chomsky had absolutely no reservations about straying well beyond the corpus. Illustrating a syntactic relation, for instance, he would simply pluck some examples (an active and a passive sentence, say) from his head. Asked, by a mildly outraged Bloomfieldian, to defend such choices as representative chunks of English, he professed bafflement at the question:

CHOMSKY: The verb *perform* cannot be used with mass-word objects: one can perform *a task,* but one cannot perform *labor.*

HATCHER: How do you know, if you don't use a corpus and have not studied the verb *perform?*

CHOMSKY: How do I know? Because I am a native speaker of the English language. (Hill, 1962c [1958]:29)

In this case, unfortunately, Chomsky was a little too rash in his *perform* generalization, but he is only mildly chastened when Anna Granville Hatcher comes up with an incisive counter-example, *perform magic* (Hill, 1962c [1958]:31), and for good reason. Though Chomsky doesn't mention it, Hatcher's counter-example also comes from a head, hers, not from a corpus. The two of them, Chomsky and Hatcher, because they are native speakers of English, could presumably come up with a more resilient generalization in fairly short order. The notion of going beyond the immediate, solicited data was not completely novel to Bloomfieldians. Hockett, in particular, had required of grammars that they go beyond the corpus to predict the structure of nonsampled utterances (1954:34). But Chomsky's attendant indifference to corpora was virtually scandalous for them, and his heavy reliance on intuition was a powerful shift in the methodological winds.

The shift toward intuition, away from corpora, had the liberating effect of opening up vast worlds of cheap data. But it also changed the nature of much of that data. In particular, the emphasis on intuition significantly raised the status (and availability) of negative data, of non-sentences.

Chomskyan linguistics hinges tightly on the notion of grammaticality, on identifying which sequences of noises are grammatical in a language and which are not. But it is very difficult to elicit such judgments reliably from an informant. Take the following siSwati sentence (siSwati is an African language spoken in Swaziland):

15 a *Umfati uteka indvodza.

SiSwati speakers unhesitatingly reject 15a, which translates quite literally as "The woman takes a husband," as a bad sentence. On the other hand, they find 15b ("The man takes a wife") and 15c ("The woman is taken by the man") both perfectly fine.

15 b Indvodza iteku umfati.
    c Umfati utsatfwa yindvodza.

So the question becomes, do siSwati speakers reject 15a because it is ungrammatical (like "Man the a takes wife"), or because it is semantically anomalous (like "The circle is square") or pragmatically infelicitous (like "The chimp is a cardiac specialist")? Is it the syntactic system of siSwati, its semantic logic, or social factors that make siSwati speakers reject 15a? Where is *langue* and where is *parole?* It is extremely difficult to tell. A linguist looking for such data is much safer looking to her own native language.

There are two issues on the table in the Hatcher-Chomsky exchange. The first is generativity. The grammar must, Chomsky says in *Syntactic Structures,* "*project* the finite and somewhat accidental corpus of observed utterances to a set (presumably infinite) of grammatical utterances" (1957a:15; Chomsky's italics). That is, the grammar determines the corpus. The second issue, more important with respect to *langue* and *parole,* is that Chomsky wants access to the data in his own head, which strictly positivist methods would prohibit.

His interest in language, remember, is profoundly psychological. The Bloom-fieldians had virtually no interest in what their grammars were models of, beyond some abstract patterns exhibited by corpora—indeed, words like *model* and *representation* and, for that matter, *langue,* were virtually nonexistent in their work. For Saussure, the job of linguistics was unequivocally to study *langue,* but, as you may have felt when we looked at the distinction earlier, the notion was left half-baked in his work, like the bit of potato Scrooge thought responsible for his night-mare. In particular, Saussure's interest in Durkheim's sociology led him to leave *langue* hanging in some social Realm of the Forms. The system underlying use for him is "out there" in society somewhere, in a vague collective unconscious of the sort that was regarded as explanatory in the early part of this century. Chomsky is concerned with something "in here," in our individual heads, in a cognitive system of the sort that is regarded as explanatory in the late part of this century.

These differences were significant enough that Chomsky coined his own terms—which, finally, is where *competence* comes in, along with its perennial traveling companion, *performance.* Competence is a refraction of the concept behind Saus-sure's *langue;* performance, a refraction of the concept behind *parole.* We will see a good deal of these terms before we're through, but for the moment the important points about them are that they identify clear cases of, respectively, knowledge and use, and that they are unambiguously psychological.

Competence refers to the familiar case of knowing the correct phonological shape of the words *two* and *martinis; performance,* to the articulation those shapes get in speech—*tee* and *martoonies* in an episode of "I Love Lucy." More generally, competence is the hard-core knowledge someone has of her language—that subjects and verbs must agree, that adjectives qualify nouns, that *easy* takes one type of complement, *eager* another, and so on. It is relatively stable after childhood acquisition, amenable to formalization, and, for Chomsky, the single proper focus of linguistics. Saussure's general, suprapersonal notion, *langue,* became Chomsky's super-personal competence.

Performance is the application of that knowledge in speech. It is relatively vari-able—subject to fatigue, exhilaration, context—more difficult to formalize in a meaningful way, and, for Chomsky, of decidedly lesser interest.

The differences between competence and performance can be subtle, as are most differences between social and psychological accounts of the same phenomena, but they are real all the same, and, even beyond the altered-states sort of example, there are a wealth of clear cases. Sticking first with the artificial instances of television, think of the situations where people exposed to some trauma lose the ability to talk and regain it later, just in time for an important dénouement. Their knowledge of English hasn't evaporated. It doesn't disappear into the ozone, and come back. The characters just have trouble getting at it for some reason. There is no loss of com-petence; rather, their performance is interrupted. Real cases are seldom so clean, but analogous things do happen, and they fall in the realm of neuropsychology.

Linguists (at least those of the Chomskyan persuasion) have virtually nothing to say about them.

Or consider the different ways people speak when they address various listeners, the different registers they use in different contexts: one type of greeting for the queen, another for a spouse, another for mates in the pub. Whether someone says "Good evening, Your Highness," "Hello, love," or "What's up, boyo?" is a matter of register, a performance phenomenon. Whether those utterances are grammatical or not, are questions or statements, are meaningful or not—those are matters of abstract knowledge, competence phenomena. Register concerns sociology (more specifically, sociolinguistics); grammaticalness, sentence type, and reference concern linguistics. Or, take a sentence, like 16a.

16   a   The horse raced past the barn fell.

Such monstrosities are called *garden-path sentences,* because they tend to lead people in the wrong processing direction. People see *raced* as taking *the horse* for its subject, a reading that makes perfect sense until *fell* shows up at the end and throws that interpretaion out of whack. Some people are so severely discombobulated by that reading that they never recover, and write the sentence off as gibberish. But another reading makes perfect sense; namely, taking *raced* as having an implied subject, like the unambiguous *ridden* in 16b.

16   b   The horse ridden past the barn fell.

The fact that people wander down a garden path trying to make sense of 16a has to do with processing mechanisms (presumably, an unwillingness to explore other syntactic potentials for *raced* once it can be linked up with *the horse* as its subject). Possibly it involves short-term memory in some way, but, in any case, it concerns something besides linguistic knowledge, and therefore belongs to performance—in this case, a fit subject for psychologists or psycholinguists, not for center-court Chomskyans. Insofar as 16a concerns competence, it is indistinguishable from 16b.

In short, it is rather easier to demonstrate what a Chomskyan grammar does *not* model than what it does, and this issue became important in the sixties because of a confusion that had dogged transformational grammar since *Syntactic Structures* (and which, in fact, still persists in some quarters). In the baldest formulation of this confusion, Carleton Hodge characterized transformational grammar as organized around the question, "How does one proceed, not to describe a sentence, but to make one?" (in Hill, 1969:38). That is, many observers thought Chomsky was marketing a production model, a representation of how speakers produce sentences.

The confusion was very natural, in large part because of the unclarity at the time about what it means to model the mind. Chomsky proposed his grammar to model a mental *state,* but many people took it to model a mental *process.* His emphasis on creativity, which has inescapably dynamic connotations, didn't help matters. For Chomsky, creativity is epitomized by a language user's ability to produce and understand novel sentences, to speak them and to figure them out when others speak them, to write them and to read them. But muddled observers saw a generative grammar as something "that explains how we produce [sentences]," illustrating how "Professor Chomsky . . . is thus interested in the creative rather than the interpretive side of grammar" (Francis, 1963:320, 321).

Even more problematically, Chomsky's model *was* a process model on another level—not that it represented a process, but that it represented a state, the condition of knowing a language, in process terms, replete with all the production-line vocabulary of *input* and *output* and *stages*. No single word is more troublesome here than *generative;* Hodge again exemplifies the confusion by using *sentence generator* (Hill, 1969:39) as a synonym for *transformational grammar*, but even Einar Haugen couldn't make heads or tails out of the difference between *generate* and *produce* (Dallaire and others, 1962:41–43). The difference is a simple one. *Generate* identifies an abstract notion, like *delineate, define*, and *enumerate*. *Produce* identifies a concrete notion, like *make, build*, and *assemble*. But the distinction was barely addressed, let alone clearly defined, before *Aspects*, and even members of the inner circle made some pretty glaring errors on this front.[22]

A production model, then—what a generative grammar *isn't*—represents how speakers formulate and spit out utterances. A generative grammar, a competence model, represents the knowledge inside a language user's head, not the techniques, strategies, and synaptic processes for getting pieces of it out of her mouth. The set {1, 2, 3, . . . } represents some mathematical knowledge that can be put to use in certain situations, but not the actual process of counting a herd of goats.

With this point sufficiently belabored, it should be clear that a good portion of transformational grammar's success outside of linguistics, and some of it within, was based on a few fundamental misunderstandings, misunderstandings that were to escalate over the next few years.

And one more thing about competence and performance. They were very convenient. Recall the Cormorant-Island sentences yet again: Katz and Postal's solution to the change-of-meaning problem they posed was essentially to claim that the conflicting readings (the same-two and different-two interpretations) were the products of performance factors; since a Chomskyan model has no direct responsibility for performance, those conflicting readings are simply disposed of. They are no longer part of the data.

Small price to pay, the Chomskyans thought. They had, in a few short and feverish years, hammered out an elegant framework which accomplished the ultimate goal of all linguistic work from at least the time of the Stoics. They had formally linked sound and meaning. The fulcrum of this beautiful theory was the underlying syntactic representation, the evocatively named deep structure. It was the direct output of the base component, a beefed-up descendant of *Syntactic Structures'* phrase structure rules. It was the direct input to the transformations, the early theory's titular technical device. Most crucially for all concerned, with the Katz-Postal principle in place, it was the locus of meaning.

Yes, the *Aspects* model was elegant. It linked sound and meaning. It represented knowledge of language. It housed the beautiful deep structure. But it leaked.

Generative Semantics 1:
The Model

Simple imitation is not enough; one should build upon the model.

Quintilian

Disciples are usually eager to improve on the master, and . . . the leader of a movement sometimes discovers he cannot or does not wish to go quite as fast to the Promised Land as those around him.

Gerald Holton

Trouble in Paradise

The Chomskyan universe was unfolding as it should in the middle of that optimistic and captious decade, the 1960s. The Bloomfieldians were driven to the margins. There was a cadre of feisty, clever, dedicated linguists and philosophers working on generative grammar. Young linguists everywhere were clamoring for their thoughts. The graduate program was up and running, generating an impressive, in-demand, soon-to-be-influential string of philosophical doctors. MIT was a stronghold of truth and wisdom in language studies, Chomsky was the uniformly acknowledged intellectual leader, *Aspects* was the new scripture.

The central, defining concern of the work codified, extended, and enriched in that scripture was to get beneath the literal surface of language and explore its subterranean logical regularities, to find its deep structures, to get at meaning. "In general," Chomsky had told the troops, "as syntactic description becomes deeper, what appear to be semantic questions fall increasingly within its scope; and it is not entirely obvious whether or where one can draw a natural bound between grammar and 'logical grammar'" (1964d [1963]:51). *Aspects* not only endorses the Katz-Postal principle which enacts this concern, and the principle's attendant innovations, like semantic interpretation rules and trigger morphemes and reevaluations of problematic data, *Aspects* strengthens it, by adding base-recursion to the model, discarding an entire class of semantic interpretation rules in the bargain, and consequently inventing deep structure.

Everybody was happy with this work, but no one was content (with the possible exception of Katz). Postal, the hands-down-sharpest syntactician, expressed his unwillingness to stay put by digging more deeply under the surface, working in a direction which rapidly came to be known as *abstract syntax,* and several other linguists joined this project—most notably George Lakoff, Háj Ross, and James McCawley. Together they pushed syntax deeper and deeper until, to the extent that semantics had substance at the time, their deep structures became virtually indistinguishable from semantic representation; indeed, their deep structures were the closest things to explicit semantic representations in mid-sixties Cambridge. They took Chomsky at his word and made a grammar in which logical form had a central place. At this point, early in 1967, the program mutated into generative semantics, and Chomsky was displeased. He voiced his displeasure, and then, having let the semantic genie out of the bottle, spent the next several years trying to stuff it back in, against the agitated resistance of the generative semanticists.

But these are just the bones of the story. There is sinew and gristle, hair and hide, wet stuff and dry, yet to tell. We can start, as in all fleshy matters, with the progenitors.

Home-brewed Dissension

As of 1965, and even later, we find in the bowels of Building 20 [the home of the MIT linguistics department] a group of dedicated co-conspirators, united by missionary zeal and shared purpose. A year or two later, the garment is unraveling, and by the end of the decade the mood is total warfare. The field was always closed off against the outside: no serpent was introduced from outside of Eden to seduce or corrupt. Any dissension had to be home-brewed.

Robin Lakoff

Paul Postal, studying under Lounsbury at Yale in the late fifties, met Chomsky on one of his Bloch-sponsored visits. Postal converted almost overnight:

I was very impressed, first with the power of his thought, but also it seemed that this stuff was from a different planet, in the sense that it was based on an entirely different way of thinking from anything I had come into contact with before.

He hitched his corrosive polemical style to the new movement, and his classmate, Jay Keyser, remembers the note-passing beginnings of his campaign against the Bloomfieldians:

I remember sitting there, listening to Bernard Bloch lecture on morphemic postulates. Bloch would be saying something, and Paul would write me a note with a counterexample. I thought to myself "Jesus. Postal's right. This is just not working."

Eagerly escaping the Bloomfieldian confines of New Haven, Postal finished his dissertation in Cambridge as part of the Research Laboratory of Electronics and joined the MIT staff, where he helped shape the first MIT generation, and, of course, hammered out most of the properties of deep structure with Katz. He also began working on a program of increasing abstraction and semantic perspicacity that excited a number of linguists; most notably, George Lakoff.

George Lakoff first encountered generative grammar as an MIT undergraduate in 1961, taking classes from Chomsky and Halle. He found it all pretty dry and uninspiring. But when he went off to do graduate work in English at Indiana, he began to read the material on his own, found it a good deal more compelling, and embarked on some unorthodox work trying to transform the principles of Propp's *Morphology of the Folktale* into a generative story grammar. Returning to Cambridge in the summer of 1963 to marry Robin Tolmach, he met Ross and McCawley, and found a job on Victor Yngve's machine translation project. Katz and Postal were down the hall, working on *Integrated Theory,* and he spoke with them frequently. Through this regular participation in the MIT community, he became more directly interested in language and returned to Indiana to work on a doctorate in linguistics, under Householder. The following summer he attended the Linguistic Institute, at which Postal was teaching, and renewed his friendship with him. So, when Householder left for a sabbatical during Lakoff's dissertation year, he naturally headed back to Cambridge, where Postal directed his dissertation, and Háj Ross became his very close associate.

Háj Ross, son of a nuclear physicist, grandson of a Nobel Peace Prize laureate, did his undergraduate work at Yale, where Postal was chafing at the taxonomic bit, and Bloch was the reigning theorist. He studied under both, played varsity football, ran a jazz show on the campus radio station, graduated, and went off to MIT to enroll in its increasingly important linguistics program. He didn't. Halle found his work singularly unimpressive and suggested he go off somewhere and "prove himself." He did. In fact, he went off to Chomsky's old stomping grounds and completed a master's thesis under Zellig Harris at the University of Pennsylvania. He returned to MIT to enroll in its now-Camelotian, brink-of-the-*Aspects*-theory linguistics program. He did. One of the shining stars in a stellar class, he went on to produce a hugely influential dissertation. He also began collaborating closely with Lakoff, particularly on Postal's abstractionist genre of analyses, and gained the friendship of James McCawley.

James McCawley, in the estimation of his teacher, colleague, friend, and opponent, Noam Chomsky, is "one of the smartest people to have gone through the field." Lees places him among "the sanest and most astute linguists of our time." Ross and Lakoff go on at great length about his intelligence, sensitivity, humor, warmth, inventiveness, pedagogical gifts, musicianship, culinary talents, . . . He is "the generative grammarians' Shiva, the many-handed one of transformational theory" (Zwicky and others, 1970:vii), an important, diverse, challenging linguist. With a background in mathematics and a thick Glasgow accent, he entered MIT in 1962, distinguishing himself rapidly for both the clarity of his thought and the deftness of his wit. He was more intrigued by phonology than syntax and produced a brilliant dissertation on Japanese tone phenomena (1968b [1965]), earning the second doctorate awarded by the new linguistics department. But he soon found himself at the University of Chicago having to teach courses in syntax. He very quickly educated himself in the area by spending a great deal of time on the phone with two friends in Cambridge who were directing a syntactic study group at Harvard, Ross and Lakoff.

These four relatively abrupt, interpenetrating paragraphs—one per progenitor—mirror the parentage of generative semantics.[1] Each of these men has been credited with engendering the theory (though Ross only, in his words, "as George's side-kick"). Susumu Kuno, a very distinguished Harvard linguist who knew all the principals, working closely with several of them, says "I have no doubt that George was the original proponent and founder of the idea." Arnold Zwicky says that the theory was the joint issue of Lakoff and Ross. Newmeyer says that it was born out of work by Lakoff and Ross, under Postal's influence. Ross, for his part, says "it's basically Postal's idea. He was basically the architect. [George and I] were sort of lieutenants and tried to flesh out the theory." But Postal says that "McCawley first started talking about generative semantics," and it was McCawley's arguments which first got him interested in the movement. McCawley says that Lakoff and Ross "turned me . . . into a generative semanticist." Lakoff says that the idea was his, but that Postal talked him out of it for a period of about three years, when McCawley then convinced him it was correct, while he was working with Ross, on some of Postal's ideas.[2]

The moral is that most ideas don't have fathers or mothers so much as they have communities. Generative semantics coalesced in mid-sixties Cambridge—partially in concert with the *Aspects* theory; partially in reaction to it. Chomsky, who is decidedly uninterested in generative semantics' parentage says, simply, "it was in the air," and there is much to recommend this account.

As transformational analyses of syntax grew more probing and more comprehensive, they increasingly involved semantics. As early as Harris (1954), there was a promise that formal syntactic work could make inroads into the jungle of meaning. *Syntactic Structures* did some important preliminary surveying, and by the Ninth International Congress, Chomsky was arguing explicitly that the deeper syntax went, the closer it got to meaning. At some indistinct point, however—or, rather, at several different indistinct points, for several people—this program began to seem exactly wrong. It began to appear that syntax should not be coopting semantics so much as exploring its own semantic roots; that syntax should not be determining semantics, semantics should be determining syntax—that semantics should be generative rather than interpretive. At some indistinct point, there was generative semantics.

But "in the air" is far too vague. There were clear leaders. As Chomsky, quite suddenly, began to show less inclination for this type of deep syntactic work, and as generative semantics began to take shape against the backdrop of *Aspects,* it was obvious who these leaders were: George Lakoff, Ross, McCawley, and Postal. Others had influence, especially Robin Lakoff, Jeffrey Gruber, Charles Fillmore, and several MIT instructors—most notably, Edward Klima, Paul Kiparsky, and the big one, Noam Chomsky.

"I sort of believed [generative semantics] myself back in the early sixties," Chomsky has said, "and in fact more or less proposed it." And the popular arguments in *Cartesian Linguistics* and *Language and Mind* support the claim. In the former, for instance, he says that "the deep structure underlying the actual utterance . . . is purely mental [and] conveys the semantic content of the sentence. [It is] a simple reflection of the forms of thought" (Chomsky, 1966a:35)—as clear as articulation of generative semantics as ever there was. But there are large differences of scale between the four horsemen of the apocalypse—as Postal, Lakoff, Ross, and McCawley became known—and everyone else.

Chomsky, in particular, certainly never adopted generative semantics, and everywhere that he comes close to endorsing something that looks like generative semantics, there is a characteristic rider attached. In *Cartesian Linguistics,* he says the relation of deep structure to thought and meaning in fact is not so clear as the above quotation suggests, that he is just expressing what the Port-Royalists held; the real connections are a "further and open question" (1966a:100n8). The horsemen expressed no such reservations. Robin Lakoff, an important horsewoman who also expressed no such reservations, and who became one of the movement's most influential teachers, was neither the proselytizer nor the large-scale theorizer that the others were. And two important theorizers, Fillmore and Gruber, had reservations; Kiparsky and Klima had even more.

The kernel of generative semantics was an obliteration of the syntax-semantics boundary at the deepest level of grammar—the axiom that the true deep structure *was* the semantic representation, not a syntactic input to the semantic component. This obliteration, in real terms, began with Postal, though George Lakoff was the first to propose it.

Lakoff's Proposal

> The approach taken by Katz, Fodor, and Postal has been to view a semantic theory as being necessarily interpretive, rather than generative. The problem, as they see it, is to take given sentences of a language and find a device to tell what they mean. A generative approach to the problem might be to find a device that could generate meanings and could map those meanings onto syntactic structures.
>
> George Lakoff

The first step toward generative semantics was a paper by George Lakoff, very early in his graduate career, which, after some preliminary hole-poking in KatznFodorian interpretive semantics, says "There are several motivations for proposing a generative semantic theory" (1976a [1963]:50). Well, okay, this is not a step. This is a hop, a skip, and a jump. In three lunging moves, Lakoff wanted to replace Katz and Fodor's just-published semantic program, to bypass Katz and Postal's still-in-the-proofs integration of that program with transformational syntax, and to preempt Chomsky's just-a-glint-in-his-eye deep structure. Whew.

He took the paper, "Toward Generative Semantics," to Chomsky, who was, Lakoff recalls, "completely opposed" to his ideas, and sent him off to Thomas Bever and Peter Rosenbaum for some semantic tutelage. Chomsky remembers nothing about the paper except that "everybody was writing papers like that" in 1963—a remark that is, at best, difficult to substantiate.[3] Bever and Rosenbaum were at MITRE corporation, an air force research lab in Cambridge where many linguistics students went in the summer to spawn transformations. Bever and Rosenbaum didn't like it either, and Lakoff remembers a huge, three-hour argument. No one budged, though Ross, another MITRE summer employee, sat in as an onlooker and thereafter began his close, collaborative friendship with Lakoff. Lakoff ran off some copies and gave one to Ross, another to McCawley, another to Postal, and sent a few more off to people he thought might be interested. It was not a success. No one held much hope for his proposals, and no one adopted them. Lakoff does not give up easily, but he respected Postal immensely and took his advice. He abandoned the notion of generative semantics (or, perhaps more accurately, suppressed it), and went back to work in the interpretive framework of the emerging *Aspects* theory.

Stepping out of the chronology for a moment to consider etymology, it's not exactly clear what a *generative* semantics is, of course, at least in the technical sense. *Interpretive* semantics is clear enough. An interpretive model has some structures respresenting sentences (deep structures, surface structures, intermediate structures, it doesn't matter), and some way of turning those structures into semantic representations, some way of interpreting the syntax to extract elements of meaning. The *Aspects* model depends, crucially, on interpretive semantics.

But Chomsky's groundbreaking work in the fifties defined a quite sweeping notion in connection with the crucial term—"generative *grammar*"—a notion which embraced phonology, morphology, syntax, and (once Katz, Fodor, and Postal came along) semantics.[4]

It's true that there was occasional talk about a "generative syntax," and syntax was unquestionably the central component of the grammar, linked critically to the creativity of the grammar (Chomsky, 1965 [1964]:136). But *generative,* in its technical sense, meant only a grammar which "specifies the infinite set of well-formed sentences and assigns to each of these one or more structural descriptions" and two or more interpretations, phonological and semantic (Chomsky, 1964b [1962]:915). Certainly it was never meant as an antonym of *interpretive;* in fact, just to confuse the issue, the phonological investigations in transformational grammar were conducted under the rubric, *generative phonology,* and they concerned an explicitly interpretive component of the grammar.

But the word, *generative,* was very highly valued at the time, with charismatic authority to bestow on any noun it abutted, and it was in this sense that there was a generative phonology (which meant simply a phonology which adhered to generative principles and which therefore belonged in a generative grammar). Partially as a result of the glory the word was gaining, and partially as a result of Chomsky's loose and influential talk of describing creativity, *generative* smudged a good deal in informal contexts, commingling with *creative* and *productive* and other dynamic terms, taking on some of their motive senses—and this is clearly part of what Lakoff meant in 1963 by *generative semantics.* He uses it as an antonym to *interpretive semantics* (1976a [1963]:50). But he also retained part of the technical sense of the term, calling for "a device that could generate meanings and could map those meanings onto syntactic structures" (1976a [1963]:44); this, in opposition to the nascent *Aspects* model which generated syntactic forms and mapped those onto phonetic and semantic structures.

The most important word in the label, it should be clear then, isn't *generative,* but *semantics,* and the opposition is to *syntax.*[5] The Katz and Fodor and Postal work, for all its inroads into meaning, under Chomsky's heavy influence, clearly left syntax in the driver's seat. As Chomsky describes his own approach, much of his work is what everyone else calls semantic; "still, I want to include all that as syntax" (1982a:114). The only aspects of meaning he has ever wanted to tackle are those that can be subsumed under (or, in some cases, redefined as) syntax.

Not so Lakoff. He wanted to put meaning behind the wheel.

But, aside from proposing the label, *generative semantics,* and raising some of the issues that engulfed that label several years later, Lakoff's paper is, as Ross puts it, "only good for historical gourmets." Nor, even though a similar model a few years later would sweep through the field like a brushfire, is it really very hard to see why Lakoff's proposal fell so flat. The paper is very inventive in places, which everyone could surely see, but it also takes a tone, an arrogant confidence, that Chomskyans were used to seeing in one another's polemics against the Bloomfieldians, but not directed at their own internal proposals. The first half of the paper, remember, is directed against Katz and Fodor and Postal's recent innovations. In particular, the Katz-Postal principle requires that sentences with the same meaning have the same deep structure, but Lakoff casually adduces counter-examples like 1–3, where the a and b sentences all mean essentially the same thing, but the deep structures are quite different.

1   a   I like the book.
    b   The book pleases me.

2   a   I made that clay into a statue.
    b   I made a statue out of that clay.

3   a   Yastrzemski hit the ball to left field for a single.
    b   Yastrzemski singled to left field.

These are problems for the *Aspects* theory, all right, but hardly incapacitating ones; there are fairly clear ways around them, none of which Lakoff explores. He says KatznFodorian semantics run into trouble here, and blithely proposes tossing out the whole approach. And the second section of the paper, the one in which he offers his replacement, has an even larger quotient of certainty, about matters which are obscure in the extreme. Lakoff talks of semantic laws, for instance, and rules for embedding thoughts into other thoughts, and even formalizes a few thoughts. Take this rule as an example, the rule which introduces subjects, predicates, and objects—presumably (though this is not specified) replacing Chomsky's iconic S → NP + VP phrase structure rule.

$$4 \quad T \rightarrow [\text{s. pred.}] \left(\text{s. subj.}, \left\{ \begin{array}{c} \text{s. obj.} \\ \varnothing \end{array} \right\} \right)$$

The *T* stands for thought, so the rule says that every thought must have a semantic predicate and a semantic subject, but need not have a semantic object. Oh, and what does a thought look like? As it turns out, a great deal like the bundles of features hanging down from the bottoms of deep structures; 5 is the thought rendered into English as "Spinach became popular";

$$
5 \quad
\begin{bmatrix}
\text{s. pred.} \\
-\text{DS} \\
+\text{change} \\
-\text{space} \\
+\text{directed} \\
+\text{direction} \\
+\text{dummy}
\end{bmatrix}
\left(
\begin{bmatrix}
\text{s. obj.} \\
-\text{p.o.} \\
+\text{state} \\
+\text{quality} \\
-\text{dummy} \\
| \\
spinach
\end{bmatrix}
\begin{bmatrix}
\text{s. subj.} \\
+\text{p.o.} \\
+\text{state} \\
+\text{quality} \\
-\text{dummy} \\
| \\
popularized
\end{bmatrix}
\right)
$$

The content of 5 is not especially important here, just that Lakoff—brainy, full of chutzpah, and taking the cognitive claims of generative modeling far more seriously than anyone else at the time—is talking confidently about thoughts, plopping representations of them down on the page, when Katz and Fodor have just started to explore what meanings might look like in Chomskyan terms, Katz and Postal just starting to etch out how they might fit into a transformational grammar, Chomsky just putting the final recursive touches on deep structure.

The final section of Lakoff's highly speculative paper is "Some Loose Ends," implying that the hard work at the grammar loom was over and a few stray threads need only be woven back into the rug of Chomskyan—or, with these advances, perhaps Lakovian—theory. The section reveals much about Lakoff, particularly the tendency to view his own work in the most grandiose and comprehensive terms, to look constantly at the big picture, but no one else was similarly moved by his case. Chomsky saw little in it, Bever and Rosenbaum ditto, Ross and McCawley quickly forgot about it, and Postal suggested Lakoff curb his speculations and try to help revise the incipient *Aspects* model, rather than drop it altogether. He did just that, shortly setting to work on his dissertation, *Irregularity in Syntax* (1970a [1965]), which not only revises the *Aspects* model, but stretches it about as far as it could go without breaking.[6]

Abstract Syntax

> As we proceed ever further with our investigations, we come of necessity to
> primitive words which can no longer be defined and to principles so clear that
> it is no longer possible to find others more clear for their definition.
>
> Blaise Pascal

George Lakoff's dissertation was "an exploration into Postal's conception of grammar" (Lakoff, 1970a[1965]:xii), and the published title reads like a diagnosis of the problems Postal saw in transformational modeling. There was too much irregularity, a diagnosis that led Postal to embark on the line of research soon known, rather loosely, as *abstract syntax*. The fact that it had its own label, and that Lakoff talks of "Postal's conception of grammar," indicates it was perceived as a separate stream, at least in some measure, but no one really felt the work to be at odds with all the other feverish research surrounding *Aspects*. Well, almost no one. The main focus of Postal's work was to reduce complexity in the base down to an axiomatic minimum of primitive categories, which entailed some moves Chomsky evidently found uncongenial (Newmeyer, 1980a:93; 1986a:82). Ross and Lakoff, on the other hand, found the work extremely congenial and began augmenting and elaborating Postal's proposals.

But again there are difficulties of demarcation; in particular, it is not at all obvious when this work became uncongenial to Chomsky. Certainly the division was one of degree, organized around a few technical proposals, rather than one of kind. There was no sharp change in the argumentation or the direciton of Postal's research. The trend in Chomskyan work from the outset was toward increasingly higher abstraction, a trend that gathered considerable momentum in the early sixties with the introduction of trigger morphemes and Δ-nodes.[7]

Postal attacked a growing problem in the early theory, which, operating under an impoverished notion of what became known as *deep regularities,* witnessed an unconstrained mushrooming of categories. Much work in early transformational research simply projected the wide variety of surface categories onto the underlying representation, and even work that winnowed off some of those categories still had an alarming number of them; Lees's exemplary *Grammar of English Nominalizations,* for instance, had dozens of underlying categories (e.g., 1968 [1960]:22–23), though it dealt with only one small corner of English syntax and used the power of transformations to reduce the surface categories, leading Schachter to complain, representatively, that the trend "was staggering to contemplate; it seems likely in fact, that each word would ultimately occupy a subcategory of its own" (1962:137). This consequence was less than appealing to an approach that prided itself on simplicity and generality.

Postal became committed to the radical reduction of these categories. He argued in classes, papers, colloquia, and at the 1964 Linguistic Institute, that adjectives are really deep structure verbs (Lakoff, 1970a [1965]:115ff); that noun phrases and prepositional phrases derive from the same underlying category (Ross, 1986 [1967]:124n9); and that pronouns weren't "real," that they were a figment of superficial grammatical processes (Postal, 1966a). All of this is good Chomskyan practice, and the titular Chomskyan was happy to use such arguments himself. Just before his as-syntactic-description-becomes-deeper remark at the International Congress, in fact, Chomsky offered this exemplary argument for adjectives like *astonishing* and *intriguing* which arose from deep structure verbs:

> The structural description of the sentence "it was an intriguing plan," as provided by a transformational grammar, will contain the terminal string underlying "the plan intrigued one (i.e., unspecified human)" just exactly as it contains the past tense morpheme; and this fact might be suggested as the explanation for the cited semantic feature [that they are connected with a specific human reaction]." (1964d [1963]:51)

As Postal began to pursue this course, though, and as Ross and Lakoff joined him so thoroughly and enthusiastically that it became something of a program, Chomsky lost his affection for it.

Ross's most effective work along these reductionist lines was to explore arguments supporting an unpublished proposal by Postal to eliminate the categorical distinction between auxiliary verbs and main verbs (more accurately, he softened the distinction on principled grounds by translating it into feature notation—Ross, 1969b [1967]). Auxiliaries, he argued, were just a species of verb, as traditional grammar usually held them to be, and therefore didn't need their own deep structure category, à la *Syntactic Structures.* The proposal may have been a source of some friction with Chomsky, since it departs markedly from one of his most celebrated analyses. Friction or not, the case is very persuasive. Ross is one of the most sensitive analysts in the transformational tradition and, at the time, one of the most dedicated to its tenets. So, while some of his conclusions are at odds with *Syntactic Structures,* the arguments are models of Chomskyan rationality. The case he offers depends on two of Chomsky's most important early themes, naturalness and simplicity, and on the descriptive power of transformations. Ross points out that a wide range of transformations must somehow refer to the complex in 6.

$$6 \quad \text{Tense} \left(\left\{ \begin{matrix} have \\ be \end{matrix} \right\} \right)$$

In particular, negative and question transformations need to refer to this complex, to account for data like the following sentences.

7   a   Pixie bought some shoes.
    b   Pixie didn't buy some shoes.
    c   Did Pixie buy some shoes?

8   a   Pixie had bought some shoes.
    b   Pixie hadn't bought some shoes.
    c   Had Pixie bought some shoes?

9   a   Pixie was buying some shoes.
    b   Pixie wasn't buying some shoes.
    c   Was Pixie buying some shoes?

The generalization for negative sentences which the transformation must express is that the negative element (*n't*) occurs after the Tense and either *have* or *be,* if either of them is present. The argument is a little subtle for nonlinguists, since it requires keeping several abstractions and the effects of several transformations straight, and since it requires taking the notion of a sequential derivation very literally, but consider the most complicated example, 7b. The deep structure string for 7b is *NEG Pixie PAST buy some shoes.* The first transformation moves the *NEG* to after the *PAST* (that is, after the tense morpheme), generating *Pixie PAST NEG buy some shoes.* A rule called D̲o̲-support now kicks in, giving any tense which doesn't immediately precede a verb something to hang on to, then Affix-hopping joins the tense and the *do* together (producing *did*), and, after a contraction rule, you get 7b. On the other hand, if there is a *have* or a *be* in the deep structure, *Do*-support doesn't apply (since the tense—as in *Pixie PAST be NEG buy some shoes,* the deep structure string for 9b—has a verb immediately following it). Similarly, the question transformation moves the tense to the beginning of the sentence, along with *have* or *be,* if either is present (as in 8c and 9c); *Do*-support gives the tense something to hang on to if there is no *have* or *be* (as in 7c).

All this frequently looks like a lot of fancy and unnecessary footwork to nonlinguists, but it is a quite elegant way to describe some of the knowledge English speakers have about negatives and questions—that the tense and the first auxiliary verb often work in close concert. The important point for Ross, however, is that the complex these rules need to refer to looks very arbitrary with relation to the rest of the grammar. In terms of the *Aspects* model, 6 is not even a constituent. As Ross puts it, a rule which refers to 6 is no more natural for *Aspects* than a rule which refers to 10.[8]

$$
10 \quad \text{Noun} \left(\left\{ \begin{array}{c} toast \\ and \end{array} \right\} \right)
$$

Ross's proposal (that all the rules in question refer to any element bearing the features [+V, + AUX], which he assigns to *have* and *be*) is far more natural than Chomsky's earlier analysis, and manifestly simplifies both the base and the transformational component. It is considerably more abstract than Chomsky's analysis, since it depends on lexical features, and there is some nose-tweaking involved in pointing out that the *Syntactic Structures* analysis is as absurd as a collection of items like any arbitrary noun and the two otherwise completely unrelated words, *toast* and *and,* but certainly Ross's proposal falls within the purview of mid-sixties transformational grammar. Indeed, it is far less abstract than much of *Aspects.*

Ross's suggestion made both the base and the transformational component simpler and more regular, but a very welcome by-product of work in abstract syntax was to make deep structure (and thereby the entire grammar) more transparently semantic. Ross's auxiliary argument, for instance, included subroutines of the following sort. Consider 11a and 11b.:

11 a Dianna doesn't need to chase the duck.
    b Dianna needn't chase the duck.

In the *Aspects* model, 11a and 11b have distinct deep structures (in 11a, *need* is a main verb; in 11b it is an auxiliary verb). If the category distinction is erased, as it is for Ross, then the two (semantically equivalent) sentences have the same deep structure.[9]

That is, the Katz-Postal hypothesis ruled virtually all of the work in abstract syntax; *Integrated Theory,* in fact, pairs the hypothesis with this heuristic, effectively a blueprint for abstract syntax:

> Given a sentence for which a syntactic derivation is needed; look for simple paraphrases of the sentence which are not paraphrases by virtue of synonymous expressions; on finding them, construct grammatical rules that relate the original sentence and its paraphrases in such a way that each of these sentences has the same sequence of underlying P-markers. (Katz and Postal, 1964:157)[10]

Nobody mined this heuristic more thoroughly, or more astutely, than Lakoff, and his Postal-sponsored thesis is an abstract syntax treasure trove. For instance, he noticed lexical gaps in English of the following sort:

12 a Bart's transgression of community standards is appalling.
    b That Bart is a transgressor against community standards is appalling.
    c That Bart transgressed community standards is appalling.

13   a   Bart's aggression toward Lisa is appalling.
     b   That Bart is an aggressor toward Lisa is appalling.
     c   *That Bart aggressed Lisa is appalling.

Now, 12a–c constitute the sort of data that early transformational grammar thrived on: there is a clear pattern, and taking 12c as basic made it easy to explain the syntactic and semantic parallels in all three sentences transformationally (the verb *transgress* would be nominalized, turned into either of the nouns *transgression* or *transgressor* depending on the structural context). But 13a and 13b were out in the cold, since *Aspects* couldn't provide a common deep structure. Lakoff proposed an abstract verb, AGGRESS, which then served as the missing link for 13a and 13b (and he prevented 13c from "surfacing" by marking AGGRESS in the lexicon to obligatorily trigger nominalization, filtering off the bad sentence in the best *Aspects* tradition—1970a [1965]:58–61).

His most celebrated abstract analysis in the thesis is related to this solution—by way of a notion that came to be very productive in generative semantics, abstract verbs—and takes the general label, *lexical decomposition*. Lakoff noticed that sentences like those in 14 are effective paraphrases of each other.

14   a   Mathew killed the bogies.
     b   Mathew caused the bogies to die.
     c   Mathew caused the bogies to become dead.
     d   Mathew caused the bogies to become not alive.

Lakoff adduced a number of strong arguments for a transformational relation holding between them—namely, that all four derive from the same deep structure, 14d; or, more properly, that all derive from 14e (since abstract verbs are semantic primitives, not English words; hence the uppercase letters):[11]

14   e   Mathew PAST CAUSE the bogies to BECOME NOT ALIVE

Most of the arguments for these abstract analyses were syntactic, at least in the expansive *Aspects* sense of *syntactic*—that the object of *kill* and the subject of *die,* for instance, have exactly the same selectional restrictions (e.g., they both must be [+alive]; *bogies* is okay, *rocks* is not)—but again the most persuasive component of the case clearly follows from its successful adherence to the Katz-Postal principle. Deriving *kill* and *cause to die* and *cause to become not alive* from the same underlying structure is a very appealing move because, as McCawley points out in the preface to Lakoff's thesis, it is "more semantically transparent" (Lakoff, 1970a [1965]:i) than treating *kill* as a lexical atom.

Chomsky had predicted in the fifties that "if we should take meaning seriously as a criterion, our analyses would be wildly complex" (1958:453), and it doesn't take much imagination to see that's exactly where these sorts of analyses were heading. Think, for instance, of the decomposition of lexical items like, just sticking with fatal ones, *slaughter, garotte,* and *assassinate;* the last would have to be, conservatively, something on the order of CAUSE X TO BECOME NOT ALIVE FOR POLITICAL REASONS IN A COVERT MANNER, WHERE X IS REASONABLY IMPORTANT. Or take a look at a real example, 15b, a proposed deep structure string for 15a (Fillmore, 1972 [1969]:15):[12]

15 a Did I give you the other book?

  b There is a set of books that both you and I know about and the car-
     dinality of that set is some number n and you and I have just had in
     mind a subset containing $n - 1$ of those books and I am now calling
     your attention to the remaining nth book. There was a time when I
     had that book in my possession and I am now asking you to tell me
     whether I did anything in the past which would count as causing that
     book to be in your possession.

The most famous of these wildly complex deep structures, which Ladoff and Ross
used frequently in their lectures on abstract syntax, and which even made *The New
York Times* (Shenker, 1972), is their analysis of 16. *Aspects* would assign a deep
structure to this sentence much like that of PM-1, but Ross and Lakoff assigned it
an underlying structure like that of PM-2 (both trees, however, are somewhat
elided).[13]

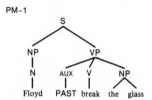

PM-1

16 Floyd broke the glass.

It is important to note that PM-2—although it looks, and has been taken, to be a
logical absurdity which vitiates the theory that spawned it—is the serious proposal
of two very good, very dedicated linguists. Lakoff and Ross realized it looked pretty
silly. Ross had a ceiling-to-floor mobile of it hanging in his office (another joke
placed their tree just after a Klee painting as the latest advance in modern art—Pop,
1970:123). But they were just loyally following the dictates of the *Aspects* model
and the Katz-Postal principle to their inevitable consequences; taking meaning seriously as a criterion is exactly what the Katz-Postal principle was all about, and
wildly complex was unavoidably what the abstract syntax analyses became. Notice
before we move on, though, that things at least as ugly as 15b and PM-2, since they
represent meanings, are inevitable in any grammar which wants to get the semantics right. If indeed a significant part of the meaning of *kill* is "cause to become not
alive," that fact has to be treated somewhere—ditto for *broken* and "cause to
become not whole"—even in interpretive semantics. Whatever a semantic representation in the *Aspects* model looked like (and they remained unspecified while
this work went ahead), it would be similarly unappetizing. Meaning is a complicated beast.

It's also worth noting that Anna Wierzbicka, who had been studying linguistics
in Warsaw with Andrzej Boguslawski, a serious semanticist if ever there was one,
visited MIT just as these complicated trees were proliferating and was somewhat
amused at how simpleminded much of the work was. She kept urging the brink-of-
generative semanticists to follow through on the implications of their work and get
really abstract.

The important point for the moment, however, is much more straightforward

PM-2

V
|
and

N
|
S

V | N
|
PRES

N
|
S

V | N | N
|
BE   I   S

V | N | N
|
CAUSE I   S

V | N
|
HAPPEN S

V | N | N
|
ASSUME I   S

V | N | N
|
NAME X Floyd

N
|
S

V | N
|
PRES

N
|
S

V | N | N
|
BE   I   S

V | N | N
|
CAUSE I   S

V | N
|
CAUSE S

V | N | N
|
ASSUME I   S

V | N
|
glass Y

N
|
S

V | N
|
PRES

N
|
S

V | N | N
|
BE   I   S

V | N | N
|
CAUSE I   S

V | N
|
HAPPEN S

V N N N
|
SAY I you S

V | N
|
PAST S

V | N | N
|
HAPPEN X S

V | N | N
|
CAUSE X S

V | N
|
COME ABOUT S

V | N | N
|
BE   Y   S

V
|
BROKEN

than whatever the appropriate semantic representation of "Floyd broke the glass" might be: many Chomskyan linguists found the arguments surrounding the Floyd tree (PM-2) extremely appealing in the context of post-*Aspects* generative grammar, and it became something of an exemplar, the abstract syntax equivalent of early transformational grammar's passive analysis.

Logic

> As the fool is to the wiseman, so is the grammarian ignorant of logic to one skilled in logic.
>
> Albertus Magnus

As another glance at the Floyd tree will show, Postal's reductionist campaign was gathering a good deal of steam—adjectives were re-analyzed as deep verbs, adjective phrases disappeared at deep structure, some nouns were also deep verbs, prepositions and conjunctions were deep verbs, prepositional phrases dissolved at deep structure, tenses were deep verbs, quantifiers were deep verbs, articles arose transformationally, the verb phrase dissolved at deep structure—and abstract syntax arrived at a convenient little core of deep categories: NPs, Vs, and Ss. There were noun phrases, verbs, and sentences at deep structure and every other category was introduced transformationally.[14]

This core was extremely attractive because, as McCawley and Lakoff began to argue, it was in very close alignment with the atomic categories of symbolic logic: arguments ($\approx$NP), predicates ($\approx$V), and propositions ($\approx$S). More: the reduction of the deep structure inventory meant a corresponding reduction of the phrase structure rules, which now fell into an equally close alignment with the formation rules of logic. More: the formalisms of symbolic logic and transformational grammar also fell nicely together. Take 17, a simple statement in symbolic logic.

17   ABUSE(x,y) & MAN(x) & DUCK(y)

ABUSE is a two-place predicate that (therefore) takes two arguments, x and y, MAN and DUCK are one-place predicates taking, respectively, the arguments x and y. The whole proposition means, pretty much (ignoring tense and other minor complications), what the sentence 18 means.

18   The man abused the duck.

These two entities, 17 and 18, are similar in many important respects, but they also look very different in others. Lakoff pointed out that the differences are actually pretty superficial. So, we know that sentences are represented in phrase structural terms as labeled trees, like PM–3.

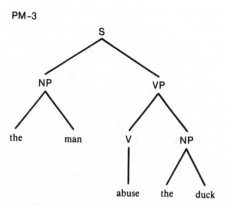

PM–3

But, as everyone realized, there was an alternate and fully equivalent formalism for representing constituent structure, namely, labeled bracketing, like 19:[15]

19   ((the man)$_{NP}$ (abuse(the duck)$_{NP}$)$_{VP}$)$_S$

Put this way—17 and 19 are both bracketed strings, and 19 is equivalent to PM-3—you can probably see the dénouement as clearly as Lakoff and McCawley did: use the tree formalism for the bracketing of 17 and you get a phrase marker like PM-4.

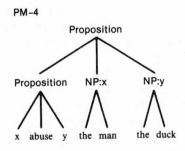

PM-4

Voilà: the formalisms of logic and transformational grammar are equivalent. And in the transformationally heady post-*Aspects* days, getting from PM-4 to PM-3 was a trivial matter. Making PM-4 the deep structure, with its great advantage in semantic clarity, seemed not only attractive to McCawley and Lakoff and their argument-buyers, but inevitable. This realization—that deep structure coincided with symbolic logic, the traditional mathematico-philosophical language of meaning—was an exhilarating confirmation that abstract syntax was on exactly the right track.[16]

In a sense, the result really shouldn't have been much of a surprise. The finding that syntactic categories could be reduced to the primitives of symbolic logic is indeed the inevitable climax of their program, but it is inevitable to the point of banality. Logic was born out of observations of words and sentences (both *word* and *sentence* are possible translations of *logos*), so it is less than astonishing that it would come to figure prominently in an approach dedicated to reducing word and sentence categories. The convergence of abstract syntax and symbolic logic should not have been much of a surprise, should not have been, as it was for many, the conversion experience that effected generative semantics.

Perhaps not, but that's the same as saying Kepler shouldn't have been euphoric over the smoother fit of the Copernican solar system to Euclid than the Ptolemaic fit, or that a whole school of theoretical physicists shouldn't currently be trumpeting the geometrical virtues of string theory, or, coming home to linguistics, that the Modistae should not have extolled their program for its fidelity to logical principles.

Logic is, in almost every way, to linguistics as geometry is to physics. Both are formal sciences which play in and around their respective empirical relatives—the sciences of understanding language and understanding matter—often diverging for long solipsistic periods, but always returning, and frequently bearing new fruit. Geometry and logic are especially compelling, as one would expect, for the more formally inclined scientists, for the theoreticians and model builders rather than the field workers and experimentalists, and the formally inclined often get the relation backwards. Hawking, for instance, calls cosmology "a dance of geometries" and Edward Witten says that "string theory at its finest is, or should be, a new branch of geometry" (Davies and Brown, 1988:95). But geometry is a tool devised for the exploration of the physical world, a tool of physics, "founded in mechanical practice and . . . nothing but that part of universal mechanics which accurately proposes and demonstrates the art of measuring" (Newton, 1960 [1686]:xvii). Geo-metry was an empirical product, the distillate of observations compiled in earth-measuring, so it is less than remarkable that it functions prominently in theories which organize and predict measurements of matter. Logic was an empirical product, the distillate of word and sentence observations, so it is less than remarkable that it came to function prominently in theories which organize and predict grammaticality. But if Hawking and Witten can be so pleased about the relation of their work to geometry, Lakoff and McCawley are not exactly in shabby company.

Geometry and logic, that is, are not just abstract and autonomous exploratory instruments. They are also extremely important confirmatory instruments, and an empirical program that converges with a related formal program is fully justified in celebrating that match as a strong indication it is on the right track. For the abstract syntacticians, the discovery that their coalescing theory was "just symbolic logic after all," was an immensely liberating experience.

A large part of its appeal is that, once again, this work looked like just what Dr. Chomsky ordered. Right back to his M.A. thesis, which bears the unmistakable imprint of Carnap's *Logical Syntax* (1937 [1934]; Newmeyer, 1988a), through his massive *Logical Structure of Linguistic Theory,* through *Aspects,* where the deep structure subject is "the logical subject," up to *Cartesian Linguistics,* in which *deep structure* and *logical form* are synonyms—through his entire career, that is— Chomsky was courting symbolic logic.[17] The abstract syntacticians thought it was time to end the courtship, and, fearing the good doctor lacked the courage to do it himself, proposed on his behalf.

There was one more factor—beyond the semantic clarity at deep structure, the convergence with a formal program, the ability to solve tricky syntactic problems, and the natural fulfillment of the Chomskyan program—which contributed enormously to the appeal of symbolic logic for generative semantics.

Exactly what logic is and what it says about the way humans acquire, manage, and perpetuate knowledge has never been entirely clear, different logicians or philosophers giving different answers. But "logic, under every view, involves frequent references to the laws and workings of the mind" (Bain, 1879.1:1), and, in the strongest views (like the Port-Royalists', for instance), logic is construed *as* the laws and workings of the mind. McCawley explicitly took this position, aligning himself with one of its strongest expressions, Booles's early logic book, *The Laws of Thought* (McCawley, 1976b [1968]:136). In short, logic brought the abstract syntacticians much closer to the mentalist goals which they had swallowed with their early transformational milk.

The Universal Base

The latent content of all languages is the same.
Edward Sapir

Along with vitamin M, mentalism, their early transformational milk included another essential nutrient, especially after Chomsky's *tour de force* linkage of his program to the goals of traditional grammar, vitamin U, universality, and in *Aspects* Chomsky associated this nutrient with one specific module of his grammar, the base component:

> To say that the formal properties of the base will provide the framework for the characterization of universal categories is to assume that much of the structure of the base is common to all languages. This is a way of stating a traditional view, whose origins can . . . be traced back at least to the [Port-Royal] *Grammaire générale et raisonée.* (1965 [1964]:117)

Somewhat more strongly, in *Cartesian Linguistics,* he added "the deep structure that expresses the meaning is common to all languages, so it is claimed, being a simple reflection of the forms of thought"—1966a:35), unambiguously bringing deep structure and meaning into the mentalist universal-base suggestion, all but proposing generative semantics. Well, alright, not so unambiguously—there is the telltale "so it is claimed"—but the early transformationalists overlooked the hedge too, and Chomsky's suggestion caught fire. It rapidly evolved into the uppercase Universal Base Hypothesis—the claim that at some deep level all languages had the same set of generative mechanisms—and became identified exclusively with abstract syntax. Lakoff endorsed it in his thesis (1970a [1965]:108–9) as one possibility for getting at the universal dimensions of language, and by 1967, at a conference in La Jolla, his enthusiasm for it had increased markedly (Bach, 1968 [1967]:114n12). Robin Lakoff also endorsed it in similarly hopeful terms in her thesis, *Abstract Syntax and Latin Complementation,* where she spells out its implications for the *Aspects* model clearly: it would radically alter Chomsky's proposed base component, reducing the phrase structure rules to only a very few (1968:168). Chomsky's base rules in *Aspects* are almost exclusively for fragments of English grammar and Lakoff points the way toward a common-denominator approach for finding base rules which can underlie both Latin and English. Betraying some of the schismatic spirit that had begun to infuse abstract syntax, she identifies this alteration with "some of the more radical transformationalists" (1968:168) and opposes it to "more conservative transformational linguists (such as Chomsky)" (1968:215n5). Emmon Bach's (1968 [1967]) contribution—an influential argument that some nouns were better analyzed as deep verbs—spells out most fully that the best hope for a universal base depends on the abstract syntax program of finding the essential core of lexical and phrasal categories.

But the two figures most closely associated with the Universal Base Hypothesis are McCawley and Ross. McCawley is important because, although he was not one of the chief marketers of the proposal—may never, indeed, have conjoined the words *universal* and *base* in print—he wrote an important paper in which many found strong support for the Universal Base Hypothesis.[18] Ross is important because the explicit claim "that the base is biologically innate" appears to have been his (Lancelot and others, 1976 [1968]:258); because of his recurrent use of the hypothesis as an appeal for generative semantics (for instance, in a paper directed at cognitive psychologists—1974b:97); and because he gave the hypothesis its most succinct, best known, formulation:

THE UNIVERSAL BASE HYPOTHESIS
The deep structures of all languages are identical, up to the ordering of constituents immediately dominated by the same node. (Ross, 1970b [1968]:260)

Ross's definition makes it inescapably clear what is being said: that the dizzying variety of linguistic expression in all known languages, in all unknown languages, in all possible human languages, derived from a common set of base rules, with the trivial exception of within-constituent ordering differences (acknowledging, for instance, that adjectives precede nouns in English noun phrases, follow them in French noun phrases). In fact, Ross and Lakoff were confident enough in the hypothesis to begin working on such a set of rules, and the confidence was infectious. Not the least of the attractions for this claim was its antithetical relation to the irredeemable Bad Guys of American linguistics: recall that one of the chief Bloomfieldians had said "languages could differ from each other without limit and in unpredictable ways" (Joos, 1957:96). The Chomskyans found this notion repugnant in the extreme, and regularly trotted out Joos's quotation as the epitome of woolly-mindedness and unscientific confusion.

But the Universal Base Hypothesis, in a typical scientific irony, actually got much of its drive from the genius that ruled Joos's comment—attending to a wide variety of languages. The overwhelming majority of transformational-generative research in its first decade was on English, and *Aspects* reflects this emphasis. The abstract syntacticians were certainly not in the Bloomfieldians' league in terms of experience with alien languages, but they began thinking more deliberately about taking generative principles beyond English. Postal's thesis was on Mohawk, McCawley's was on Japanese, Robin Lakoff's on Latin, Ross's was widely cross-linguistic, and most of them studied under G. H. Mathews at MIT, who taught portions of his influential generative grammar of Hidatsa (1965). Perhaps most importantly, though the full repercussions of his work were still several years off, Joseph Greenberg had just published the second edition of his typology of linguistic universals (1966), which surveyed the morphological and syntactic patterns of a great many, quite diverse languages.[19]

Now, since the base component sketched out in *Aspects* depended on English, attempts to universalize it inevitably led to serious changes. The *Aspects* base included adjectives, for instance, but Postal's work on Mohawk had shown him that not all languages have a separate category of adjectives, distinct from verbs, leading directly to his adjectives-are-deep-verbs arguments. The *Aspects* base included a verb phrase of the form V + NP, but Japanese, Latin, and Hidatsa can't easily accommodate such a VP; other languages seem to have no VP at all. The *Aspects* base included auxiliary verbs; not all languages do. The *Aspects* base included prepositions; not all languages do. The *Aspects* base included articles; not all languages do. The *Aspects* base ignores causatives and inchoatives (CAUSE and BECOME, respectively, in the conventions of abstract syntax), which are often covert in English; in many languages, they play very overt roles. (For instance, in Ainu there is a causative suffix, -*re*. So, *arpa* is "to go," *arpare* is "to send (or, cause to go);" *e* is "to eat," *ere* is "to feed.") The *Aspects* base adopted the basic order of English; Greenberg demonstrated very convincingly that there were several distinct basic orders.

The rub, then: while following on some of Chomsky's general comments, abstract syntax was forced to reject or modify many of his specific analyses.

Where there is a rub, there is friction.

Filters and Constraints

> We would like to put as many constraints as possible on the form of a possible transformational rule.
>
> George Lakoff

The abstract syntax move toward a universal grammar underneath the literal skin of languages had another reflex: the shift in emphasis from phenomena-specific and language-particular rules to general grammatical principles.[20] Early transformational grammar was rule-obsessed. The motto seemed to be, "Find a phenomenon, write a rule; write two, if you have time," and the search for phenomena was heavily biased toward English. Once again, Lees's *Grammar* provides the best illustration. A brief monograph, focusing on a small neighborhood in the rambling metropolis of English, it posits over a hundred transformations (while noting frequently along the way the need for many other rules that Lees doesn't have the time or the mandate to get to), most of which are highly specific. Some rules refer, for instance, to particular verb or adjective classes, some to individual words or morphemes, some to certain stress patterns or juncture types, and a good many of them have special conditions of application attached to them. Take this pair of transformations (rules 20 and 21), which handle two specific deletions (Lees, 1968 [1960]:103–4):

20   Nom + X − Nom′ + Pron − Y ⇒ Nom + X + Y

[where Nom = Nom′, Nom is subject of a complex sentence, and Nom′ + Pron is inside a *for*-phrase Infinitival Nominal within that sentence]

$$21 \quad X - P - \begin{bmatrix} \text{that} \\ \text{for} \\ \text{to} \end{bmatrix} Y \Rightarrow X \begin{bmatrix} \text{that} \\ \text{for} \\ \text{to} \end{bmatrix} Y$$

[where *that* introduces a Factive Nominal, *for* and *to* introduce Infinitival Nominals]

What these rules are up to isn't particularly important at the moment (though it's not nearly so easy to make fun of them once you gain an appreciation of the sort of machinery and assumptions in which they are embedded,)[21] but there is one point we do need to notice: how incredibly mucky these sort of analyses are for anyone interested in general grammatical principles—how sticky it is to move from a rule that depends on the presence of specific English words to a principle holding of all languages—and the defining move of abstract syntax was toward general grammatical principles.

A principle-oriented approach to grammar was clearly not going to achieve a very convincing universal grammar with transformations as detailed as 20 and 21. In fact, one of the axioms of "Postal's conception of grammar" was that "transformations may not mention individual lexical items" (Lakoff, 1970a [1965]:xii) like *that* and *for*, an axiom easy to see at work in Ross's smoothing out of the auxiliary system, for instance. Ross indicts the use of individual lexical items like *have* and *be* and *toast,* and the central virtue he cites in favor of his analysis is that it permits the generalizing of several transformations to refer to a natural class of items rather than to a seemingly arbitrary list of words. A transformation that refers to a class of items with the features [+verb, +auxiliary] has a much better chance of getting a job in a universal grammar than a transformation that refers to a class of items like the set {*have, be*}. Similarly, Lakoff's use of hypothetical verbs like AGGRESS allowed nominalizing transformations to apply more widely—and don't be fooled by the English spelling, a mnemonic convention. Hypothetical verbs were abstract meanings, pure and simple, with no phonological structure, not language-specific words.

The first truly important work on general transformational principles was—no surprise here—Chomsky's. In his International Congress talk, he proposed a constraint on movement rules to help illustrate explanatory adequacy. Describing languages, Chomsky had said of the Bloomfieldian descriptive mandate, is only part of the linguistic battle, the least interesting part at that, and one can describe languages with myriad specific rules like 20 and 21. The tough linguistic work was in explaining Language, and explaining Language, in Chomsky's terms, involved curbing the power of transformations. One way to do this was by fiat, such as Postal's stipulation that transformations couldn't refer to individual words, and a more famous stipulation from the period, that transformations could only delete "recoverable" constituents,[22] and a more famous one yet, that transformations couldn't change meaning. All of these came from formal considerations. Certain stipulations (like, in another arena, "parallel lines cannot meet"), just made the system work better. But another way was by empirical investigation: look for activities that transformations can do, by virtue of their definition, but which they don't appear to do.

Transformations can easily move any constituent anywhere in a sentence, for instance, so Chomsky looked for movement possibilities that don't seem to be exploited.

The constraint Chomsky proposed at the International Congress (later called the *A-over-A-principle* by Ross) was a tentative suggestion about something transformations don't seem to do—more constituents from certain locations[23]—and, therefore, something that should be built-in as a prohibition in universal grammar. The reasoning may look peculiar: if transformations don't do X, why does X need to be prohibited? The answer has to do with the very important notion of descriptive power. Transformations are extremely powerful descriptive devices. They can be used to describe virtually any conceivable sequence of symbols, including a lot of sequences that never show up in human languages. Let's say that languages are analogous to the system of whole numbers; if so, then transformational grammar would be a theory capable of describing the rational numbers. You have a theory which includes fractions describing a system which doesn't contain fractions. And there is more at stake here than economy. If a transformational grammar is what children acquire when they become language-users, and if Chomsky's poverty of stimulus argument goes through, there have to be some innate guidelines that direct the acquisition process and restrict the grammar. There have to be some genetic principles which ensure the specific grammar a child ends up with is one that describes a human language: that it describes only whole numbers, so that the child does not begin spouting fractions. A principle that says "whatever else transformations can do, they can't move constituents out of the location specified by the A-over-A formula" restricts the descriptive power of the grammar—it excludes fractions—and therefore gives the theory housing that grammar a better shot at explaining language acquisition.

Unfortunately, the A-over-A principle was a bust. When Chomsky discovered that it made the wrong predictions in certain cases, he—quickly, quietly, and reluctantly—dropped it, adding a prophetic and hopeful "there is certainly much more to be said about this matter" (1964c [1963]:74n16a; 1964d [1963]:46n10). The one who said it was Ross.

It is the concern with general principles of grammar, in fact, that makes Ross's (Chomsky-supervised) thesis such a landmark in grammatical theory. With a cautiousness and modesty uncommon to the mid-sixties MIT community, with a cross-linguistic sensitivity equally uncommon to transformational grammar of the period, with an eye for abstract universals unprecedented in syntactic work, Ross plays a remarkable sequence of variations on the theme of Chomsky's A-over-A principle, coming up with his theory of syntactic islands. His definition of *islands* is quite abstract, but the metaphorical long and short of them is that certain constituents are surrounded by water, and transformations can't swim. So, for instance, 22a is fine (from 22b), but *23a isn't, because transformations can't move constituents off a "complex noun phrase" island; 24a is fine (from 24b), but *25a isn't, because transformations can't move constituents off a "coordinate structure" island; and *26a is bad because transformations can't move constituents off a "sentential subject" island (there is no grammatical equivalent, but notice that 26b and 26c are effectively the same type of deep structure string, and that the phrase structure rules have to generate 26c in order to get grammatical sentences like 26d).

22   a   Which spud do you believe Bud peeled?
     b   you believe Bud peeled which spud

23   a   *which spud do you believe the claim that Bud peeled
     b   you believe the claim that Bud peeled which spud
         [where *the claim that Bud peeled which spud* is a complex noun phrase]

24   a   What was Bud eating?
     b   Bud was eating what

25   a   *what was Bud eating spuds and
     b   Bud was eating spuds and what
         [where *spuds and what* is a coordinate structure]

26 a *what that Bud will eat is likely
    b   that Bud will eat what is likely
        [where *that Bud will eat what* is a sentential subject]
    c   that Bud will eat spuds is likely
    d   That Bud will eat spuds is likely.

Ross suggested his syntactic-island theory as an integral part of the innate mechanism guiding language acquisition, and his work is volcano-and-palm-trees above
similar efforts at the time. But other abstract syntacticians were exploring similar
restrictions. In particular, Postal worked out his Crossover principle (1971a
[1968]), so-named because it prohibited the transformational movement of a noun
phrase in a way that "crosses over" another noun phrase which has the same real-
world referent, in order to explain a number of grammaticality facts. So, for
instance, 27a is fine, but *27b and *27c are bad because *Jeff* and *himself* cross over
one another when the rule Passive moves them; 28a is fine, but *28b and *28c are
bad because *Jeff* and *himself* cross over one another by the rule *Tough*-movement.

27 a Jeff shaved himself.
    b *Jeff was shaved by himself
    c *himself was shaved by Jeff
28 a It is tough for Jeff to shave himself.
    b *Jeff is tough for himself to shave
    c *himself is tough Jeff for to shave

Considering the process graphically (figure 5.1), the crossover in example A does
not violate Postal's principle, since *Jeff* and *Andrew* have different real-world referents. The crossover in example B, which corresponds to 27b and 27c, does violate
the principle, since *Jeff* and *himself* have the same real-world referents, and is
therefore prohibited.

Again, the attractive element of Postal's principle is that it allows much more
general formulations of the individual rules: rather than individual conditions on
Passive and *Tough*-movement (and, in fact, several other rules), a single condition
holding true of all movement rules can be stated once, and the transformations
themselves become much more general. More importantly—for Chomsky's proposal, for Ross's theory, for Postal's principle—rule-specific conditions of the sort

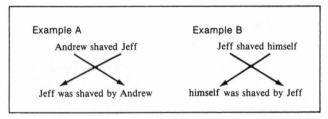

Figure 5.1. Illustrations of Postal's Crossover principle.

illustrated by Lees's work just missed something critical about language. A bunch of individual statements of the same condition tacked onto the bottoms of a bunch of distinct rules makes everything look coincidental. It makes the grammar into a list of rules just like a long catalog of mammals which includes, after each entry, "this one lactates." What makes biologists happy is a theory of mammals that includes lactation; what makes transformational grammarians happy is a theory of movement rules that includes an overarching account of where they can't apply.

There is another way of looking at these proposals, as performing the function Chomsky called *filtering* in *Aspects*. In that testament, Chomsky had said that transformations, in addition to their regular job describing grammatical sentences, also had the job of weeding out structures that the base component erroneously generated, of filtering off deep structures that couldn't achieve grammatically. The notion of filtering can be tough to grasp, but the *Aspects* base component generated structures like 29a and 29b:

29   a   the cat the cat chased the dog had a nap
     b   the cat the kangaroo chased the dog had a nap

Since these structures come from the base component, they have to pass through the transformational component, which has no trouble with 29a. A transformational rule of Relativization changes the second occurrence of *the cat* into the relative pronoun *which,* yielding a grammatical sentence containing the relative clause *which chased the dog,* 29c.

29   c   The cat which chased the dog had a nap.

Not so with 29b. Relativization requires two identical noun phrases: it doesn't find them in 29b, so it can't apply and filters the offending structure off.

The island constraints and the Crossover principle do exactly the same job, but more explicitly. Movement rules could do whatever they wanted, in this view, but if they pulled something off an island or crossed-over same-referential noun phrases, Ross and Postal's conditions would filter off the mess by declaring any violation to be ungrammatical.

Filtering was an important line of post-*Aspects* research which went beyond movement constraints. Lakoff's thesis was the first significant investigation here, proposing a theory of exceptions that gave some of the filtering work over to the lexicon. He was worried about situations like 30a and *30b, where an otherwise normal transitive verb refuses to do what normal, red-blooded transitive verbs usually do, undergo Passive.

30   a   Jan resembles Mick.
     b   *Mick is resembled by Jan.

Recall that Lakoff kept AGGRESS from showing up in the surface as a verb (filtered it off) by marking it to obligatorily undergo Nominalization. Well, that was only a small part of a more general solution in which he developed a system of rule features—adapting a proposal from Chomsky and Halle's at-the-time-still-underground phonological tome, *The Sound Pattern of English* (1968)—which permitted lexical items to "govern" the application of transformations. In effect, AGGRESS is marked [+Nominalization] in the lexicon, *resemble* is marked [−Passive], allowing the good sentences to make it through the transformational component just as they should, but filtering off the bad ones.[24] Lakoff's rule features were more general than the sort of conditions Lees proposed, but they were still language specific, tied to individual English words like *resemble,* and for that reason not especially attractive. Their important advantage, though, is that they helped to make the transformations more general.

One other development bears mention, by a student of Ross and Postal, David Perlmutter, whose (yes, Chomsky-supervised) thesis explored *Deep and Surface Structure Constraints in Syntax* (1971 [1968]). The title explains pretty much what Perlmutter was up to—investigating the need for filters operating at either end of a derivation, in English. These filters had their own name, *output conditions,* and since they were language specific they were viewed with a tolerant distaste, a necessary evil of exactly the same type as Lakoff's exceptions. Perlmutter was an abstract syntactician who trotted alongside without ever hopping aboard the generative semantics bandwagon, but his output conditions contributed significantly to the general growth of extra-transformational mechanisms which all generative semanticists endorsed. Several generative semanticists proposed output conditions of their own, including Ross (who apparently gave them their name—1970a [1967], 1986 [1967], 1972a) and Lakoff (1968a).

The Performative Analysis

> "I do (sc. take this woman to be my lawful wedded wife)"—as uttered in the course of the marriage ceremony.
>
> "I name this ship the Queen Elizabeth"—as uttered when smashing the bottle against the stem.
>
> "I give and bequeath my watch to my brother"—as according to a will.
>
> "I bet you sixpence it will rain tomorrow."
>
> examples cited by J. L. Austin

Abstract syntax was, in large part, an effort to increase the scope of the *Aspects* model, and some of it was true to the label and focused exclusively on syntax (in particular, the category reduction research and the filters-and-constraint research). In back of most other developments in abstract syntax, just as it was in back of most developments leading up to the *Aspects* theory, was meaning. Nowhere is this clearer than in another development closely associated with Ross, the performative analysis.

Linguists were just beginning to quit their game of ostrich with meaning in the fifties and sixties, but philosophers had maintained a very strong interest in meaning all along. So, when linguists pulled their heads out of the sand, they saw a few others who had been exploring the terrain awhile. One group of explorers were logicians, and Lakoff and McCawley looked over their shoulders. Another group, members of a school called *ordinary language philosophy,* intrigued Ross, and he began importing some of its insights.

Those insights go at least as far back as the Greek Sophists. One of the first things the Greeks noticed when they started looking at sentences, rather than just at sounds and words, is that they come in different types—assertions, questions, commands, and so on—and Protagoras is credited with the first typology of sentence functions. But in the intervening millennia very little research went beyond identifying a few of these types. When transformational grammar took notice of different sentence types in the fifties, it did no more. It just cataloged a few of the basic types. Harris, Chomsky, Katz and Postal, all took brief looks at a few of the more basic sentence functions (asserting, asking, and commanding), but it was very far from a central concern, which is where ordinary language philosophy comes in.

Ordinary language philosophy is something of a parallel movement; indeed, a parallel revolution, if one of its founding voices is to be trusted. J. L. Austin not only unequivocally calls the movement a revolution, but hints rather broadly that it is "the greatest and most salutary" revolution in the history of philosophy (1962 [1955]:3); he adds, however, that given the history of philosophy, this is not a large claim. It develops out of Wittgenstein, but really took wing in the fifties, especially through Austin's William James lectures at Harvard in 1955. The lectures came out as *How To Do Things with Words,* a title which gets right at the driving theme of ordinary language philosophy—that people do things with language—a ridiculously simple but remarkably productive notion. Austin's starting point is the observation that philosophy had generally only noticed one of the things that people do with words—namely, assert propositions—and had consequently paid virtually all of its attention to truth conditions. But people also inquire, Austin says, and order, and warn, and threaten, and promise, and christen, and bequeath, and bet, and generally perform a wide variety of actions when they talk, actions in which truth conditions are either subservient or wholly irrelevant. Austin calls the effective intentions of these speech acts *illocutionary forces,* and the sentences that perform them, *performative sentences.*[25]

Early transformational grammar and Austin's speech act theory had very little to do with one another, although there were some close brushes—Chomsky was writing *Logical Structure* at Harvard when Austin gave the lectures that became *How To Do Things with Words,* and a few years later Austin was planning a discussion group on *Syntactic Structures* when he fell fatally ill.[26] Speech act notions also made a brief appearance in *Integrated Theory,* when Katz and Postal footnote their proposal for the imperative trigger morpheme with "a case can be made for deriving imperatives from sentences of the form *I Verb$_{request}$ that you will Main Verb* by dropping at least the first three elements" (1964:149n9).[27] After listing some of the virtues this proposal has over the course they actually took, they add "although we do not adopt this description here, it certainly deserves further study."

Like so much of their work, Katz and Postal's buried peformative suggestion did get further study a few years later. Robin Lakoff's dissertation proposes and analyzes a number of abstract performative verbs for Latin (1968:170–217), and McCawley also outlines the hypothesis in an important paper for the Texas Universals conference (1976b [1967–68]:84–85). But the "performative proposal," as it has become known, clearly belongs to Ross, and to his paper "On Declarative Sentences" (1970b [1968]).[28] The principal claim of the paper is that the deep struc-

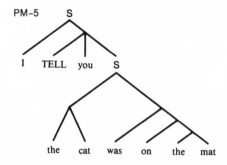

tures of simple declarative sentences have a topmost clause of the form "I TELL you," as in PM–5, which was usually excised by a rule of Performative deletion.[29]

By extension, the same mechanism would take care of questions, imperatives, promises, and the entire remaining panoply of speech acts—giving topmost clauses like "I ASK you" and "I COMMAND you" to questions, imperatives, and so on.[30] Ross offers a battery of arguments on behalf of the analysis, but consider just one of them, based on reflexive pronouns. In English, reflexive pronouns require antecedents which agree in person and number (and sometimes gender). Within the mechanisms of transformational grammar, the most efficient way to explain the grammaticality of 31a and 31b but the ungrammaticality of *31c is to assume underlying antecedents that get deleted (as in 31d):

31   a   Jery thinks that the sandcastle was built by Debbie and myself.
     b   Jery thinks that the sandcastle was built by Debbie and yourself.
     c   *Jery thinks that the sandcastle was built by Debbie and themselves.
     d   I TELL you that Jery thinks that the sandcastle was built by Debbie and myself.

With 31d as the deep structure, 31a is okay (since *I* and *myself* agree in person and number). Ditto for 31b (since *you* and *yourself* agree in person and number). But *31c is out (since there is no third-person plural antecedent, like *they,* anywhere in 31d). Simple and clean. And without the underlying "I TELL you" clause, the grammar has to come up with some other explanation for why 31a and 31b are grammatical sentences, but 31c is not.

Bolstered by a phalanx of other arguments, the case for an underlying "I TELL you" clause has a certain force to it, and all of the facts that Ross adduced are like the ones illustrated by 31a–d, syntactic facts—facts about the distribution and co-occurrence of words in sentences—in the best, most rigorous *Aspects* tradition. But it escaped no one's notice that Ross's proposal increased the semantic clarity of the deep structure. Ross likely wouldn't even have noticed the facts in 31a–d if he wasn't trying to explore Austin's insights about meaning. Ross's proposal pushed deep structure ever deeper, toward abstraction, and toward meaning: the defining trend of abstract syntax. In McCawley's terms,

> Since the introduction of the notion of 'deep structure' by Chomsky, virtually every 'deep structure' which has been postulated (excluding those which have been demonstrated simply to be wrong) has turned out not really to be a deep structure but to have

underlying it a more abstract structure which could more appropriately be called the 'deep structure' of the sentence in question. (McCawley, 1976b [1967]:105)

"This," McCawley adds, "raises the question of whether there is indeed such a thing as 'deep structure'."

The Opening Salvo

> We believe semantics may be generative.
>
> George Lakoff and Háj Ross

The next step (and *step* is not exactly the right word; these are all more or less overlapping developments, not sequential ones) in the emergence of generative semantics, the one cued by McCawley's question, may seem like a peculiar one. We have just been looking at arguments that deep structure has only a few essential categories, that symbolic logic is the language of deep structure, that there is a universal base which implies all languages share a common core of deep structures, that deep structure calls for abstract verbs, including abstract performatives, that deep structure is very deep indeed. And the next step is the abandonment of deep structure.

The signal document, the birth announcement, the first important gauntlet, of generative semantics emerging from abstract syntax is a slight mimeographed paper—a letter, really—by Lakoff and Ross, entitled "Is Deep Structure Necessary?" (1976 [1967]) which says no.

In the spring of 1967, Ross wrote a letter to Arnold Zwicky outlining the work he and Lakoff had been doing, in telephonic collaboration with McCawley, and the conclusions the three of them had all come to about the relations of syntax and semantics in Chomskyan linguistics. Zwicky recalls being "very impressed" by the letter, and Ross decided to circulate their conclusions more widely. By this point Ross's collaboration with Lakoff had become a sort of Lennon and McCartney affair, where it didn't really matter who wrote what; both names went on the letter, the important passages were mimeographed, and copies very quickly made the rounds. The force of the letter is that there can be no such a thing as deep structure, at least as it is defined in *Aspects.*

Ross and Lakoff's case was sketchy at best, but the letter was very effective, for Zwicky and for most of its secondary readers—McCawley says it's what turned him "from a revisionist interpretive semanticist into a generative semanticist" (1976b:159), and he was not alone. Where Lakoff's "Toward Generative Semantics" sputtered, the joint letter sparked, kindling a brushfire not so much because of the immediate force of Lakoff and Ross's arguments, but because of all the promising work done in the name of abstract syntax. The letter plugged directly into a feeling that had begun to pervade generative linguistics, that deep structure was just a way station. Chomsky had said, after all, that the deeper syntax got the closer it came to meaning and the abstract syntacticians were getting awfully deep. Look at how many fathoms down McCawley was in the spring of 1967, just weeks before Ross licked the stamp and posted his letter to Zwicky:

On any page of a large dictionary one finds words with incredibly specific selectional restrictions, involving an apparently unlimited range of semantic properties; for example, the verb *diagonalize* requires as its object a noun phrase denoting a matrix (in the mathematical sense), the adjective *benign* in the sense 'noncancerous' requires a subject denoting a tumor, and the verb *devein* as used in cookery requires an object denoting a shrimp or a prawn. (1976a [1967]:67)

Selectional restrictions in the *Aspects* model were considered syntactic, but, clearly, calling [±tumor] or [±prawn] *syntactic features* parallel to [±transitive] or [± plural] rebels against any traditional notion of syntax. Lakoff and Ross, in fact, cite this argument as damning evidence against deep structure.

Once McCawley was pushed over the abstract-syntax-to-generative-semantics brink, or nudged over it, he began to develop more compelling arguments against deep structure, including a rather notorious and ingenious one revolving around the word *respectively* that follows Halle's anti-phoneme lead very closely. But his most damaging argument for deep structure was not a negative argument that it had to go. It was a positive proposal that one could get along just fine without it.

McCawley's positive argument built on some of the lexical decomposition ideas in Lakoff's thesis. Lakoff had argued that "*kill, die,* and *dead* could be represented as having the same lexical reading and lexical base, but different lexical extensions" (1970 [1965]:100). That is, they would all involve the primitive definition for *dead* (something like NOT ALIVE), but *die* would additionally be marked to undergo the transformation, Inchoative, and *kill* would be further marked to undergo Causative, capturing rather smoothly that *dead* means *not alive*, that *die* means *become not alive*, and *kill* means *cause to become not alive*. Among the attractive features of this suggestion was an increase in scope. The transformations were necessary for the grammar anyway, Lakoff argued, to account for the range of semantic and syntactic properties in words like *hard* (as in 32); he just increased their workload.

32   a   The metal is hard.
      b   The metal hardened. [i.e., became hard]
      c   Tyler hardened the metal. [i.e., caused it to become hard]

McCawley went a step further, proposing a new rule which "includes as special cases the inchoative and causative transformations of Lakoff" (1976b [1967]:159), and collects atomic predicates into a subtree to provide for lexical insertion. The new rule, Predicate-raising, was about as simple as transformations come. It simply moves a predicate up the tree and adjoins it to another predicate, as in phrase markers 6, 7, and 8.[31]

Lexical insertion could take place on any of these phrase markers, yielding any of the synonymous sentences in 33.

33   a   Stalin caused Trotsky to become not alive.
      b   Stalin caused Trotsky to become dead.
      c   Stalin caused Trotsky to die.
      d   Stalin killed Trotsky.

Moreover, the dictionary entry for *kill* no longer needs the markerese of the *Aspects* model; it could be expressed simply "as a transformation which replaces [a] sub-tree" (McCawley, 1976b [1967]:158), like the conglomeration of abstract verbs in PM–8. The entry for *kill* could be the transformation in figure 5.2.[32]

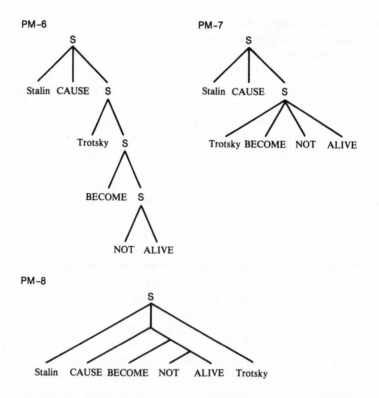

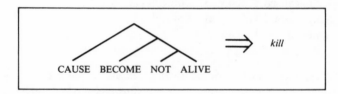

The implications of this proposal are very sweeping, and many transformation-alists found it extremely attractive. In particular, as the title of the paper in which McCawley proposed it promises, "Lexical Insertion in a Transformational Grammar without Deep Structure," it showed linguists how to make do without *Aspects'* biggest drawing card.[33] Lakoff and Ross's mimeographed assault on deep structure was important, but it was almost exclusively corrosive. Their arguments ate away at deep structure without a solid proposal for what to do once it had completely dissolved. One of the criteria they cite and then denounce for deep structure is its role as the locus of lexical insertion, the place where the words show up in a derivation, but their dismissal is remarkably curt. "We think we can show," they say, that "lexical items are inserted at many points of a derivation" (1976 [1967]:160); ergo, that there is no one specific location, no deep structure, where all lexical items enter a derivation. This bold claim is followed by a one-sentence wave in the direc-

Figure 5.2. McCawley's lexical insertion transformation for *kill* (From McCawley, 1976b [1968]:158).

tion that such a proof might take, and they're on to other matters. At best, this move looked like chucking out the lexical-insertion baby with the deep-structure bathwater. This apparent recklessness and the lack of positive substance are the main reasons that Lakoff and Ross's letter had virtually no impact on Postal at all. Even if he agreed that deep structure was compromised by their arguments (a conclusion he could not have been eager to embrace, since he was instrumental in developing deep structure), there was nothing for him to sink his teeth into.

When Postal says that "the fomenter of [generative semantics] was McCawley; I've always considered it to be his," it is primarily the lexical insertion arguments he has in mind.

Postal was off at the IBM John Watson Research Center in Yorktown Heights by this point—closer to Cambridge than McCawley, actually, but further out of the academic loop. He had stayed in contact with Lakoff and Ross, but not closely enough to find their patchy claims very persuasive. His response to their arguments, in fact, was much the same as it had been to Lakoff's solo arguments four years earlier: the case was suggestive, promising even, but much too sketchy to warrant abandoning the interpretive assumptions that had grounded his research from the outset. It was the more explicit, more closely reasoned, and more positive arguments of McCawley that eventually persuaded him to dump deep structure and work in (and on) the emerging generative semantics framework.

With the dissolution of deep structure, and the four leading figures all on their semantic horses, there is still one more point to make before we can get to the generative semantics model directly: the campaign against deep structure was in many ways a campaign against Chomsky. Ross's letter doesn't even mention him, or *Aspects* for that matter, but look at what McCawley says when he questions the existence of deep structure:

> As an alternative to Chomsky's conception of linguistic structure, one could propose that in each language there is simply a single system of processes which convert the semantic representation of each sentence into its surface syntactic representation and that none of the intermediate stages in the conversion of semantic representation into surface syntactic representation is entitled to any special status such as that which Chomsky ascribes to 'deep structure.' To decide whether Chomsky's conception of language or a conception without a level of deep structure is correct, it is necessary to determine at least in rough outlines what semantic representations must consist of, and on the basis of that knowledge to answer the two following questions, which are crucial for the choice between these two conceptions of language. (1) Are semantic representations objects of a fundamentally different nature than syntactic representations or can syntactic and semantic representations more fruitfully be considered to be basically objects of the same type? (2) Does the relationship between semantic representation and surface syntactic representation involve processes which are of two fundamentally different types and are organized into two separate systems, corresponding to what Chomsky called 'transformations' and 'semantic interpretation rules', or is there in fact no such division of the processes which link the meaning of an utterance with its superficial form? (McCawley, 1976b [1967]:105–6)

This passage, about the clearest and most succinct expression of the central issues that dominated the generative semantics debates, mentions Chomsky four times—

no Katz, no Fodor, no Postal, rather critical members of the team which designed deep structure and semantic interpretation rules to come up with the *Aspects* model—and spells out explicitly, in the language of binary choices, that there is a new kid on the block.

The Model

In syntax meaning is everything.

Otto Jespersen

The new kid on the block is the grammatical model, simple in the extreme, given as figure 5.3.

Leaving aside the *Homogeneous I* label for now, it isn't hard to see where the model came from, or why it was so appealing. The defining allegiances in the historical flow of science—call them *movements, paradigms, programs, schools,* call them by any of the hatful of overlapping collective terms of the trade—are all to conglomerations, to knots of ideas, procedures, instruments, and desires. There are almost always a few leading notions, a few themes head and shoulders above the pack, but it is aggregation that makes the school, and when the aggregate begins to surge as one in a single direction, the pull is, for many, irresistible. Witness James's enthusiasm over the manifest destiny of the philosophical school of pragmatism in 1907:

> The pragmatic movement, so-called—I do not like the name, but apparently it is too late to change it—seems to have rather suddenly precipitated itself out of the air. A number of tendencies that have always existed in philosophy have all at once become conscious of themselves collectively, and of their combined mission. (1981 [1907]:3)

Precisely this teleological sense pervaded the abstract syntacticians in the midsixties, the sense that they were witnessing a conspiracy of ideas marching ineluctably toward the position that meaning and form were directly related through the interative interplay of a small group of transformations: simplicity and generality argued for a few atomic categories; these categories coincided almost exactly with

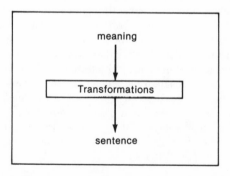

Figure 5.3. Generative semantics (Homogeneous I) (adapted from Postal, 1972a [1969]:134).

the atomic categories of symbolic logic; symbolic logic reflected the laws of thought; thought was the universal base underlying language. Meaning, everyone had felt from the beginning of the program, was the pot of gold at the end of the transformational rainbow, and generative semantics, if it hadn't quite arrived at the pot, seemed to offer the most promising map for getting there. Listen to the unbridled excitement of someone we haven't heard from for a while, someone who was there at the beginning of the transformational program, Robert Lees:

> In the most recent studies of syntactic structures using [transformational] methods, an interesting, though not very surprising, fact has gradually emerged: as underlying grammatical representations are made more and more abstract, and as they are required to relate more and more diverse expressions, the deep structures come to resemble more and more a direct picture of the meaning of those expressions! (Lees, 1970a [1967]:136)[34]

There are a number of noteworthy features to the directionality-of-abstractness appeal—as it was awkwardly dubbed by Postal (1971b [1969])—all of them illustrated by Lees's passage. First, the word *fact* appears prominently in them; that transformational analysis led to semantic clarity was an undoubted phenomenon, in need of explanation the way the fact that English passives have more morphology than English actives is in need of explanation. Second, there is usually a combined expression of naturalness to the finding (as in Lees's "not very surprising") and enthusiasm for it (as in his exclamatory ending). The naturalness follows from the conviction shared by most transformationalists (1) that they were on the right track, and (2) that the point of doing linguistics was to provide a principled link between form and meaning. The enthusiasm follows from the immense promise of the result.

If the *Aspects* model was beautiful, generative semantics was gorgeous. The focus of language scholars, as long as there have been language scholars, has always been to provide a link between sound and meaning. In the *Aspects* model, that link was deep structure. To the generative semanticists, deep structure no longer looked like a link. It looked like a barrier. As Háj Ross expresses it, still somewhat rapturously, once you break down the deep structure fence, "you have semantic representations, which are tree-like, and you have surface structures, which are trees, and you have a fairly homogenous set of rules which converts one set of trees into another one." The link between sound and meaning becomes the entire grammar. *Aspects'* semantic component was grafted onto the hip of the *Syntactic Structures* grammar, having only limited access to a derivation, and extracting a distinct semantic representation. Generative semantics started with that representation ("the meaning") and the entire machinery of the theory was dedicated to encoding it into a configuration of words and phrases, ultimately into an acoustic waveform. As an added bonus, the semantic interpretation rules could be completely discarded, their role assumed by transformations.

Chomsky's claims to Cartesian ancestry, too, as everyone could see, only fed the appeal of generative semantics. Robin Lakoff hinted broadly in a review article of the *Grammaire générale et raisonée* that when placed cheek-by-jowl with generative semantics the *Aspects* model looked like a very pale shadow of the Port-Royal work. She dropped terms like *abstract syntax* about the Port-Royal program, and drew attention to its notions of language and the mind as fundamentally logical, and to its use of highly abstract deep structures, even to a somewhat parallel discussion of inchoatives (1969b:347–50).

But she needn't have bothered. The case is even more persuasively offered in Chomsky's own book, *Cartesian Linguistics.*

Generative semantics was inevitable. In the tone of utter, unassailable conviction that he had brought to the Bloomfieldian rout, Postal put it this way:

> because of its *a priori* logical and conceptual properties, this theory of grammar [generative semantics, or, as the paper terms it, Homogeneous I] . . . is the basic one which generative linguists should operate from as an investigatory framework, and that it should be abandoned, if at all, only under the strongest pressures of empirical disconfirmation. In short, I suggest that the Homogeneous I framework has a rather special logical position *vis-à-vis* its possible competitors within the generative framework, a position which makes the choice of this theory obligatory in the absence of direct empirical disconfirmation. (1972a [1969]:135)

But, as Homogeneous *I* suggests, the story is far from over. Generative semantics leaked. The beautiful model has some mutation ahead of it, but that will have to wait a chapter or so. Even figure 5.3 wasn't entirely accurate in 1972, since, as we've seen, there were also a few other devices beyond transformations. Ross's and Postal's constraints don't change the model, because they are external to it, part of the overarching general theory that defines the model. But there was something else, which Ross called a *conditions box,* an additional trunk full of odd bits of Perlmutterian tubing and pieces of cheesecloth to filter off the grammatical effluvia the rest of the grammar couldn't control. As it turned out, *Pandora's box* would have been a better label. The filtering problem in generative semantics led to several mutations.

Before we look at those mutations, though, we should check in with Chomsky and see what his reaction was to all of these developments. It looked at the time to be a flat rejection of generative semantics, and it still looks to many people in retrospect to have been a flat rejection. But this is Chomsky, don't forget. Things were not so simple.

CHAPTER 6

Generative Semantics 2: The Heresy

Beliefs are most clearly and systematically articulated when they are formed *via negativa*. The boundaries of what is true and acceptable are marked through a systematic identification of what is false and unacceptable. . . . It is through battles with heresies and heretics that orthodoxy is most sharply delineated.

Lester Kurz

The transformationalist position is adopted in much recent work, for example Lakoff (1965) . . .

This solution is proposed by Lakoff (1965, p. A-15f), but on the transformationalist grounds that he adopts there, there is no motivation for it. . . .

The scope of the existing subregularities, I believe, has been considerably exaggerated in work that takes the transformationalist position. For example, Lakoff (1965) . . .

Noam Chomsky

The George and Háj Show

Almost everyone who talks about the beginnings of generative semantics uses the same image about Lakoff and Ross, even if only to downplay its applicability; in Lees's phrasing, "it may well not have been so important that the cat was away while those two mice played, but only that the two had come together." That is, almost everyone involved sees a significant correlation between Chomsky's 1966 sabbatical visit to Berkeley and the dramatic rise in the fortunes of abstract syntax as it seguéd into generative semantics. The reasons for this correlation are clear, and by this point, very familiar. I'll let Ross say it this time: "Chomsky is extremely intellectual, brilliant, amazingly quick in debate, and hard working, and reads everything, at huge speed, and retains everything: just an overwhelming intellect"—qualities whose effect on the MIT community is to make it "very much a one-man show." Chomsky sets the agenda whenever he is there, and setting the agenda at MIT in the first half of the sixties was tantamount to setting the agenda for transformational grammar. When he left, there was a vacuum.

Lakoff, who had a lectureship at Harvard, and Ross, who had just been appointed at MIT, had the energy, the ideas, and the charisma to fill that vacuum, at least for a while. They had conducted an informal seminar at Harvard the previous year, in which they mined the abstract syntax program to unprecedented depths—articulating, elaborating, and augmenting Postal's proposals, with constant telephonic input from McCawley. Now they were both teaching similar paradigm-stretching courses. Their students were cross-registering. And their ideas were getting airtime outside of Cambridge.

There is a fair amount of ambiguity about all of this work, the Lakoff-Ross work that pushed abstract syntax over the generative semantics brink. Unquestionably, there was a driving attempt to be, in Lakoff's phrase, "good little Chomskyans," to greet the founder of the movement, on his return from Berkeley, with "Hey Noam! [Your theory] is *even greater* than you already think it is!" But the role of a good little Chomskyan was an uncertain one:

> We were always perplexed. I could never figure out, for instance, in my work, which things [Chomsky] would like, and which things he thought were just completely off-the-wall, that he would fight against tooth and nail. Going right back to my thesis, we had terrible fights. (Ross)

One aspect of the ambiguity, that is, concerns the difficulty of knowing exactly what is required of a good little Chomskyan, a notion about as easy to grasp as an eel in Jell-O. Almost any linguist in 1966 willing to bet on Chomsky's reactions and the future of transformational research would have given Ross and Lakoff's extensions pretty good odds, though certainly not huge ones, of success. (On the slippery other hand, the modifications to the *Aspects* model Chomsky did endorse, chiefly articulated by his student, Ray Jackendoff, would have received miniscule odds; they flew in the face of the defining core of transformational research of the previous five years.)

But another, more subtle, aspect of Lakoff and Ross's ambiguity about good-little-Chomskyanism concerns knowing very clearly one of the requirements, evident in virtually every corner of transformational-generative grammar: the willingness—indeed, the drive—to wrangle fiercely about everything from the grammatical nuts and bolts of specific analyses to the overall architecture of linguistic theory. Chomsky has a very agonistic view of science, which his students lap up year after year. One of the highlights of the early transformational student's reading list, recall, was the record of Chomsky's battles with the Bloomfieldians at the 1958 Texas conference; another was his fiercely polemical dismissal of Bloomfieldianism at the 1962 ICL; another was his and Halle's brutalization of Householder.

Encouraged strongly to participate in the clash of adverse opinions, Lakoff and Ross naturally generated some notions that were pointedly antithetical to some of Chomsky's notions, and there was clearly a sense of challenge to a few of their specific proposals. Chomsky was known, for instance, to be unsympathetic toward Postal's suggestions that verbs and adjectives were members of the same category, and Lakoff argued the case forcefully. Ross's re-analysis of English auxiliaries went after a set piece of *Syntactic Structures* and an exemplar of the first decade of Chomskyan linguistics. "Is Deep Structure Necessary?" attacked the *Aspects* model's most spectacular success.

Ambiguity or no, with Chomsky gone their proposals became more daring, more original. Ross advanced his performative analysis. Lakoff elaborated his abstract verb proposals to include quantifiers (*many, all,* etc.), and explored arguments against deep structure. And their classes were as exciting as their proposals. Dwight Bolinger, then a Romance professor at Harvard, remembers the "sparkle and fireworks" of the classes vividly:

> I attended quite a few sessions, watching those two throw ideas back and forth with each other, and listening to the always bright, sometimes brilliant comments from the class. One had the feeling of grammar in the making, as if we were out exploring a freshly turned field, with everyone retrieving specimens before the bulldozers moved in.

Bolinger also views this period as "the time of greatest TG ferment, out of which challenges to the Master were bound to emerge."[1]

As Lees's use of the cat's-away metaphor suggests, though, the vacuum may have been a sufficient but not a necessary condition for the growth of generative semantics. There was something particularly vital in the combination of Ross and Lakoff. Some found their catalytic relationship extremely grating, others found it inspiring, but there is a common thread to the reactions: that Ross is a tremendously gifted naïf, while Lakoff's gifts are more sophistical, more superficial. A thinly veiled *dialogue à clef* which played on these perceptions made the mimeograph rounds at the time:

G: H, I'm going to prove that your eyes are purple.

H: How will you do that?

G: You see this pencil? It's not purple, and it's not your eyes. You see this book? It's not purple, and it's not your eyes. [Twelve more arguments on this template.] So, I have fourteen arguments that your eyes are purple.

H: Great data, but when I look into the mirror, I see that my eyes aren't purple.

G: Yes, but that's a fact about mirrors. There are many strange and wonderful facts about mirrors that we don't yet understand.

H: Thank you, G, for explaining this to me.

This dialogue is astute about some of the elements of generative semantics—it parodies the argument style very well—but it misses the give-and-take dynamism of the collaboration rather widely. The dialogue does Lakoff a rather severe, if common, injustice; G[eorge] is far more of a shyster than a scientist. But Ross clearly comes off worse; H[áj] is a linguistic rube being taken for a ride by the big-city con man. In fact, Ross is both an incisive judge of argumentation and (almost) Lakoff's equal in volubility. You can take Jerrold Sadock's word for it. He became a significant generative semantic contributor, and he describes his initiation in the terms of a classic conversion story, Ross breezing into Urbana like Elmer Gantry:

> In 1966 Háj Ross stopped by the University of Illinois. He was actually on a big tour across the country. He spent four days in Urbana, and he lectured almost nonstop. It

was absolutely awesome. He was so good and this work he was doing was so intricate and so neat that I immediately fell in love with it. I decided then and there [about my dissertation topic]. I was going to find more arguments and push that theory.

Ross was not the only one on tour; he and Lakoff began a dedicated round of conference appearances—arguing the virtues of abstract syntax and insinuating the virtues of a generative semantics—from California to Texas to Illinois. They used the lectern like a pulpit. Of the two, Lakoff was the most forceful, and in this sense the purple-eyes dialogue reflects his professional relationship with Ross. The parody goes way too far—G just leads H around by the nose—but Lakoff was clearly the driving energy in the alliance. Bolinger calls their Harvard sessions "a kind of duet. George was the lead voice, but Haj harmonized at intervals in a lively exchange that kept the class on its toes" (1991 [1974]:29), and Ross is remembered from this period principally for a few technical innovations; in particular, for his performative hypothesis and, something we will see a little later, his"squishoids." (If you're wondering why his magnificent island work isn't one of the accomplishments for which his activity in the growth of generative semantics is best remembered, it's because subsequent ideological developments detached that work from generative semantics, and even, to some extent, from Ross.) Lakoff's contributions are recalled on another scale. He was the major prophet of abstract verbs, lexical decomposition, rule government, and generative semantics' most notorious innovation—also reserved for a later appearance—global rules. He wrote the major policy statement of the movement, "On Generative Semantics" (1971b), and confronted Chomsky at every turn. He was the most tireless boundary-crosser, swelling the movement with material he brought back from forays into philosophy and psychology.

They were, in short, a gifted duo. Their performances, built on their extravagant analysis of "Floyd broke the glass," were tireless and inspiring. The preface to the proceedings of the 1967 Texas Conference on Universals, for which they were invited commentators, gives some indication of the impression they were cutting:

> We would give a totally false picture of the symposium if we did not mention the contribution of Ross and Lakoff, who not only played a role as discussants but were kind enough to remain in Austin to devote approximately six hours to a presentation of some of their recent work [followed by an epitome of the "Floyd" analysis]. (Bach and Harms, 1968:viif)

Generative semantics was well on its way; lo, in the east, Chomsky.

The Backlash

> Should they, urging the minds of their listeners into error, ardently exhort them, moving them by speech so that they terrify, sadden, and exhilarate them, while the defenders of the truth are sluggish, cold and somnolent?
>
> Augustine

Of his first venture into generative semantics, Lakoff says "Chomsky didn't like it from the beginning" (compromising, to some degree, his remarks that the Ross-Lakoff seminars were all about being good little Chomskyans; Lakoff knew Chomsky's position), and although he means in this remark only to date Chomsky's antipathy from his 1963 paper, "from the beginning" holds more widely of Chomsky, all the way back to the start of his career. As early as 1955, in a paper that Sledd (1962 [1958]) takes to represent the hardest Bloomfieldian hard line on meaning, Chomsky was very explicit on this point—"if it can be shown that meaning and related notions do play a role in linguistic analysis, then . . . a serious blow is struck at the foundations of linguistic theory" (1955a:141)—and he has only rarely softened this position, never abandoned it. Indeed, Chomsky's position can be traced even earlier, to Goodman and Harris, Chomsky's teachers, certainly to Bloomfield, who had a deep influence on Harris, back to Whitney, an influence on Bloomfield, and earlier. Chomsky comes to what looks like the brink of generative semantics in *Cartesian Linguistics,* true, when he describes deep structure as something "that is purely mental, that conveys the semantic content of the sentence" (1966a:35), but he is careful to distance himself from the characterization. He offers it only as his interpretation of the Port-Royal stance, and accompanies it with phrases like "so it is claimed." Chomsky has always had a sidelong interest in meaning, but he is a deep and abiding syntactic fundamentalist.

Phrases like "so it is claimed" were easily overlooked in all the commingle of meaning and mentalism in *Cartesian Linguistics* and Chomsky's other works of the mid-sixties, especially since they came hard on the heels of the promising work of Katz and Fodor, and Postal, and Chomsky himself, that seemed to tame meaning. Too, his early hyper-Bloomfieldian arguments about keeping structure and meaning separate were read much less frequently than works suggesting a new semantic expansiveness. So it was, in Ray Jackendoff's phrase, "a dreadful surprise" when Chomsky returned to MIT from his Berkeley sabbatical in 1967, and launched a series of lectures that completely reversed the abstract syntax trend of deepening deep structure.[2] His students, after some initial shock and puzzlement, found these lectures invigorating; Jackendoff, a second-year doctoral student, was particularly thrilled that his own recent research, research which had no home at all in generative semantics and compromised the *Aspects* model considerably, was resonating with Chomsky.

There was no puzzlement about where these lectures—the "Remarks" lectures, named after the famous paper that came out of them, "Remarks on Nominalization"(1972b [1967])—were aimed. Everyone immediately perceived them as an attack on generative semantics, a reactionary attempt to cut the abstract legs out from underneath the upstart model. The best term for the lectures is Newmeyer's. He calls them a "counteroffensive" (1980a:114; 1986a:107), which captures the air of reaction, assault, and upping-the-ante in which they were received. Chomsky, though—here the story gets particularly bizarre—says he wasn't much interested in generative semantics or in abstract syntax at the time, that he "knew virtually nothing about" either, that he barely noticed the work Postal, Lakoff, Ross, and McCawley were up to. His 1967 MIT lectures, he says, were just a delayed reaction to Lees's *Grammar of English Nominalizations* (written in the very late fifties with considerable input from Chomsky).

I have depicted Lees's *Grammar* as something of a negative example in the growth of generative semantics, and aspects of that growth were certainly a reaction to the welter of overly specific transformations and the panoply of syntactic categories in the early sixties work Lees epitomizes. But another aspect in the growth of generative semantics continued a trend that Lees was pioneering. A key part of Lees's project was to account for a wide number of nominalizations with a smaller number of underlying categories, transformationally deriving, for instance, the nouns *transgressor, transgression,* and *transgressing* (the gerundive) from the verb, *transgress.* Abstract syntax took over this category-reduction work with a vengeance, thinning out the deep lexical categories to just nouns and verbs, deriving everything else (and a good many of the nouns to boot) transformationally. To the man who has only a hammer, runs the adage, everything looks like a nail. To Lees, who had only the transformation, everything must have looked like a case of movement.

Chomsky now recalls thinking that Lees "was way overdoing the use of transformations," an understandable concern, though Lees was religiously following the program laid out in Chomsky's own work (in particular, 1975a [1955], 1957a, 1962a [1958]). This concern with Lees's use of transformations, Chomsky says, preyed on him for several years, simmering on some cortical back burner for seven or eight years, when it boiled over into his work, coincidentally taking a completely inverse approach to the one that abstract syntax had adopted and coincidentally coming just months after Lakoff and Ross had declared generative semantics open for business.

The big puzzle of the "Remarks" lectures, the dreadful surprise Jackendoff mentions, was their theoretical direction. One of the flagships of the transformational program had always been what *Syntactic Structures* called the "very interesting and ramified set of nominalizing transformations" (1957a:72). The transformational analysis of nominalizations dated back to Harris's earliest work, and Lees's *Grammar* was widely regarded as a how-to lesson in transformational grammar (it had already been through five printings). But Chomsky organized his 1967 MIT lectures around the proposal that some nominalizations are better treated without transformations. In particular, where earlier work had derived 1a from 1b transformationally, Chomsky now argued that words like *refusal* should be represented fully in the lexicon.

1   a   George's refusal to beat his rug.
    b   George refused for George to beat his rug.

For everyone who sat in on the lectures, or read the paper abstracted from them, the immediate motive of this approach seemed clear: Chomsky's proposal struck squarely at the kidneys of generative semantics.

The implications of his arguments went considerably beyond the procedural matter of how to account for nominalization. He proposed an approach—called the *lexicalist hypothesis*—which greatly reduced the heretofore divine right of transformations to change syntactic categories, a right which virtually defined the abstract syntax program underlying generative semantics, and he systematically stood this approach in direct opposition to the *transformationalist* approach, which, although it was really the only approach around (and, in any case, Lees was its most obvious proponent), he connected closely and repeatedly with George Lakoff's unpublished thesis.

Chomsky's formulation of the lexicalist hypothesis proposes that "a great many items appear in the lexicon with fixed selectional and strict subcategorization features, but with a choice as to the features associated with the lexical categories noun, verb, adjective" (1972b [1967]:22), but it rapidly came to mean, in Lakoff's practical rephrasing, that lexical items "may not change category in the course of a transformational derivation" (1967:8).[3] Once a category, that is, always a category. This move declares much of the abstract syntax program illegitimate. One of *its* flagships was the boiling down of lexical variation to just a few underlying categories—a distillation that depended heavily on the ability of established transformational mechanisms to change categories; to, for instance, change a deep verb into a surface adjective.

Now the founder of the field proclaimed the arguments behind all such derivations misguided. In the process, he disowned much of the work elaborated under his specific endorsement (he had virtually directed Lees's thesis on nominalizations). He advocated a position he specifically dismissed in the New Testament: "Clearly, the words *destruction, refusal,* etc., will not be entered into the lexicon as such. Rather, *destroy* and *refuse* will be entered into the lexicon . . . [and] a nominalization transformation will apply at the appropriate stage in the derivation" (1965 [1964]:184). In the bargain—perhaps the most dreadful surprise of them all—he cast suspicion on one of the cornerstones of the *Aspects* model, the Katz-Postal hypothesis.[4] As Chomsky's student Susan Fischer recalls these lectures, "the shit hit the syntactic fan."

Lexicalist Wranglings

> One reads "Remarks on Nominalization" without a clue that the description there of the lexicon and of deep structure existed nowhere before and was developed only because GS [generative semantics] pushed Chomsky to redefine his position—quite radically.
>
> Robin Lakoff

Almost everyone outside of MIT, and some inside, took the "Remarks" lectures to be little more than crackpot revisionism. There are actually a few more indications in *Aspects* that the lexicalist hypothesis is one potential future for Chomsky than Robin Lakoff prefers to admit, and there are a few more clues in "Remarks on Nominalization" that it is a reaction to abstract syntax and generative semantics than she admits. But, in general, her assessment is right on the mark: *Aspects* certainly looks to endorse abstract syntax much more than lexicalism, and "Remarks" plays its cards pretty close to the vest. It dresses up surprising new proposals as natural extensions of the *Aspects* model, and characterizes the natural extensions pursued by the "transformationalists" as a misguided detour.

In support of these partly conservative, partly radical, moves Chomsky's arguments are vague, half-baked, and ad hoc. To get out of one pickle, for instance, he appealed to analogy—the common-sense notion that a speaker might coin the word *prioritize* on analogy to an existing word like *civilize,* or *oxidize*—said to be a property of performance rather than competence. Well and good, you might say, except that his dismissive views on the role of analogy were well known,[5] and except that the use he puts it to, supporting lexicalism, is rather peculiar. For reasons we needn't go into, the following two grammaticality claims serve Chomsky's argument for the lexicalist hypothesis:

2   a   His criticizing the book before he read it annoyed me.

     b   *His criticism of the book before he read it was strangely insightful.

"Suppose that we discover, however, that some speakers find [expressions like 2b] quite acceptable," Chomsky adds (and, indeed, some people do), then "we might propose that [2b is] formed by analogy to the gerundive nominals [*criticizing* in 2a], say by a rule that converts X-*ing* to the noun X *nom* (where *nom* is the element that determines the morphological form of the derived nominal) in certain cases" (1972b [1967]:27–8). If this analysis is correct, he goes on, then it "indicates that speakers who fail to distinguish [2b] from [2a] are not aware of a property of their internalized grammar" (1972b [1967]:28); in short, for speakers who do not conform to the grammaticality judgments required by the lexicalist hypothesis, 2b is *still* ungrammatical, but acceptable due to analogic rules of performance.

Even more problematic at the time than such offhand appeals was the schema he proposed for phrase structure rules which has become known as *x̄-syntax* (pronounced "x-bar syntax"). It completely reoriented the phrase-structure component, critically required an amorphous and undefended entity called *specifier,* harkened back in some uninvestigated way to earlier work by Harris—and its only role in the paper is to explain certain correspondences (the structural similarities between sentences like *Chomsky criticized Lakoff* and noun phrases like *Chomsky's criticism of Lakoff*) that were already explained under the transformational account but not under the new lexicalist hypothesis. That is, the only purpose of x̄-syntax seemed to be to prop up the lexicalist hypothesis, and the only purpose of the lexicalist hypothesis seemed to be to undermine abstract syntax.

Chomsky repudiated successful early work, proposed radical changes to the *Aspects* model, and opened ad hoc escape channels for those changes—all on the basis of quite meager evidence—with no more motivation, as far as anyone could see, than to cripple the work of his most productive colleague and of some of the most promising former students they shared.

Lakoff and Ross were shaken to the bone. They were expecting some sort of brouhaha when Chomsky returned, but it was taking a completely different route than anyone could have predicted, and, from their perspective, a very peculiar one. They sat in on the lectures (which included specific attacks on many of the abstract analyses presented in Lakoff's dissertation), and argued with Chomsky every step of the way. They also tried to talk about these issues with Chomsky outside of class, but had very little success. Chomsky was getting thickly involved in political activism at the time, and his popularity was spreading rapidly inside and outside of academia. He had very little free room in his schedule. Lakoff, in particular, felt snubbed, but Postal says that Chomsky probably just didn't see any potential profit in the meetings:

> Viewed from his perspective, I don't think he had any great desire to talk to them. It certainly was no urgent matter for him. They viewed it with some urgency. They felt that they probably had something important to tell him, but I doubt if he felt they had anything important to tell him.

I doubt that he really avoided them. I suspect he didn't care one way or the other, and if he had more pressing things to do, he would do them. He would certainly much rather talk to some guy from *Time* magazine than them, so they probably got shunted off, and got pissed, and gave up.[6]

Lakoff's interpretation is far less generous, and he dates the animosity that characterized much of the next ten years from exactly this point: "[Since we couldn't talk to him in private] we figured that the only way to talk to him was to start bringing up counter-examples in class, and that infuriated him." Chomsky's classes became increasingly contentious, heated arguments breaking out frequently, and Lakoff was usually at the center of the storm.[7] But Chomsky is a virtuoso debater, and he handled Lakoff's objections to the satisfaction of virtually the entire audience. His own graduate students were especially impressed. In addition to the distinctively anti–generative semantics bent of the lectures, they were also incredibly rich theoretically, and set off a flurry of activity among his students, activity which rapidly bloomed into the interpretive semantics model.

But this was only the beginning. Chomsky followed the "Remarks" lectures with an absolutely unprecedented period: several years of almost exclusively negative rhetoric, documented in two long papers (1972b [1968]:62–119, [1969]:120–202), whose express aim was to eviscerate generative semantics. Chomsky has certainly devoted considerable time and energy to assaulting other people's work. His "Bad Guys" courses in the watershed years of MIT linguistics are notorious, and he lets very little criticism escape his withering notice, if only to condemn it as beneath his notice. Take, for instance, this typical dismissal:

> I will not consider Reichling's criticisms of generative grammar here. The cited remark is just one illustration of his complete lack of comprehension of the goals, concerns, and specific content of the work he was discussing, and his discussion is based on such gross misrepresentations of this work that comment is hardly called for. (1966b [1964]:9)

But, even given his penchant for returning the fire of his critics and for attacking views he regards as intellectually dangerous, this period is anomalous. Aside from his paper to the Ninth International Congress, most of his arguments against the Bloomfieldian program are scattered among original proposals. And B. F. Skinner, the apostle of everything Chomsky considers evil in twentieth-century psychology, received only two papers, over a decade apart (1959 [1957]; 1970), his review of *Verbal Behavior,* and a more general, more clearly ideological assault in *The New York Review of Books.* Attacking generative semantics, a development within his own framework, by colleagues and former students, occupied virtually all of his linguistic energies for several years. Lecture after lecture, he took up, dissected, and discarded generative semantics proposals: one lecture to McCawley, one to Postal, three to Lakoff, another one to McCawley . . .

Students and opponents alike were alarmed. One of the most fecund minds of the century was doggedly stamping the life out of a heresy, producing little (by his standards, extremely little) original research. Even the "Remarks" work, rich and original as it was (both lexicalism and x̄-syntax eventually became immensely successful), was too sketchy to make much impact, especially on people who weren't present for the lectures and had only the mimeographed version to go by. The evidence he presented for the lexicalist hypothesis was inconclusive at best, spliced in with unconvincing appeals like the one to analogy, and came with a vaguely adumbrated proposal (x̄-syntax) for completely overhauling the phrase structure rules. As one of his most sympathetic students at the time puts it, apropos of Chomsky's seminal influence: "We all thought he had shot his wad."

The Extended Standard Theory

> By making the enemy soldier doubt the well-foundedness of his cause, and giving to confused and bewildered minds reason for doubting the wisdom of their choice, you bring to your side all chances of success.
>
> Chairman Mao

Chomsky, in a deft capture of the nominal high ground, promptly dubbed the *Aspects* model "the standard theory," and his lexicalist revisions to it, "the extended standard theory." Most linguists thought that Postal's program and the generative semantics that it spawned were the most natural extensions to *Aspects*. Indeed, Lakoff's public-letter answer to "Remarks" speaks repeatedly of abstract syntax as "extending transformational theory" (1967:7) and of mechanisms like filters and constraints as constituting an "extended version of the assumed theory of grammar" (1967:4). More importantly, the convergence of transformations and semantic interpretation rules seemed like the inevitable fulfillment of the Katz-Postal principle's promise, virtually the inevitable fulfillment of linguistics: "We know what we want to say," Lakoff said in the first argument offered for generative semantics, and then we "find a way of saying it" (1976a [1963]:50); after the innovations leading to *Aspects,* it seemed apparent that transformations were the way of saying it.

There had always been a problem with transformations moving blithely from meaning to form, sentences like the familiar 3a and 3b.

3   a   Everyone on Cormorant Island speaks two languages.
    b   Two languages are spoken by everyone on Cormorant Island.

For the Chomsky of *Syntactic Structures,* 3a and 3b illustrate "that not even the weakest semantic relation (factual equivalence) holds in general between active and passive" (1957a:101); therefore the Passive transformation alters meaning; therefore transformations won't take you from an underlying meaning to a surface form. Katz and Postal tried to patch this hole, and for them 3a and 3b "are ambiguous in the same way and so are full paraphrases of one another" (1964:72)—they both have the same-two and the different-two readings. For the increasingly cagey Chomsky of *Aspects,* 3a and 3b are a minor puzzle. They seem to suggest that the Katz-Postal hypothesis is "too strong," but "we might maintain that in such examples both interpretations are latent," a maintenance that he finds attractive (1965 [1964]:224n9), and in his most authoritative statement on the subject in *Aspects* he says

> It is clear, as Katz and Fodor have emphasized, that the meaning of a sentence is based on the meaning of its elementary parts and the manner of their combination. It is also

clear that the manner of combination provided by the surface (immediate constituent) structure is in general almost totally irrelevant to semantic interpretation. (1965 [1964]:162)

But, in 1967, with the "Remarks" lectures, when the matter seemed completely settled and people were pretty happy with it, Chomsky shifted his position again, in ways that almost everyone but Jackendoff found alarming.

It's impossible to recover what Chomsky said in the lectures, but in the accompanying paper, he hints about the renewed relevance of such examples as our Cormorant-Island sentences, describing the *Aspects* grammar as one which contained "semantic rules that assign each paired deep and surface structure generated by the syntax a semantic interpretation" (1972b [1967]:12). This statement was strictly true within the *Aspects* framework, of course, as would be an assertion that the grammar assigned a semantic interpretation to the phonological representation, or the derivation as a whole, but in *Aspects* Chomsky clearly says that deep structure is where the semantic action takes place, and this new phrasing signaled a change in the winds. More explicitly, but still with a coyness that many found exasperating, he also said in print that year:

> I think that a reasonable explication of the term "semantic interpretation" would lead to the conclusion that surface structure also contributes in a restricted but important way to semantic interpretation, but I will say no more about this matter here. (1967b:407)

Such remarks were surely elaborated in his lectures. His students likely knew his reasons and some of his arguments. Everyone else was puzzled and disconcerted.

In 1968, he was more forthcoming, saying that "surface structure determines (at least in part) the scope of logical elements [like the quantifier *every* of *everyone* in 3a and 3b]" and consequently plays "a role in determining semantic interpretation" (1972b [1968]:209).[8] And by this point his students had begun circulating papers, over and under the ground, with analyses that invoked surface structure interpretation. There goes the Katz-Postal hypothesis: if surface structure plays a role in meaning, then transformations can't be semantically neutral. By 1969, for Chomsky, sentences like 3a and 3b were hopelessly problematic for both the *Aspects* model and, *a fortiori,* generative semantics, requiring a major overhaul of the former and a wholesale rejection of the latter.

Much of the work that took the name *extended standard theory,* that is, extended nothing. It was completely negative—looking not for solutions to specific problems but only for data that compromised or disconfirmed the Katz-Postal hypothesis, and, therefore, generative semantics.

Why the shift? Chomsky, again suggesting his move was wholly incidental to generative semantics, says it was occasioned by his student, Ray Jackendoff:

> The first person who offered a substantial critique of the Standard Theory, and the best as far as I can recall, was Ray Jackendoff—that must have been in 1964 or 1965. He showed that surface structure played a much more important role in semantic interpretation than had been supposed; if so, then the Standard hypothesis [the Katz-Postal hypothesis], according to which it was the deep structure that completely determined [meaning], is false. (1979 [1976]:151)

Chomsky's chronology is off a bit here. Jackendoff didn't even show up at MIT until the fall of 1965.[9] And (although this depends on how he chooses to define "a substantial critique of the Standard Theory") Chomsky's notions of priority are similarly off. Lakoff's thesis, which many people regarded as a (constructive) critique of *Aspects*, was in circulation before Jackendoff finished his first semester, and several other linguists were revamping the *Aspects* model substantially before Jackendoff—including Gruber, Postal, Ross, and Fillmore. The only point we can grant Chomsky here is *best*, since it is a subjective evaluation and he's entitled to his opinion.

Too, the development of this surface-structure critique seems not to be as simple as Chomsky recalls. In particular, the student remembers that Chomsky was entertaining this critique of the *Aspects* model at least as early as he was:

> During my second year, [when] Chomsky was away, writing *Cartesian Linguistics*, . . . I started thinking about rules of interpretation, and in an independent study under Halle during the fall semester of 1966 wrote three papers—one arguing for base-generated pronouns whose antecedents are determined by rules of interpretation; one arguing that selectional restrictions are enforced in the semantic component, not as part of lexical insertion (as in the *Aspects* position); and one trying to work out a theory of base-generated PRO and relative pronouns instead of the deletion rules then in fashion. Ross pointed out to me that this would require surface structure interpretation, in contradiction to the then-dogmatic assumption that deep structure determines interpretation. This worried me a lot. I remember a discussion with Chomsky on one of his brief reappearances on campus during that year, in which I brought up this worry, and he said calmly, Of course surface structure plays a role in interpretation. So evidently he was thinking along those lines too by at least early in 1967.

In one way, Chomsky's calm "Of course surface structure plays a role" looks like just another example of Lees' observation that he "is so smart that any idea you came up with he had already thought of, and thought over long and deeply." Chomsky's quantifier-scope quandaries date at least to *Syntactic Structures*, and his lexicalist developments evidently germinate from 1960 or so, in partial reaction to *The Grammar of English Nominalizations*. In another way, though, the issue has less to do with Chomsky's vaunted depth of analysis than with his enigmatic choice of which particular thread of which analysis to follow at any given time. The quantifier-scope, semantic-neutrality-of-transformations question, raised very early, seemed to be solidly resolved by 1965; raising it again not only seemed very peculiar, it seemed like a step backward, and a betrayal of everything the transformational community had been working toward. Even after several years of MIT arguments against the Katz-Postal hypothesis, Barbara Partee expressed her bafflement over the effort in terms that harken back to the Bloomfieldian *bête noire* of the entire Chomskyan community:

> The position that transformational rules *don't* preserve meaning is of much less inherent interest than the contrary position, since it amounts simply to the position that a certain strong hypothesis is false. It may of course turn out to be the correct position, but it doesn't seem like anything one could rationally *want* to champion—it is analogous to the position that synchronic rules don't reflect historical development, or that not all languages use the same stock of phonological features, and so forth, the sum of

all such positions being [in the notorious Joos quotation] that "languages can differ from each other without limit and in unpredictable ways." (1971 [1969]:8; Partee's italics)

Now, add to this program against the Katz-Postal hypothesis Chomsky's arguments on the use, or over-use, of category-changing transformations, and his efforts looked odder yet. Why swim against the current of the most productive post-*Aspects* line of research in transformational grammar, particularly on the basis of the sketchy and vaguely adumbrated arguments of "Remarks"? And, in retrospect, why not raise his concerns about category-changing transformations seven years earlier, since they clearly had ramifications for the whole movement toward deep structure that characterized Chomskyan linguistics in the early sixties?

Who knows?

But it seemed to many linguists that Chomsky had simply lost the courage of his convictions, that he couldn't follow through on the natural implications of his own work, that he was retrenching rather than advancing, that he was, sadly, being left behind by more vigorous, more visionary linguists.

The Best Theory

The fact is that Homogeneous I is . . . the best grammatical theory *a priori* possible.

Paul Postal

Meanwhile, generative semantics continued to gather momentum, the four horsemen of the apocalypse—Postal, Lakoff, Ross, and McCawley—each playing somewhat distinct roles in its propagation, as they had in its genesis. Certainly there was much scatter and overlap in these roles, and an element of caricature shows up if the tendencies are pushed too far, but their respective impacts on the movement are quite clear. Lakoff and Ross, we have seen, were the evangelical salesmen, taking generative semantics to the consumers. McCawley, as we will see shortly, was the den mother, rearing a second generation of generative semanticists in Chicago. Postal was the big gun.

Though just into his thirties, Postal was a worthy veteran, below only Chomsky in stature, and his endorsement of generative semantics meant a great deal to its success. He did not proselytize for the model with anywhere near the fervor with which he sold the earlier versions of transformational grammar; he went to conferences to present and sometimes to dispute, but not to attack. He was, as always, prolific during the dispute, with almost all of his publications organized around reasons to pursue generative semantics, with most containing asides on why not to pursue its "almost completely open ended" rival (Postal, 1972b [1970]:37n4), and with some even carrying the resolutely polemical, doggedly certain tone of his anti-Bloomfieldian campaign. But his mere presence in the camp was almost enough on its own. Postal gave the movement a great deal of necessary credibility.

He looked, in fact, like its inventor. His spearheading of the movement toward deeper, more abstract syntactic analyses, everyone could see, grew into generative semantics. So, even though he didn't publicly ally himself with the theory until 1968, his influence suffused and supported it from the beginning. His and Katz's formulation of their meaning-preservation hypothesis, and their performative prefiguration, and their abstract trigger morphemes gave generative semantics a developmental warrant. His sponsorship of Lakoff's dissertation—a text at the blurry borders of abstract syntax and generative semantics—went a long way toward its validation, and therefore toward the validation of many arguments underlying those theories. His offstage (but acknowledged) participation in Lakoff's deep verb arguments, in Ross's auxiliary analysis, in McCawley's arguments, all gave them increased authority. The progression from abstract syntax to generative semantics seemed so inevitable to many linguists at the time (both supporters and opponents) that they assumed Postal was the latter's "real" progenitor.

And many people found his arguments the most challenging. Katz, for instance, says that Postal's work "was the sort of thing that I found I could take seriously. I felt that I had to come to terms with his work in some way. From an outsider's point of view, it seemed to me that he was the brains behind the outfit." Keyser says, simply, "He was the guru." Much of this perception had to do with style of presentation; Katz, Keyser, and Postal spoke the same language. The other three generative semantics principals are given to more informal and breezy presentations. Postal's work is—uniformly—very detailed, very rigorous, and the reasoning is very explicit. Part of the perception also had to do with ethos. Postal worked with Katz, studied with Keyser, went to battle on behalf of Chomsky. He was their colleague. Lakoff, Ross, and McCawley were their students. For the brood of younger interpretivists Chomsky was rearing at MIT, the situation was similar. Postal had been a teacher, in addition to being something of a legend; the other three were little more than older students. Postal's arguments are also among the strongest for generative semantics; his *remind* paper in particular (1971b [1969]) is a model of generative argumentation, and contains, for Ross "the best articulation of what generative semantics is about."

His most remembered contribution, though, is probably "The Best Theory." In part its reputation comes from its thoroughly polemical nature—polemical in terms definitive enough to be called *Postalian*—the best theory of the title referring, of course, to generative semantics, with the discussion proceeding as if anyone choosing not to work on the model was clearly in need of psychiatric counseling. In part it is remembered because it was the losing entry in a rhetorical showdown with Chomsky at the most heated conference of the dispute, the 1969 Texas Conference on the Goals of Linguistic Theory. In part it is remembered because Chomsky has a penchant for invoking it before describing generative semantics as "the worst theory," in fact as "the worst possible theory . . . the worst imaginable theory." In part it is remembered because it's a great paper.

The Texas Goals conference was full of "bad vibes and yelling and shrieking and shit like that," Ross recalls: "By then it was clear there was going to be no quarter asked nor given." And Postal's paper was in the middle of it all, full of bad vibes and provocations, neither asking nor giving any quarter. The title makes it clear intransigence was the order of the day, not compromise, and it makes for a nice contrast with Chomsky. The paper he gave at the conference was published under the title, "Some Empirical Issues in the Theory of Transformational Grammar," as neutral and mild-mannered as one could get, in sharp relief with Postal's titular claim to inherent superiority, but the content was more intransigent that anything in Postal's paper, the tone is bullying, and the whole paper is as far from compromise as possible.[10]

Postal's paper makes an unbudging but straightforward Ockham's-razor case for the strong theoretical priority of generative semantics over more complex models, especially all versions of interpretive semantics, with a few aspersions and digressive counter-arguments thrown in. Chomsky goes after every generative semantics argument he can squeeze into the paper (taking the opportunity to attack Fillmore's related grammar to boot), and offers nothing but vague promissory notes for an alternative. As always, the performance is impressive. He begins and ends with conciliatory statements, adopts a removed, disinterested persona, and exhibits interest only in the good of linguistics. But he attacks relentlessly, constantly characterizing generative semantics as "uninteresting" and "vacuous" and "totally obscure" and "[having] no substance" and "permitting any rule imaginable" and "[incorporating] at best dubious rules" and "[constituting only] a terminological proposal of an extremely unclear sort" and "not only unmotivated but in fact unacceptable" and much, much more. Interpretive semantics, though, "is probably correct, in essence" and "very interesting" and "more natural" and "somewhat more careful" and "well-supported" and "to be preferred" and "again to be preferred" and "more restrictive, hence preferable" and, of course, considerably more.[11] Postal's belligerence pales in comparison.

McCawley in Chicago

This paper is concerned with claims about the applicability of transformations to idioms made in a recent article by the notorious war criminal, U. S. Air Force Lt. Bruce Fraser. Fraser maintains that 'conjunction reduction will never be applicable' within an idiom, that 'no noun phrase in an idiom may ever be pronominalized [or] take a restrictive relative clause,' and that 'gapping never occurs' within an idiom; he in addition states that he has 'been able to find no idioms in which a noun phrase may be clefted.' Field work which I and my colleagues Yuck Foo and Tri Bung Quim have carried out, using as informants other U.S. Air Force war criminals who were undergoing political reeducation, has demonstrated the existence of a class of counterexamples to all of Fraser's claims of supposed inapplicability of transformations to idioms. I refer to the idioms *take a piss, take a shit,* and *blow a fart.*

Quang Phuc Dong

McCawley was something of a proselytizer, more than Postal but less than Lakoff and Ross, and he published a number of important foundational papers. In conferences, he was especially compelling when he showed (gently) how a given interpretivist account couldn't pull the weight its author claimed for it, coupling the demonstration with a suggestion that generative semantics was better suited to the task. On paper, his arguments were more persuasive yet, and, in hindsight, his publications are more definitive even than Postal's.[12] McCawley also lectured winningly and participated in several crucial gatherings. He was at the important 1967 Texas Universals conference, arguing about the need for deeper, more semantically transparent deep structures; he was at the 1968 Urbana Linguistic Institute; he was at all the major Chicago Linguistic Society meetings; he taught semantics at the 1969 First Scandinavian Summer School of Linguistics and the 1970 Tokyo International Seminar in Linguistic Theory, while Ross taught syntax and Kiparsky taught abstract phonology. He got around. And he took generative semantics with him. But his principal role was to stabilize the movement, give it a home, and nurture a second generation of generative semanticists in and around Chicago.

Postal was at IBM, with no students, by the time generative semantics began to mushroom, though he also took occasional teaching assignments. Ross was at the only interpretive enclave, MIT, under a formidable shadow which he says (much too gloomily) kept him from ever "influencing a single mind there." Neither had any students who contributed to generative semantics. Lakoff moved from Cambridge, Massachusetts, to Michigan to Palo Alto to Berkeley, preventing him from cultivating a stable base of graduate students. He had a few notable students who made an imprint on the model—especially Guy Carden at Harvard, and later, at Michigan, John Lawler. But McCawley's impact was substantially greater. He was comfortably ensconced in the University of Chicago, attracting a group of bright young students, many of whom made significant contributions to generative semantics (among the more notable, Robert Binnick, Alice Davison, Georgia Green, Judith Levi, and Jerry Morgan). Moreover, Lees was at the nearby University of Illinois (Urbana) and very tolerant of the new trend, as was Zwicky, an inspiring teacher who also attracted bright young graduate students to the theory.

When McCawley's (and Lees's, and Zwicky's) students encountered the inspirational fervor of Lakoff and Ross, generative semantics made the transition from an alternative version of transformational grammar to a movement. Two events were especially important for the growth of generative semantics, for its roots in Illinois, and for defining its genius: the 1968 Linguistic Institute at Urbana and the fifth meeting of the Chicago Linguistic Society the following spring.

Lakoff and Ross taught at the Institute (officially, the course was "Abstract Syntax"), McCawley also taught, and brought many of his more promising students over from Chicago. Lees, the legendary polemicist and articulator of the romantic pre-*Aspects* days, was the Institute's director. The mix was, in the argot of the period, mind-blowing. The spirit of the Institute has been immortalized by several of the participants in "Camelot, 1968" (Lancelot of Benwick and others, 1976 [1968]), a paper so recklessly enthusiastic for generative semantics and hostile to interpretive semantics that even McCawley, a free-speech advocate and itinerant pornographer, felt compelled to bowdlerize it (1976b:249). The paper, in its own words, is "an account of some of the linguistic Events of that Year: wherein are detailed the Declarations of the New Court and the Weapons used in the Awefull Battle to repeal Certain Decrees of the Old Court" (1976 [1968]:249). Newmeyer, a University of Illinois student who attended the Institute, has somewhat harsher words for the Events of that Year, saying the Institute

stands out not only as the high-water mark in the ascendant tide of generative semantics, but also as the epitome of mixing reasoned argument with pure showmanship and pure salesmanship. (1980a:152; 1986a:102)

There was a strong thread of comaraderie binding the Institute participants, a reincarnation of the spirit that pervaded MIT's Research Laboratory of Electronics less than a decade earlier. But, as the tone of "Camelot, 1968" suggests, the genius presiding over Urbana was not the earnestness of Noam Chomsky; it was the irreverent gusto of Lakoff, Ross, and, especially, McCawley. The latter, for instance, had written under the pseudonym of Quang Phuc Dong, from the South Hanoi Institute of Technology, and the Institute participants wore their S.H.I.T. T-shirts with great pride. (Quang's other position in the relevant period was at the Free University of Central Quebec, but that affiliation apparently had no sartorial impact.)

By the following spring, says Ross, "the forces of light and truth had gotten their act together." The fifth meeting of the Chicago Linguistic Society included crucial generative semantics papers by Postal and Lakoff; significant contributions to the theory by Ross and Robin Lakoff; and a wide range of papers investigating the implications of the model by members of the second generation—Green, Morgan, Newmeyer, Laurence Horn, and Lauri Karttunen. (McCawley's paper that year, on Tübatulabal phonology, was uncharacteristically out of the fray.) The proceedings of the conference (Binnick and others, 1969) constitute, if not quite a manifesto— generative semantics never spawned one—a much more thoroughly explored, carefully reasoned, and challenging articulation of the theory than any other document in its brief history. Certainly it represented a more comprehensive invitation to linguists outside the central group than Lakoff and Ross's "Is Deep Structure Necessary?" or even than McCawley's several, isolated early papers.

And, of course, all the participants had a great deal of fun. One of the generative semanticists' most distinctive qualities was the ability to combine probing scholarship with thoroughgoing levity. This feature is reflected in, among other mirrors, the movement's papers—in frivolous titles, funny data constructions, and a rather loose-jointed style. Ross's paper at the conference, for instance, was titled "Guess Who?," and Postal's included a whole range of sentences like 4.

4 Irma's a blonde and she got it caught in the fan.

Ross's paper explored a rule which operates under very sophisticated conditions and phenomena which create havoc for interpretive theories. Postal's example (4) illustrates that words like *blonde,* although they are referential for constructions like *has hair which is blonde* cannot sponsor a pronoun that refers to *hair.*

Generative semantics was out in the light, out from under the shadow of *Aspects,* and well ahead of Chomsky's new, ill-specified, lexicalist, x̄-syntax, surface-semantics, interpretive model in the transformational sweepstakes.

Bad Blood

> The enmity of one's kinfolk is far worse than that of strangers.
>
> Democritus

Postal's label for this period—"The Linguistic Wars"—looks extreme to anyone outside the field, and to many linguists who have entered the field in its wake, but *wars* is the only fitting term. It was a vicious, aggressive, frequently ignominious period. On one side, there were the spirited and abrasive generative semanticists, who had already captured much of the high ground. On the other side, there was Chomsky with a growing number of supporters, young and eager for the types of giant-killing episodes that had made legends out of Lees and Postal. Toward the middle, Katz was buttressing the *Aspects* model against both hordes. And around them all was a vast wash of ill-formed issues, underdetermined proposals, and amorphous data. Like Verdun, there were no dry footholds.

Chomsky's papers were severe but relatively well mannered, with only occasional jibes—especially in the notes, and especially directed at Lakoff—about "misleading" (1972b [1969]:141n19) and "totally obscure" (1972b [1969]:148n22) arguments. The lectures, though, were considerably more agonistic. He argued in these lectures and throughout this period as he always argues—calmly, dispassionately, even convivially. When his voice rose (again the target was usually Lakoff) it was out of exasperation with his interlocutor's vehemence, or outright rudeness. And a little testiness is understandable. It was, after all, Chomsky's class.

But even with his polite and assured tone—or rather, partially *because* of his polite and assured tone—Chomsky's polemics are almost always very hard for their targets to take. Lakoff is quoted in *The New York Times* as saying "Chomsky . . . fights dirty" (Shenker, 1972), and the impression is quite widespread. In particular, very few people recognize themselves in the representations Chomsky gives of their arguments. The Bloomfieldians, of course, were completely mystified by many of his versions of their work, and Skinner complained that "Chomsky simply does not understand what I am talking about."[13] But, since Chomsky's impact on these two sources was so devastating that they no longer carry a lot of authority at the moment—maybe they were just cranky because they were so badly roughed up—listen to The Venerable Quine's reaction to one of Chomsky's representations of his work:

> Chomsky's remarks leave me with feelings at once of reassurance and frustration. What I find reassuring is that he nowhere clearly disagrees with my position. What I find frustrating is that he expresses much disagreement with what he thinks to be my position. (1969:302)

This is several steps removed from "Chomsky fights dirty," but where Quine is frustrated by the funhouse mirror Chomsky holds up to his work—stretching his ideas here, squashing them here—and where Skinner simply refused to look, others are enraged. Certainly Lakoff was. He recalls Chomsky's principal tactic during the

"Remarks" lectures as systematic misrepresentation, and his own reaction as furious umbrage:

> [In one lecture] he took up McCawley's paper on *respectively,* and he put forth the argument that McCawley was arguing against as McCawley's position, and he himself put forth the position that McCawley was arguing for, and he said "See how dumb McCawley is."
> There were a hundred people present, or so. And I raised my hand and I said "No. You've got it wrong. You've got the positions reversed."
> He said "No. No. I'll read you the text." And he read a bunch of things out of context. Of course, nobody in the room had the text, except him. I just got furious about this.
> The following week, he took a paper of mine and did the same thing. He put forth as my position a position I specifically avoided saying. And then he argued, correctly, against this position. And I said "No. I specifically did not give that position in the paper." Then he began, with me in the room, to take quotes out of context from my own paper, with no one else having the text. It was mortifying.

Lakoff's memory is surely blurring and smudging things a bit here, and he has called upon his poetic license to render the dialogue, but it is the kernel in his story that is important: the generative semanticists found Chomsky's presentation of their positions to be willful, mean-spirited distortions, and the result was a series of heated exchanges, and lasting bitterness.

Others remember the lectures a little differently, particularly about the source of aggression:

- Lakoff would sit in the back row and say "Noam! Noam! You're wrong! You're wrong! Everything you say is wrong!" He just basically made a pest of himself.
- The typical scenario at Chomsky's course would be that Chomsky would say something, then Robin [Lakoff], who was pathologically shy but smart, would whisper into George's ear, then George would raise his hand and in his smarmy pseudo-civil way make some comment.
- [Lakoff] was combative and impolite . . . [He] would go to Noam's class and sit at the front of the room. One time Noam said something, and George said "I have been saying the same thing." Noam asked "Where did you write about it?" And George responded "I have been lecturing about these things, and if you are interested, you should come to my class."

The level of gall required for anyone, let alone a junior lecturer, to tell the inventor of the field to attend his classes if he wanted to stay current goes right off the chutzpah meter. It could well go right off the accuracy meter as well (though perhaps not—it is not dreadfully out of character, and several people report the same incident); there have to be some distortions in a collection of quotations that have Lakoff exclaiming "Everything you say is wrong!" one minute and being smarmy the next, sitting in the back row and at the front of the room. And the animosity was probably more evenly distributed than most of the participants recall. Most tend to lay the blame for all evils and misunderstandings squarely on one side or the other, rarely seeing fault with both. But two constants emerge from all the reports: Chomsky was allergic to Lakoff, and Lakoff was pushy with Chomsky.

The encounters galvanized both sides. The generative semanticists found Chomsky intractable, if not perverse, and consequently felt that much of his criticism was empty. Chomsky's students found him to be eminently reasonable, and his opposition—mostly embodied in Lakoff, whose fuse was easily the shortest, but also in Ross, and in scattered sympathizers, like Carden—to be intractable, if not perverse.

Katz was there too, and he played an especially interesting role in all this, one much like the role Herbert Butterfield describes for Marin Mersenne, an important presider at the birth of modern science, "a man who provoked enquiries, collected results, set one scientist against another, and incited his colleagues to controversy" (1957:83). Katz certainly made valuable contributions of his own, but his most prominent role was that of provocateur, fanning the fires on every side. One student of the period, for instance, remembers his participation in Chomsky's lectures thusly: "Jerry Katz was very contentious—not just about generative semantics-interpretive issues, about all kinds of stuff." He was contentious about all kinds of stuff, but Katz's particular bugaboo was any analysis that threatened the *Integrated Theory–Aspects* notion of deep structure, what he called at the time, "the CKP thesis about deep structure" (1970:221)—an ironically appropriate designation, since the initials stand for *Chomsky-Katz-Postal,* and Katz now stood midway between his old allies on the question of deep structure. For Katz, Chomsky was Momma Bear, whose deep structure was too shallow, Postal was Poppa Bear, whose deep structure was too deep. His own, the one everybody had agreed upon only a year ago, was juuust right.

In Chomsky's class, defending the CKP thesis largely meant fighting Chomsky, but away from MIT Chomsky's revisions to the *Aspects* model were not getting a very serious hearing, and the main opposition was the burgeoning program that officially dated itself from the discovery of a negative answer to the question, "Is [the CKP] Deep Structure Necessary?" Katz's chief polemical writings of the period are all attacks on generative semantics, and, although he differed sharply from the general trend in Cambridge at the time, Katz's allegiance to the *Aspects* model put him in Chomsky's camp on the central issue at the outset of the hostilities, the nature of semantic rules. He was one of the staunchest bearers of the interpretive semantics banner.

The friction and rancor at Chomsky's lectures set the tone for the several years of bickering that followed. There were a few frothing publications, like Ray Dougherty's (1974) "Generative Semantics Methods: A Bloomfieldian Counterrevolution," and an exchange between Katz and McCawley that saw titles like "Interpretive Semantics Meets Frankenstein" (McCawley, 1976b [1971]:333–42) and "Interpretive Semantics Meets the Zombies" (Katz, 1972a), but the more blatantly abusive papers were generally confined to mimeograph circulation—"Camelot, 1968," for instance, and Dougherty's "Generative Semantics: Galileo Died for Your Sins," and a best-kept-anonymous parody of Jackendoff, in which he is portrayed as a Nazi scientist attacking a Jew.

There was certainly a lot of bickering in the publications, but little of it was overt. It usually took one of two general forms, neither of which look like bickering to the uninitiated. One form is best represented by Chomsky's many attacks on generative semantics—explicit in its disagreement with theoretical stances or proposed mechanisms, but not betraying much overt hostility. The other form is best represented by many of the generative semantics papers—for instance, Lakoff's syntactic amalgams paper (1974). It contains implicit attacks on two foundations of the Chomskyan program (transformational derivations and the competence-performance dichotomy), but does not, except in passing, mention either of these issues, or even Chomsky, or, for that matter, even appear to be participating in a controversy of any sort.[14]

Most of the explicit enmity was not in journals or proceedings. As in the first round of Chomskyan hostilities, against the Bloomfieldians, the majority of invective was oral. There was, for instance, a celebrated flare-up in 1967 at a small conference in La Jolla which was something of a coming-out party for generative semantics. Jackendoff rode west, as he recalls, essentially "as Chomsky's point man," and the generative semanticists regarded him much as Chomsky regarded Lees on his first appearance at MIT, a gunslinger sent to quell the insurrection. Unlike Lees, though, Jackendoff was not converted, and there were a number of shrill exchanges, including the loud swapping of graphic imprecations between him and another renowned Chomskyan gunslinger, Postal. There was an equally vitriolic, and even more public exchange between Jackendoff and Lakoff a few years later, at a plenary session of the 1969 Linguistic Society of America conference, "when for several minutes [they] hurled amplified obscenities at each other before 200 embarrassed onlookers" (Newmeyer, 1980a:162; 1986a:126).

There was more, much more. Every linguist even close to the mainstream at the time has a host of anecdotes about individual firefights, in colloquia, job interviews, classrooms, chance hallway meetings . . . seemingly wherever and whenever representatives of the two sides came into contact. The stories range quite widely in color and credibility, and it is difficult to take many of them at face value. They have grown epic in the retelling: the stupidity of the antagonists, the forbearance of the protagonists, the simplicity and clarity of the point under dispute, are all surely exaggerated. One account even has Chomsky and Lakoff, very unlikely in the roles of Achilles and Hector, physically wrestling for a microphone.

But, in concert, these tales have the general cultural value any body of folklore has. Take the several versions of a generative semantics story which has Jackendoff (or sometimes Dougherty), up against thorny opposition and unable to think for himself, thumping *Aspects* like the Bible and intoning plaintively "but *Noam* says . . ." On the other side of the schism, there are several versions of a story which has some interpretive semanticist attempting to use an *argumentum ad absurdum* of abstract syntax (say, arguing that *the* is a verb), which a generative semanticist falls for, hook, line, and theoretical sinker. ("By golly, *the* is a verb. We'll have to change the theory right away to accommodate this new fact. I'll get on the phone this afternoon.") Some approximations of these stories may have played themselves out, but their accuracy is less important than the attitudes they reveal. The generative semanticists felt that the other camp consisted largely of Chomsky and his puppets. The interpretive semanticists felt the other camp far too willing to overhaul its general theory on whatever scanty evidence they heard over lunch that day. Each side thought the other uncritical and obtuse to self-evident points.

Although the general level of mutual tolerance was pretty low, and many tempers ran high in individual encounters, and there is lasting bitterness in some mouths, and there are clear casualties of the debate, it is important to note that courtesy, friendship, and respect for one another's work were not completely absent between antagonists. Postal, for instance, at the height of the wars, dedicated his through-and-through-generative-semantics book, *On Raising* (1974), to Chomsky. Jackendoff saluted the generative semanticists in the preface to his *Semantic Interpretation*, for motivating his research, though the compliment does have an undercurrent—he says much of his interpretive model is "due to my being in sufficiently violent disagreement with their work to want to do something about it" (1972:xii). Chomsky himself was unflaggingly supportive for the careers of at least some of his opponents, writing letters on their behalf, supporting Ross's appointment and advancement at MIT, doing the same for Perlmutter, trying to keep Postal there. McCawley and Chomsky corresponded about their respective arguments, with warmth and mutual respect, albeit with little agreement.

Still, the battles went on, boiling over into more public forums, first to include linguists outside the Chomskyan tradition, and then to include scholars in neighboring disciplines. Chomsky had acquired a great deal of prominence by this point—psychologists and philosophers routinely debated his work, his books were reaching a fairly broad public, there were profiles and articles in magazines like *The New York Review of Books, Horizon, Time, The New Yorker,* even an appearance in Norman Mailer's book *The Armies of the Night,* and he was at the forefront of protests against the U.S. invasion of Vietnam. Additionally, the generative semanticists became very wide-ranging in their interests, drawing on the work of logicians like Zadeh, philosophers like Grice and Searle, psychologists like Rosch, and sociologists like Goffman. They shopped far and wide for data and mechanisms to build their theory around, and they were never reticent about letting their discontent with Chomsky be known. Inevitably, the sectarian squabble within Chomsky's framework found a wide audience.

A mildly notorious symptom of the public dimensions of the dispute is an exchange of letters by Lakoff and Chomsky in *The New York Review of Books.* The *Review* had just published an article by Searle (1972) on Chomsky's revolution. While revealing the familiar tendency of outsiders to be several years behind the work in any field, it is fairly evenhanded and offers a good précis of Chomsky's major contributions to that point, their implications for linguistics, and their ramifications for other fields. It certainly includes some lionizing, but the praise is neither undue nor excessive. Still, it got under Lakoff's skin. He felt compelled to write a letter to the editor (1973c), a letter with two purposes. The stated motive is to spell out recent transformational-generative developments more clearly. Searle, for instance, had said that the interpretive-generative semantics dispute was occurring "entirely within a conceptual system that Chomsky created" (1972:20), which by the early seventies was no longer strictly true; as we will see, the grounds of the debate had shifted considerably. It is the barely hidden motive, though, that is of principal interest at the moment: to diminish Chomsky. Lakoff accuses Chomsky of dishonesty both about the Bloomfieldians and about the Cartesian parallels of his own work. He says that the important results about Chomsky's transformational work are its disconfirmations, mostly at the hands of generative semanticists. He shoos away psychologists who might be interested in examining Chomsky's hypotheses. He discounts the importance of Chomsky's early work, and depicts his current work as misguided. Several of Lakoff's individual points have merit, but the constraints of a brief, nontechnical letter led to many conflations and simplifications, and, in any case, much of it is completely irrelevant to Searle's article. The letter serves as little more than a catalog of reasons to suspect and devalue Chomsky, if not to disregard him completely. It is simple invective, of which the kindest characterization is *gratuitous*.

Chomsky, of course, responded (1973b). His letter tries to clarify his position on a number of the issues Lakoff raised, and to vindicate his general approach to the study of language, but he faces exactly the same constraints of space and audience that cause Lakoff problems, giving rise to similar conflations. He also returns invective for invective, and raises the pitch of the attack considerably. As a result, the strongest message that comes across in the letter is his extreme disregard for Lakoff's intellect, if not for Lakoff's person. The letter is littered with unsubtle *ad hominems,* and far outstrips Lakoff's in the category of devaluing its object:

> Lakoff presents a very confused picture . . .
> is completely wrong . . .
> misrepresents my account . . .
> has thoroughly misunderstood the references he cites . . .
> is confused beyond repair . . .
> presents a hopelessly garbled version . . .
> [has] discussed views that do not exist on issues that have not been raised, confused beyond recognition the issues that have been raised and severely distorted the contents of virtually every source he cites.

This collection is only a representative sample, not a comprehensive list, of the aspersions. Lakoff's letter is invective in the technical sense of epideictic rhetoric which aims to devalue its object, though it shades toward the ordinary language sense in some passages; Chomsky's letter does not bother much with shading, and its invective fits the common definition of venting venom much more closely. Just as he and Halle had spent more time abusing Lakoff's teacher, Householder, than addressing his arguments years earlier, Chomsky now focused on Lakoff, rather than on his case. Lakoff was naturally incensed, though by this time in the debate he had very little love left to lose for Chomsky, and tried to publish a longer and more vituperative response to the response. The editors wisely decided the matter had gone far enough.

The generative-interpretive dispute reached an even more public forum the same year, getting aired in over forty column-inches of the Sunday *New York Times.* The story, by a childhood schoolmate of Chomsky's, Israel Shenker, skews the dispute somewhat, and oversimplifies it drastically, but it is a reasonable epitome of how things looked to a good many people at the time, including many linguists: that Chomsky had, rather sadly, stopped short of fulfilling his own revolution, which another generation of Young Turks was now bringing to fruition. By this time the theme was getting familiar in popular articles. *The New Yorker,* for instance, had quoted Janet Dean Fodor saying "Chomsky was once a revolutionary. Now he has been forced into the position of a conservative" (Mehta, 1971:218), and Searle's article that spawned the Lakoff-Chomsky exchange had commented on the irony that "the author of the revolution now occupied a minority position in the movement he created. Most of the active people in generative grammar regard Chomsky's position as having been rendered obsolete" (1972:20). But Shenker's piece was the first devoted to the dispute alone, and its verdict is the same as the others.

The debate is rendered nicely by Shenker's quotations from all the key players:

ROSS: There's no question that Chomsky is a genius and has revolutionized
            linguistics, and created a field of mathematics which didn't exist
            before, and helped in the revolution of psychology and the rebirth of
            interest in problems like cognition and perception. But I think that
            he's so committed to the truth of this view he grew up in that he can't
            see where it's inadequate.

POSTAL: By ignoring meaning [as Chomsky does] you get an artifact.

LAKOFF: Since Chomsky's syntax does not and cannot admit context, he can't
            even account for the word *please.* . . . But it's virtually impossible to
            talk to Chomsky about these things. He's a genius, and he fights dirty
            when he argues. He uses every trick in the book.

MCCAWLEY: Chomsky assumes that there are sentences which belong to the lan-
            guage and other sequences of words which don't—and the grammar-
            ian's task is to write rules which will determine which belong and
            which don't. Postal and Lakoff and I say this isn't a coherent notion.

And, getting the last word,

CHOMSKY: I'd gladly accept parts of their work if these turn out to be right, and all
            of it if it is right, and I assume they'd do the same.

Lakoff feels he was misquoted here,[15] though it is clearly his opinion that Chomsky's eristic techniques are not always aboveboard. The others may also feel poorly represented. But the point here is not whether these quotations are accurate. They could be blatantly wrong, although all fit the temper and the issues of the debate quite well. The important point for the moment is just that they indicate the hullabaloo was loud enough for the neighbors to hear.

Chomsky is clearly outgunned in the piece. The only quotations in the article that could even remotely be ascribed to "his side" in the schism are from Katz, who dismisses it all as an inconsequential squabble over formalisms, and from Bever, who simply fills in some historical detail. Indeed, there is really no impression that he has a side at all. All the defining criteria are negative: he can't see certain implications because of his Bloomfieldian blinders, he doesn't attend to meaning, he doesn't attend to context, his central methodology lacks coherence. He looks isolated and stuck in the past. At best, in the remarks of Ross, he looks something like Einstein, who did pioneering work in quantum physics, but whose classical prejudices wouldn't allow him to follow its implications to the necessary conclusions. At worst, in the remarks of Lakoff, he looks like a petulant old Freud, using every trick in the book to ostracize and marginalize his erstwhile followers—and it is this image that Shenker fosters. The headline is "Former Chomsky Disciples Hurl Harsh

Words at the Master," though only Lakoff's remarks are truly harsh, and the article is framed by a discussion of Freud and his heretical disciples, particularly Tausk, whom Freud treated the most shamefully. Although Chomsky is just into his forties, and he gets the last word, and the word is very politic, he comes off pretty much as a sorry old crank.

The Vicissitudes of War

And when men did engage in debate about their deepest concerns, they found
that each man could say unto his brother, Racca, thou fool.

Wayne Booth

Nyaah, nyaah!

George Lakoff

Weapons of Choice

"For base men," Empedocles warns us, "it is indeed possible to withhold belief
from strong proofs" (fragment 55), but baseness, like beauty and contact lenses, is
in the eye of the beholder. When scientists across the way refuse to grant the force
of an argument that its sponsor finds compelling, they are base by definition, which
is why the most common accusations in science are forms of *ad hominems*—implications of personal failings, like stupidity, sloppy scholarship, and often, dishonesty. Personal attacks are far more common in science than is generally thought
(suggesting, among other things, that thin veils such as *x is confused,* or *y fails to
understand the issue,* or *z misrepresents my position* actually work). Even the
parched and stolid pages of professional journals are full of them, and they cluster
fructiferously around paradigm disputes. It isn't difficult to see why. Most immediately, mudslinging is easy. If you find an observation disagreeable, says Hawking,
"you can always question the competence of the person who carried the observation out" (1988:10). But there is of course a more emotional reason. Personal vituperation is a very natural response when you see something clearly, believe it
strongly, propound it fervently, and find your colleagues looking at you like a cheap
snake-oil merchant. Listen to Alvarez's reasoning about Oppenheimer, coming to
the most damning conclusion one could offer of an atomic physicist in the fifties:

> Oppenheimer and I often have the same facts on a question and come to opposing deci-
> sions—he to one, I to another. Oppenheimer has high intelligence. He can't be analyz-
> ing and interpreting the facts wrong. I have high intelligence. I can't be wrong. So with
> Oppenheimer it must be insincerity, bad faith—perhaps treason.[1]

160

There weren't any accusations of political treason in the generative-interpretive debates (though such charges would have had a much different impact in the anti-establishment sixties, possibly even conferring honor), but there were plenty of charges of scientific treason—especially of the x-is-just-a-Bloomfieldian-in-Chomskyan-clothing variety. Since Bloomfield had become the absolute Bogey Man of linguistics, and since the generative and the interpretive semanticists seemed to be working within the same general framework, it took a while for these accusations to surface. They marked the death throes of the dispute, the obvious sign that reconciliation was completely hopeless, that each side was on its own, that neither side thought the other was doing Real Linguistics.

When the differences still revolved around a few technical questions under the same Chomskyan umbrella, a wide variety of *ad hominems* ruled the day. The most common of them, as with most debates, was the straw man charge: "x is arguing not against my real position, but against a caricature of my position." Thickly implicit, and occasionally explicit, in the charge was willful, deceitful distortion, but building straw men is a natural side effect of conviction. The practice of reading (or listening) only for weaknesses inevitably leads to distortions of one another's arguments—McCawley's wonderful title, "Interpretive Semantics Meets Frankenstein," jumps on Katz for his attack not on a living, breathing generative semantics, but on the spare parts he has stitched and bolted together to attack.

Another frequent and related charge, in science generally and the generative semantics dispute particularly, is vagueness. Witness Chomsky's troubles—a man who says "I never had the slightest problem reading papers in [generative semantics]"—trying to decode generative semantic arguments:

At best, the logic of [Lakoff's] argument is unclear . . . this is hardly clear enough even to be a speculation. (1972b [1967]:35)

Until these matters are cleared up, I see no force to McCawley's contention. (1972b [1967]:48n30)

I do not see how these questions can be resolved without undertaking an analysis of these structures which does propose rules as well as underlying structures, and in this sense, goes well beyond the approach to these questions that Lakoff presents. (1972b [1968]:82)

I fail to see what more can be said, at the level of generality at which McCawley develops his critique. (1972b [1968]:80)

The charge, naturally, returned to Chomsky and his camp in quadruplicate:

[To Chomsky:] You ought to make explicit just what new devices you would need in the lexicon. . . . Throughout your ["Remarks"] lectures you remarked in an offhand fashion that you would need certain principles of word-formation in the lexicon, without stating just what these principles would be, how many, what kind, [or] how they would differ from one another. Without fairly precise claims along these lines and without fairly precise claims about what new rules of semantic interpretation you would need, it is impossible to figure just what the lexicalist hypothesis is claiming. (Lakoff, 1967:6)

[An] interpretive or 'Surface Structure' approach to semantic interpretation . . . has been posited, not very clearly in my opinion, in a number of papers by various authors at MIT over the last year or so, most notably Chomsky, and Jackendoff. (Postal, 1988a [1969]:79)

The concept of 'independent motivation' is highly obscure in ["Remarks"]. (Ross, 1973c:212)

I can speak only vaguely of the bad guy [interpretive semantics] conception of semantic structure, since the papers by bad guys which I have seen generally give very little clue as to what they think a semantic structure looks like or a semantic interpretation rule does. (McCawley, 1973a [1970]:276–77)

Some motivation for this charge is once again simple disparagement, implying woolly-mindedness or deliberate obfuscation. Some of it also has to do with a combative spirit that prevents one from tolerating the same level of informality from opponents as allies: a confederate writing vaguely is *suggestive* or *promising* or *intriguing;* enemies are *sketchy* or *obscure* or *sloppy.* And much of it has to do with the high levels of vagueness which are in fact present in such disputes. "Is Deep Structure Necessary?" is a vague paper, which Ross and Lakoff's colleagues found promising and programmatic and their opponents found very thin. "Remarks" is a vague paper, which Chomsky's students found promising and programmatic and his opponents found very thin.

There were also, of course, issues—some of them genuine, some smokescreens, some of them lasting, some ephemeral. Some of them require a little attention, some don't.

Chomsky's first direct move against generative semantics, for instance, was to charge that it was just a hollow imitation of the *Aspects* model, and the charge is a whiff of smoke we can dispel pretty quickly. "It is easy to be misled," Chomsky said, "into assuming that differently formulated theories actually do differ in empirical consequences, when in fact they are intertranslatable—in a sense, mere notational variants" (1972b [1968]:69).[2] Moving past the aspersions latent in *misled* and *mere,* what Chomsky is saying is quite clear: no matter what two theories look like, in terms of formalisms, architecture, what-have-you, if they make the same empirical predictions, they are just different ways of saying the same thing. It is around this question, in fact, that Chomsky coined the term, *standard theory,* and then argued repeatedly that generative semantics was a mere notational variant of the standard theory.

Katz picked up this stick and used it to beat generative semantics awhile, but almost everyone else was completely baffled by the argument; in particular, by how it could conceivably count as a reason to dismiss generative semantics. The most immediate implication of the notational-variants position is that there is just no point in squabbling, that the two theories should go off arm-in-arm, like Tweedledum and Tweedledee after their pillow fight. But, if so, generative semantics wins. Everyone agreed, Chomsky included, that generative semantics was by far the prettiest and simplest theory—two criteria that have traditionally been extremely successful in theory-marketing.

So—since Chomsky's motivation wasn't to endorse generative semantics—he devoted a fair amount of attention to showing not only (1) that generative semantics says the same thing as the standard theory, but (2) that generative semantics is wrong. Almost everyone, friend and foe alike, found the combination of these claims incoherent, and the notational variants charge had a fairly short shelf life. It was widely treated as a joke (one that linguists were still chuckling about ten years later).[3] After a few more developments, after Chomsky had coined the term *extended standard theory,* after Jackendoff had put some flesh to it, after the claims about surface-structure interpretation were developed enough to be taken seriously, Chomsky's argument was a little clearer: the standard theory is wrong, generative semantics compounds these wrongs, and the extended standard theory remediates them. But these issues were too fuzzy at the time to get anywhere. By the time they were sharp enough to make sense, the debate had moved on to other stages. We can safely ignore the argument here.[4]

The next set of arguments (taking *next* metaphorically; we are, as always, in overlapping domains), though, does require some attention. These are the exchanges over deep structure and the Katz-Postal hypothesis that formally opened the dispute—the issues the debate is most remembered for, and the issues over which generative semantics is said to have lost its empirical shirt. They are certainly real issues, but they are ephemeral—for the very good reason that they had the clearest resolutions: deep structure was thrown out, and so was (the strong version of) the Katz-Postal principle.

Chomsky's next tack was restrictiveness—generative semantics was said to be descriptively wanton, interpretive semantics responsibly restricted—and it was hugely successful. The argument was never resolved; in fact, it *could* never be resolved (Gazdar, in Longuet-Higgins and others, 1981:[690]). But Chomsky won the day completely. Restrictiveness was the principal issue which led to generative semantics being laughed from the scene for irrationality and error, while Chomsky's post-*Aspects* approach triumphantly defined a new vein of constraint-based research that determined much of the linguistics of the seventies and eighties. The issue is real enough, though its inability to be meaningfully resolved also lent considerable smoke to the exchanges, and its effects are still widespread.

For their part, the generative semanticists countered with charges that the interpretive camp was playing fast and loose with crucial notions like grammaticality and sweeping data under any rug they could find. What Chomsky and Jackendoff couldn't handle directly, they said, was simply banished to some netherworld of ill-behaved phenomena, in a shady shell-and-pea game with the facts.

Bloomfield might have been invoked at this point. The generative semanticists might easily have accused the interpretivists of his border-shuffling gambit—recalcitrant data, Bloomfield had said, should "properly be disposed of by merely naming them as belonging to the domain of other sciences" (1926:154; 1970:129). It might look like linguistics, and it might smell like linguistics, Bloomfield said, but if we call it "psychology" or "sociology" we don't have to look at it or smell it any more. Chomsky and his followers did not invoke other sciences, but they did shift the borders within their model—chiefly, the competence-performance border and the syntax-semantic border—and conveniently left the troublesome facts on the other side.

The Bloomfieldian charge, in fact, came first from the interpretive side. Dougherty raised it in an embarrassing so's-your-old-teacher diatribe that really wasn't very clear on what it was alleging. He just called up a convenient specter and, insofar as he went beyond name-calling, attended narrowly to certain structuralist methodologies he mistakenly took to represent the forces of darkness associated with that specter. But the case was actually very strong, particularly in connection with those flip-side epistemologies, empiricism and rationalism. Bloomfield was an unapologetic empiricist, as was the generation he invigorated and defined; Chomsky was a champion of rationalism, but the first generation he invigorated and defined became increasingly empiricist.

At this point, the two sides could do little more than wave good-bye and walk away from each other, with some muttering of "Nyaah, Nyaah!" over their shoulders.

The Decline and Fall of Deep Structure

> He who sets to work on a different strand destroys the whole fabric.
>
> Confucius

The earliest concerted argument against deep structure is McCawley's *respectively* argument, coming in a curious paper delivered at the 1967 Texas Conference on Universals and published in the proceedings of that conference with a postscript substantially modifying his position.[5] The main body of the paper, "The Role of Semantics in a Grammar" (McCawley, 1968a), attempts to stretch the *Aspects* model in several directions, particularly in the use of indices in syntax and the use of logic as a tool of semantic representation. Most of the discussion involves interesting and subtle facts about plurality and subject-verb agreement, such as in 1 (in which the subject is a two-member conjunction, hence plural, and the two verbs agree with the subject, although the predication involves only one member at a time; Howard likes the movie, but it disappoints Jenny).

1   a   Howard and Jenny like the movie and are disappointed by it respectively.
    b   *Howard and Jenny likes the movie and is disappointed by it respectively.

For all the stretching, however, the discussion falls clearly within the (rather broad) scope of the *Aspects* model, in the genre of abstract syntax. But between delivering the paper and submitting the manuscript for the proceedings, McCawley turned "from a revisionist interpretive semanticist into a generative semanticist" (1976a:159), and the postscript has the whiff of manifesto about it. Like any recent convert in a rhetorical enterprise like science, he went quickly to work uncovering arguments to justify the conversion, focusing on the titular question of Lakoff and Ross's "Is Deep Structure Necessary?" The principal argument he came up with (and, actually, Postal had a hand in it as well) is based on indices, quantifiers, and the word *respectively*. Chomsky's manhandling of this argument is one of the things that caused Lakoff to blow his cool last chapter, and we'll get to that manhandling in a moment. First, the argument. Taking the venerable Hallean syllogism as a template, McCawley argued from sentences like these three

2 a Larry and Tom love their respective wives.
  b Those men love Susan and Dot respectively.
  c That man loves Susan and Dot.

Aspects says that these sentences need at least three levels of representation: surface structures (which, for our purposes, we can represent with 2a–c), deep structures (3a–c) and semantic representations (4a–c). The last two sentences are especially interesting since both of them come from very similar deep structures, different only in the distinguishing indices which identify whether the two occurrences of *that man* refer to the same guy or to different guys, as in 3b (underlying 2b) and 3c (underlying 2c). The deep structures (terminal strings) are:

3 a Larry loves Larry's wife and Tom loves Tom's wife
  b that man$_i$ loves Susan and that man$_j$ loves Dot
  c that man$_i$ loves Susan and that man$_i$ loves Dot

And McCawley gives their respective semantic representations as:

4 a $(\forall x)\, x \in \{Larry, Tom\}$ & $(x$ loves $f(x))$, where $f(x) = x$'s wife
  b $(\forall x)\, x \in \{x_i, x_j\}$ & man(x) & $(x$ loves $f(x))$, where $f(x_i) = $ Susan, $f(x_j) = $ Dot
  c $(\forall x)\, x \in \{x_i, x_i\}$ & man(x) & $(x$ loves $f(x))$, where $f(x) \in \{$Susan, Dot$\}$

So far, so good; now comes the tricky part. According to the sketch of a rule McCawley offers (but never formulates), his *Respectively* transformation, all of these phenomena can be handled in a unitary way. The trouble is that the rule can't operate on structures like 3a–c (*Aspects*-type deep structures). It needs access to universal quantifiers ($\forall$) and set indices. That is—and this is the key point—for McCawley's *Respectively* transformation to work, it has to go directly from the semantic representations (4a–c) to the surface structures (2a–c). Generative semantics therefore can do the same work with one rule that takes the *Aspects* model at least two rules—operating between two different levels—a semantic interpretation rule which extracts the quantification facts, functions, and predicate relations from 3a–c, and a transformation (Conjunction-reduction) which crunches the set indices to yield *those men* from *that man$_i$ and that man$_j$* and *that man* from *that man$_i$ and that man$_i$*.

The argument may be pretty dense for nonlinguists, and for linguists without some training in logic, but structurally it is almost identical to Halle's galvanizing argument about the phoneme. Halle had said there are three levels in a Bloomfieldian grammar, with a box of rules separating each one, and he uncovered some data from Russian that could only be handled by this three-level grammar in a clumsy way, with two identical rules, one per box. But if the middle level—the phonemic level—was tossed out and the grammar restructured, one rule by itself would do the trick. For reasons of simplicity, therefore, the middle level had to go.

McCawley said that there are three levels in an *Aspects* grammar, with a box of rules separating each one, and he uncovered some data from English that could only be handled by this three-level grammar in a clumsy way, with two rules, one per box. But if the middle level—the level of deep structure—was tossed out, and the grammar restructured, one rule would do the trick (see figure 7.1). For reasons of simplicity, therefore, deep structure had to go.

Chomsky predictably and rightly objected to the argument, but not on any of the grounds which seem fairly obvious—such as "Let's see the rule before we decide if anything follows from it" or "Sure, but do the facts really show that there is a unitary phenomenon here? (And, if so, prove it.)" He objected by reconstructing McCawley's argument in a peculiar and ugly way; in fact, in a way that lends credence to Lakoff's charge of willful distortion.[6] But there is a more bizarre element in the story: McCawley had written Chomsky to repudiate his reconstruction long before it was published. Chomsky was nonplussed. He wrote McCawley back, maintaining that his reconstruction was accurate and, in fact, that McCawley was confused about what his own paper says. McCawley tried again a few times before it went to press, with similar results. Additionally, there was also a more explicit version of McCawley's argument (1976b [1967]:121–32) available several years before Chomsky's criticisms went to press.

Perhaps Chomsky genuinely could not see what McCawley was saying—stranger things happened in the course of the dispute—but his treatment of it suggests how completely unwilling Chomsky was to view his former students' work with any charity, and his insistence on publishing a misconstrual that McCawley explicitly rejected shows, in the most generous interpretation, a distinct lack of interest in harmoniously resolving the debate. A less gracious interpretation—that is, an interpretation of the sort which, by this point, was second nature to many generative semanticists—has more deceit than disinterest about it. And, deceit or disinterest, there was (at minimum) a remarkable arrogance in Chomsky's insistence that he was a better judge of what McCawley's argument said than McCawley himself was.

The short version, however, is just that this use of the Hallean syllogism, though formally much weaker, had largely the same effect as the original. It stirred up animosities. Those who were prepared to believe it, did (though not without some res-

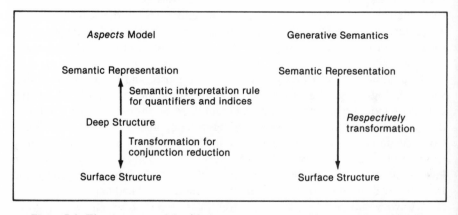

Figure 7.1. The *Aspects* model, with deep structure; generative semantics, without.

ervation). Those who were not prepared to believe it, did not. In particular, the interpretive camp felt pretty much that "the [*respectively*] argument has been effectively refuted by Chomsky" (Wassow, 1976:288), and considered the case closed. No one but McCawley ever paid too much attention to it after the fireworks from Chomsky's misconstrual died down, and for good reason.[7] There was a much stronger class of arguments about deep structure available, again, with McCawley at the helm—the lexical-insertion-without-deep-structure arguments.

One of the principal tactics of the generative semanticists in their attack on deep structure was just to claim they had no particular responsibility to construct counter-arguments against it, that the burden of proof fell on its proponents, not on its foes. A familiar generative semantics argument (or, perhaps, *aspersion* is more accurate) against deep structure was that "it was simply assumed in *Aspects* that [deep structure] contained all lexical items and preceded all transformations; no arguments were given" (Lakoff, 1971b:281). McCawley connects this point explicitly to the Hallean syllogism and the existence of deep structure:

> Chomsky's remark [1966c:48] that "the burden of proof is on the linguist who believes
> . . . that there is . . . a linguistically significant level of representation meeting the con-
> ditions of taxonomic phonemics and provided by the phonological rules of the gram-
> mar" applies equally well to the linguist who believes that a level such as 'deep structure'
> exists intermediate between the semantic and surface syntactic representation. (1976b
> [1968]:170; McCawley's elisions; see also 1976b [1967]:92–93)

But Chomsky (and, in his footsteps, McCawley) has it exactly backwards here—the burden of proof *was* on Halle. And he met it, by providing a model of phonology that worked efficiently without the phoneme (or "the phonemic level"). It's true that *Aspects* does not offer specific justifications for deep structure, but since deep structure was the linchpin of a theory that the entire community—including, of course, all of the budding generative semanticists—found very compelling, the *Aspects* model itself was quite literally an extended argument for deep structure. More importantly, historical developments made specific justifications moot: deep structure was simply what the *Syntactic Structures* model ended up with after incorporating sentence morphemes, doing away with generalized transformations, adding a semantic component, and so on. In the terms of the good Bishop Whately, a stuffy cleric and astute rhetorician of the nineteenth century, deep structure pre-occupied the ground of transformational grammar—just as the taxonomic phoneme pre-occupied the ground of structuralist phonology—and generative semantics was obliged to dislodge it:

> According to the most correct use of the term a "Presumption" in favor of any suppo-
> sition, means, not (as has been sometimes erroneously imagined) a preponderance of
> probability in its favour, but a *pre-occupation* of the ground, as implies that it must stand
> good till some sufficient reason is adduced against it; in short, that the *Burden of Proof*
> lies on the side of him who would dispute it. (1963 [1846]:112; Whately's italics)

In Chomsky's blunter, more concise terms: "There is no burden of proof on the person who provides the only theory that exists."[8] Halle's anti-phoneme argument came "embedded in a comprehensive theory of phonology, whose elegance was illustrated by the accompanying analysis of highly complex patterns of phonological relationships in Russian" (Anderson, 1985:321). McCawley's *respectively* argument came only with a postscripted IOU, and the choice between a reasonably well articulated theory (the *Aspects* model) and a promissory note for a theory (generative semantics), much the same but without its most glamorous element, is no choice at all.

McCawley, whatever he said at the time about burden of proof, knew better, and accordingly set right to work on developing a grammatical model that could get by without deep structure, rapidly formulating the Predicate-raising and lexical-insertion proposals that Postal, and a lot of others, found so winning.[9] This work marked the first real step toward the articulation of a genuine alternative to the *Aspects* model. Recall that the relation between the lexicon and deep structure in *Aspects* is such that (1) virtually all lexical items show up at deep structure, and (2) all transformations occur subsequently (as in figure 4.3). McCawley's Predicate-raising necessarily precedes at least some lexical insertion rules (since it collects deep predicates together and turns underlying strings like *Jones* STRIKES *me as* LIKE *a monkey* into sentences like *Jones reminds me of a monkey* by a lexical insertion rule that replaces STRIKES LIKE with *reminds*). The resulting grammar, with some transformations preceding some lexical insertions, could not, therefore, include an *Aspects* level of deep structure.

Generative semanticists took to McCawley's predicate-collecting proposal like a thought to a word. Not only were parallel deep-predicate analyses proposed in fairly short order for a number of other words, several ingenious support arguments for lexical decomposition quickly sprang up.[10] The interpretive semanticists, of course, reached exactly the opposite conclusion, and quickly gathered counter-evidence. Many papers questioned the semantic implications of lexical decomposition, the most renowned being Jerry Fodor's "Three Reasons for Not Deriving *Kill* from *Cause to Die*" (1970), which hinged on distinctions between such structures as 5a and 5b:

5   a   Lucretia caused them to die on Sunday by stabbing them on Saturday.
      b   *Lucretia killed them on Sunday by stabbing them on Saturday.

Other interpretive semantics arguments, mostly by Chomsky, contested the implications of the supporting case for lexical decomposition.[11] Chomsky also dismissed Predicate-raising as an upside-down semantic interpretation rule, and unmotivated to boot, and attacked the general consequences of lexical decomposition, which, he said, would lead to such ugly conglomerations as CAUSE TO DIE BY UNLAWFUL MEANS AND WITH MALICE AFORETHOUGHT for the word *murder* (1972b [1968]:72). But the most recurrent objection to McCawley's proposals was that they had no force because the only job Predicate-raising did was to make things easier for a theory which embraced lexical decomposition, and the interpretive semanticists rejected lexical decomposition. Chomsky denounced Predicate-raising in the most stinging term of transformational grammar, as *ad hoc,* since "the unit that is replaced by *kill* is not a constituent . . . it becomes one by the otherwise quite unnecessary rule of predicate raising" (1972b [1969]:142).[12]

The argument, that is, got nowhere. The generative semanticists saw enough in

McCawley's proposals to warrant their disposal of deep structure; the interpretive semanticists saw nothing to make them budge an inch. The interpretive reaction was pretty odd, given that nobody (except Katz) wanted deep structure anyway. The generative semanticists wanted to do away with the *term,* and the interpretive semanticists wanted to hang on to it. But nobody (except Katz) wanted the syntactic level that actually wore that label in the *Aspects* model. The generative semanticists wanted it much deeper, the interpretive semanticists wanted it shallower, which brings us to the Katz-Postal principle.

The Katz-Postal Principle

> The great tragedy of science—the slaying of a beautiful hypothesis by an ugly fact.
>
> T. H. Huxley

The Katz-Postal principle is a different story: argument and counter-argument over the principle were very successful, two-ways successful, in modifying linguistic theory. First, as virtually everyone would agree, the hypothesis was disconfirmed. Generative semantics retained it, but in a clearly attenuated form; interpretive semantics rejected it (while retaining some of its implications). Second, both theories changed markedly as a result of the issues raised by the question of meaning preservation and transformations.[13]

By 1969—and 1969 was an important year, the time of the contentious Texas Conference on the Goals of Linguistic Theory at which all the major players played—Chomsky was saying that the Katz-Postal hypothesis was false, and his students were coming up daily with new data they could fashion into arguments to exactly the same end. "Death to the Katz-Postal principle!" echoed in the halls of MIT.

"Long live the Katz-Postal principle!" continued to echo in other hallways, but by 1969 it was not the same old Katz-Postal principle. It had hardened into dogma.[14] The step to dogma—neither unusual nor necessarily unwholesome in science (see, e.g., Popper, 1970 [1965]:55, Feyerabend, 1978 [1975]:42)—was a very natural one in the transformational milieu. From Harris's earliest efforts to explore synonymy relations, through trigger morphemes, the discarding of generalized transformations, and the design of deep structure, the entire transformational program had grown toward, if not assumed tacitly, the Katz-Postal principle. Chomsky had upgraded it from a heuristic to a hypothesis in works like *Aspects, Cartesian Linguistics,* and *Language and Mind;* the next reasonable step was to see what conception of grammar one got by taking it very seriously, and the answer was generative semantics.

The earliest position of generative semantics on apparent violations of their defining axiom was the *Integrated Theory* position; namely, that seeming counterexamples to it were merely the result of bad analyses. Either the offending transformation had been wrongly formulated or the data was insufficiently understood. *Integrated Theory* and *Aspects* had done such a thorough job banishing violations, in fact, that there seemed only to be one major class of troublesome data left, the relative deep and surface structure locations of quantifiers, and the generative semanticists made a good deal of progress taming them with these faulty-analyses sorts of arguments. For instance, the conventional derivation for 6a ran into trouble, since it would derive from 6b, resulting in a nonsynonymous relation and a Katz-Postal violation.

6 a Everyone expects to live forever.
    b Everyone expects everyone to live forever.

But Carden argued (1968) that the deep structure sources for 6a and 6b should be, respectively, something like 7a and 7b:[15]

7 a every one$_i$ expects one$_i$ to live forever
    b every one expects every one to live forever

Lakoff wove similar re-analyses around sentences like 8a, which *Aspects* would derive from (the Katz-Postal violating) 8b, but which he said ought to come from (the Katz-Postal maintaining) 8c.

8 a Few lawyers are both popular and successful.
    b few lawyers are popular and few lawyers are successful
    c lawyers who are both popular and successful are few

But Chomsky's busy interpretivists kept finding more and more examples of this sort, where the deep structure position of quantifiers supported one reading and, after certain transformational derangements, the surface structure position supported another. Interpretive semanticists used these arguments to justify a major renovation of the *Aspects* semantic component, albeit a very vaguely adumbrated renovation: the deep structure continued to feed the semantic component information about grammatical relations and lexical content, but the surface structure now fed it information about logical elements (like quantifiers and negatives). This change made for a much more powerful semantic component, which most linguists regarded with a good deal of nervousness. But there were more changes to come.

Chomsky (1972b [1969]:180) says, with his familiar casualness, that semantic interpretation rules must apply "to deep and surface (perhaps also shallow) structure." Just as he had hinted earlier that surface structure semantic rules were on their way, he was now hinting that a new syntactic level, shallow structure, might be getting its own complement of semantic rules.[16] These rules, too, were left unspecified.

The task of putting meat on these bony suggestions, and many others, fell to Chomsky's conscience, Ray Jackendoff.

Semantic Interpretation in Generative Grammar

> Jackendoff is quite candid about the aesthetic appeal of [interpretive semantics]
> —or lack of it. He admits that the picture would be prettier if all semantic interpretation took place at a single level, but maintains that the facts militate against such an assumption.
>
> Michael Kac

Lees was called "Chomsky's Huxley" in the early years, with a certain appropriateness, but the phrase holds at least as well of Jackendoff. Like T. H. Huxley and natural selection, Jackendoff had the "Remarks" material as part of a small privileged group, directly from the source, and at length, before most of the field had even heard of it:

> I heard the lectures. That's different from reading the paper. They were more fleshed out. It took him probably the better part of a semester to cover that material.

In both cases, the presentation was also dynamic. Jackendoff did not begin with as much resistance as Huxley did, but he similarly raised objections to the work, which Chomsky settled to his satisfaction, and he witnessed the considerably more hostile objections of Lakoff and others, which Chomsky also settled to his satisfaction. (Indeed, having the recalcitrant, baiting, generative semanticists constantly raising objections to the "Remarks" proposals likely gave the entire interpretive semantics community a sense of shared purpose, rather than undermining their resolve. The same sense of conspiratorial camaraderie shows up in the letters among Darwin, Huxley, and Hooker, discussing some of their opponents' blockheadedness.) And, while Chomsky is not as shy of combat as Darwin was, he is reluctant to become involved in nasty public disputations.[17] Jackendoff showed all the reserve of Huxley in the area of disputation, tearing into not just Ross and Lakoff and McCawley (who were his professional seniors by only a few years), but into the considerably more formidable Postal as well. He also showed, repeatedly, the subtle ability to play partisan science with the best of them—such as calling a constraint he endorses "the Complex NP Constraint" (*not* "Ross's Complex NP Constraint"), and one he doesn't endorse "the Postal Crossover Condition" (1968). The favored proposals are just proposals, the unfavored ones are errors by specific bad guys.

There is a major difference between Huxley's role and Jackendoff's, however, which is where "conscience" comes in. While Huxley largely explicated and rephrased the meticulously detailed arguments of Darwin, Jackendoff elaborated and expanded Chomsky, putting considerable formal flesh on suggestions that even the unflaggingly faithful Jackendoff called "sketchy and programmatic" (1977:xi).[18] It is as if Darwin had quit after his 1858 paper to the Linnean Society and Huxley had written *Origin* and *The Descent of Man* himself. Or, fishing for a better analogy than Huxley to Darwin on this front, we might settle briefly on Kepler to Copernicus. Like Kepler's elliptical orbits, the form that Jackendoff gave to some of Chomsky's beautiful airy proposals violated his own sense of aesthetics; it was ugly, but it worked.

Whatever the analogy, Jackendoff is a hero in the tale of interpretive semantics. There were others—especially Adrian Akmajian and Joseph Emonds (Dougherty was a polemicist and little more)—but aside from Chomsky, no one else came anywhere close to him in terms of a contribution to the interpretivist side of the debate, both positively (in supporting lexicalism, x̄-syntax, and post–deep structure semantic interpretation) and negatively (in discrediting the Katz-Postal principle). Indeed, in some ways, his contribution was more substantial than Chomsky's; certainly it was more sustained, more comprehensive, and considerably more rigorous. The level of his heroism is clear if we step back into the historical long view for a moment. Chomsky's theoretical development is generally said to be punctuated by four main grammatical models (or, sometimes, "four main theories," though the number of theories associated with Chomsky is exponentially higher than four). They generally go by the names *early transformational theory, the standard theory, the extended standard theory,* and *government-and-binding theory.* Each of these models but one is associated with a major Chomsky text: *Syntactic Structures* (or, more properly, *Logical Structure of Linguistic Theory*) for the early theory, *Aspects* for the standard theory, and *Lectures on Government and Binding* for his most recent model (respectively, 1957a, 1975a [1955], 1965 [1964], and 1981a [1979]). The odd model out is the extended standard theory, and the major text associated with it is not by Chomsky. It is Jackendoff's *Semantic Interpretation in Generative Grammar* (1972).[19]

If Chomsky's central "Remarks" proposals had stayed as he left them in 1967, it is a very good bet that many fewer linguists would have been drawn to them; the other two papers in Chomsky's anti–generative semantics trilogy—"Deep Structure" and "Some Empirical Issues"—are a great deal shorter on details even than "Remarks." They are both important collections of arguments, which interpretive semanticists, virtually en bloc, regarded as lethal to any form of generative semantics; they are extremely effective works of rhetoric, completely shifting the agenda of the debate; and they touch on a number of issues that have been central topics in the field ever since. All the same, they are little more than catalogs of negative criticisms, attended by only the most allusive positive suggestions. Most disturbingly to the generative semanticists, they carry the strong implication that Chomsky has no responsibility to provide positive accounts of the phenomena he introduces. For instance, he brings the notions of focus and presupposition into the debate, arguing that generative semantics can't adequately account for them, but where his own model is concerned, he goes little further than the remark "these notions seem to involve surface structure in an essential way" (1972b [1968]:101). Jackendoff was the one (1972:229–78) who offered concrete proposals for how surface structure might be involved, and how the semantics might be able to make sense of it, just as he did for virtually every other thorny issue of the day—grammatical relations, pronouns, modals, negation, quantifiers—even venturing with some success into the very murky regions of intonational meaning.

While he was building the Rube Goldberg contraption that could accomplish these daunting tasks, the major ideas of "Remarks" were largely neglected. Chomsky (except for very cursory discussions—e.g., 1972b [1969]:158–62) had not revisited them, and his other students, despite a totemic identification with the name *lexicalists,* had taken them no further. Even Jackendoff appeared reluctant to embrace them at first.[20]

But his reluctance had less to do with embarrassment than with a lack of time. Both the lexicalist hypothesis and x̄-syntax are very ramified ideas, and exploring them with any seriousness would have meant dropping the project Jackendoff found more urgent, justifying post–deep structure semantics—more urgent, of course, because more devastating to generative semantics. Once he had offered that justification, in *Semantic Interpretation,* he turned first to the redundancy rules necessary to give the lexical hypothesis some formal substance, then to x̄-notation, necessary to give the hypothesis some explanatory capacity—all the while maintaining a fervent opposition to generative semantics. Some of these digressive attacks were annoying to readers on the sidelines—Michael Kac's review of *Semantic Interpretation* indicted Jackendoff for squandering his considerable skill "in meaningless polemical exercises" and with "deliberate parochialism" and with "the desire to score points in a sectarian debate" (Kac, 1975:30)—but those sectarian energies are what kept his imagination fired.

The model lovingly detailed in *Semantic Interpretation* is a pretty homely beast, particularly in light of the *Aspects* model and generative semantics, both of which look very good on a blackboard. The *Aspects* model had an orderly semantic component that looked in on a derivation at deep structure, and only at deep structure (figure 4.1). Its semantic rules were all of the same type, and they produced one semantic representation per derivation. The generative semantics model had a semantic component that, to all effects, constituted deep structure (figure 5.3). Its semantic rules were not only all the same, they were also the same as its syntactic rules; namely, transformations. There was only one semantic representation per derivation. Jackendoff's model, though, looked in on a derivation virtually at will. And it did away with a guiding principle of transformational grammar from the outset, that each derivation has a single semantic representation; Jackendoff's derivations had four distinct semantic representations.[21]

Moreover, *Aspects* had reduced the complexity of the Katz-Fodor semantic model by eliminating one of their rule classes. Generative semantics had eliminated semantic interpretation rules altogether, giving their job over to transformations. *Semantic Interpretation* added three new classes of semantic rules, each of which produced its own semantic representation. To make things worse yet, Jackendoff's model also had several other bits and pieces of theoretical paraphernalia that had entered the field since the mid-sixties—lexical redundancy rules, output constraints, conditions on transformations, and the like. A relatively conservative diagram of the *Semantic Interpretation* model (the one that Jackendoff himself presents) is given in figure 7.2. Jackendoff now says of this model, "I'm sure everybody thought that it was off-the-wall and weird, although nobody complained to my face."

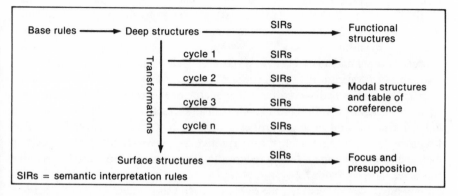

Figure 7.2. Jackendoff's interpretivist grammar (adapted from Jackendoff, 1972:4).

174   *The Linguistics Wars*

[They must have thought] 'that is really wacko. If semantic interpretation is like that, forget it.'"[22]

On the contrary, many people thought exactly the opposite: if semantic interpretation was like that, it needed close attention; it needed to be cleaned up, not discarded. *Semantic Interpretation in Generative Grammar* is a rhetoric of assent which rarely fails to put its money where its mouth is. Virtually every negative criticism is balanced by a positive proposal.[23]

The Katz-Postal Principle Again

> It has become clear over the past five years that Transformational Generative Grammar is nowhere near being an adequate theory of human language. Those of us who have tried to make Transformational Grammar work have attempted to patch up the classical theory with one ad hoc device after another: my theory of exceptions, Ross's constraints on movement transformations, the Ross-Perlmutter output conditions, Postal's Crossover principle and anaphoric island constraints, Jackendoff's surface interpretation rules, Chomsky's lexical redundancy rules and his analogy component, and so on. . . . Most, if not all, of these ad hoc patching attempts [are] special cases of a single general phenomenon: global derivational constraints.
>
> George Lakoff[24]

Interpretive semantics was a very different beast, and a much lumpier one, once Chomsky's proposals got a little meat on their bones. But there was uglification going on in the generative semantics camp, for much the same reasons. The sort of data that the interpretivists' research kept turning up against the Katz-Postal principle proved too much for arguments modeled on the *Integrated Theory* approach of reconfiguring analyses. Jackendoff and Chomsky had no trouble abandoning the Katz-Postal principle by enriching the semantic component substantially. But generative semantics was incomprehensible without some form of the principle, and it had no concept of a semantic component, distinct from syntax, which it could enrich. There were only two conceivable options, abandon the theory, or enrich the homogenized rule system in a way that would preserve a weaker version of the Katz-Postal principle. Aside from the dialectical pressures of the dispute, which would have made surrendering to Chomsky's new vision impossible, the generative semanticists also saw that vision as essentially abandoning any hope for a realistic account of meaning.

Most of them had a good deal of respect for Jackendoff's ingenuity, and for his willingness to confront the implications of data that Chomsky apparently had no interest in once it had served his anti–generative semantics purposes. But they regarded Jackendoff's efforts as an endless, fruitless, series of patches in a wall built to keep meaning and structure artificially apart; when the *Aspects* version of deep structure wasn't strong enough, the interpretive crew invoked surface structure, and then shallow structure, and then, recurrently, the nameless structural levels at the end of each cycle (figure 7.2).

Generative semanticists felt the right approach was simply to admit the artificiality of that wall, to acknowledge semantics and syntax intermingled so thoroughly as to make autonomous accounts of either futile. Having made this admission, the real task of linguistics was then to find the order in this gumbo; the Katz-Postal principle still looked to be the best bet on this front, even in a compromised form, and one new rule type looked a small price to pay for its maintenance.

This new rule type was introduced by Lakoff, who was rapidly becoming the most prominent generative semanticist, and was immediately endorsed by the other three leaders; the interpretive camp, led by Chomsky, threw up its hands in horror. Lakoff proposed (1970b; 1971b:238ff) to incorporate devices he called *global derivational constraints* (*global rules*, for short), moving the theory into what Postal called its Homogeneous II phase. In brief, global rules recognize that some transformations can alter the relations of words such that deep and surface structures of the same derivation could support different semantic readings; however, they outlaw such derivations. Derivations in which transformations change meaning were legislated out of the theory.

To take an analogy from *Logical Structure* (Chomsky, 1975a [1955]:146), early transformational theory allows the generation of sentences like the famous 9:

9   Colorless green ideas sleep furiously.

But the grammar rejects it for not achieving "the highest degree (first order) of grammaticality" (1975a [1955]:154). Katz and Fodor adopted the same sort of approach, by having the semantic component fail to return any semantic reading for it. This general plan of attack, which we looked at a little earlier, was known by the *Aspects* term, *filtering*.

Lakoff simply extended the filtering approach to sentences like 10a when they derive from underlying structures like 10b (and thereby violate the Katz-Postal principle); in order to be legitimate, the global rule stipulates, 10a must arise from 10c.

10   a   Few books are read by many men.
     b   Many men read few books.
     c   Few are the books that many men read
         (Lakoff, 1971b:238–39)

Transformational rules alone permit the derivation of 10a from 10b or 10c, but a global rule makes the 10b ⟹ 10a derivation illegitimate, and the Katz-Postal principle is maintained. Essentially, what we have is a semantic output condition. But there's a problem. The principle is maintained in an uncomfortably artificial way: indeed, by entirely circular means. The Katz-Postal principle says that transformations can't change meaning, and Lakoff argues that derivations like 10b ⟹ 10a are illegitimate because they involve a transformational change in meaning (see especially 1971b:240).

The Katz-Postal principle cannot be violated because the Katz-Postal principle cannot be violated.

From one perspective, then, global rules admit the falsification of the Katz-Postal hypothesis, since only stipulation prevents a violation, and, therefore, they vitiate the entire generative semantics program. Certainly that's the way most interpretive semanticists, as well as a few generative semanticists, saw it, and Chomsky still has quantifiers and global rules in mind when he says that generative semantics "was proven wrong, very early."

But, Lakoff argues compellingly, global rules are necessary for completely non-semantic reasons. In particular, he points out that a number of non-transformational rule types which both sides of the schism had already adopted (like Ross's island constraints, Postal's Crossover principle, Perlmutter's output conditions) have similar descriptive powers. And then there is the whole panoply of general conditions on rules that had mushroomed in transformational grammar and was codified in *Aspects*—ordering, cyclicity, recoverability of deletion, the Katz-Postal principle—the specifics of which aren't important, but which formed a knobby bag of extra-transformational goodies. Therefore, Lakoff asks, since "global rules are necessary, whatever position one takes on the relative merits of generative and interpretive semantics" (1970b:638n9), why not use them to save the semantic phenonema necessary to maintain the Katz-Postal principle?

Global rules, because they involve a sort of direct communication between non-contiguous phrase markers in a derivation (like deep and surface structure in the *many men–few books* example above), raise a number of complications that contributed heavily to the downfall of generative semantics, which we will investigate shortly, and it's not clear that Lakoff did any more than stencil a name onto the knobby bag. But, for the moment, we can give him the last word, pointing to a definite advantage of global rules over the aesthetic and conceptual messiness of incorporating several distinct classes of semantic rules:

> For each different case [Chomsky] would propose not a different rule, but a different KIND of rule, adding a new type of theoretical apparatus to the theory of grammar for each new global rule discovered.
>
> It is sad and strange to encounter such [an attitude]. (1970b:637; Lakoff's emphasis)

Restrictiveness

> May 27, 1969: George Lakoff discovers the global rule. Supermarkets in Cambridge, Mass., are struck by frenzied buying of canned goods.
>
> entry in James McCawley's
> "Dates in the Month of May That Are of Interest to Linguists"

Chomsky's anti–generative semantics campaign had several stages, each one defining a new direction for its own model. First, he undermined abstract syntax with the "Remarks" proposals, then he attacked the Katz-Postal principle by promoting the surface-structure-impinges-on-meaning arguments, and then, in his response to Postal's "The Best Theory," he attacked generative semantics for its descriptive wantonness. The first two approaches—lexicalism and post–deep structure semantics—failed to resonate with anyone beyond his immediate students. But the worst-imaginable-theory argument hit home.

"The great weakness of the theory of transformational grammar," Chomsky said, "is its enormous descriptive power, and this deficiency becomes more pronounced to the extent that we permit other rules beyond transformations (i.e., other sorts of 'derivational constraints' [global rules]). . . . Any imaginable rule can be described as a 'constraint on derivations'. The question is: what kinds of rules ('derivational constraints') are needed, if any, beyond those permitted by the standard theory?" (1972b [1969]:133–34)[25] Chomsky cannot even bear to use the term *derivational constraint* without the sanitizing effect of quotation marks (sometimes even double-bagging it by adding a set of parentheses), but the general point is reasonably clear: unless some rigor is brought to the notion, it drives linguistic theory away from what he regards as its primary goal, restrictiveness.

Ordinarily, one thinks of descriptive power as a virtue in science, and enormous descriptive power as the mother of all scientific virtues. Within certain parameters, this is indeed the case. Big-time descriptive sciences like astronomy and biology earn their bacon by having enormous descriptive ranges, from pin-size black holes to pulsars, amoebas to elephants. But the parameters are extremely important, because they represent the limits of the science. Astronomy is not the science of all objects with mass and weight and velocity. The objects it describes do not include '56 Chevies. Biology is not the science of all objects that consume and excrete and have inherited characteristics. The objects it describes do not include '56 Chevies. Linguistics is not the science of all possible symbols or symbol systems. The objects it describes do not include '56 Chevies.

Linguistics is the science of natural languages, and there are lots of symbol systems which are not natural languages. It is extremely easy, in fact, to come up with symbol systems that operate in ways that natural languages don't. Sentences might be ordered by word length, or vowel frequency, or chronological occurrence. Questions might be formed by reversing the order of words in a declarative sentence, or rearranging them alphabetically, or transposing every second pair of consonants, or only being uttered when the speaker is leaning against a '56 Chevy. An unconstrained transformational grammar—say, the one outlined in Lees's *English Nominalizations*—can describe all of these systems, and many, many more. Transformational grammar is so powerful, Emmon Bach said, that "a not too far-fetched analogy" to the way it describes language "would be a biological theory which failed to characterize the difference between raccoons and lightbulbs" (1974:158). This ugly situation is made all the beastlier in generative grammar because of its cognitive mandate. At its core, remember, transformational-generative grammar is supposed to be psychologically plausible, describing what is between the ears of a language user, and it is supposed to be particularly attuned to the problems of language acquisition. How is a child to acquire a language if she can't even know what one is, if she is in the same position as a biologist trying to learn about raccoons who is unable to distinguish one from a lightbulb or a '56 Chevy?

Hence, the opposition to enormous descriptive power. Hence, Chomsky's work on the A-over-A principle, Ross's on island constraints, and Postal's crossover research. Hence, Chomsky's burning question, the one that interpretivists set out to answer in the seventies—"What kinds of rules ('derivational constraints') are needed, if any, beyond those permitted by the standard theory?"

Hence—though very few people noticed it at the time—Lakoff's "Global Rules."

Chomsky's great rhetorical triumph was that, in very short order, he managed to turn the words *global rule* into a synonym for "any imaginable rule," completely reversing the thrust of Lakoff's argument, and the words *generative semantics* into a synonym for "enormous descriptive power," and—here is where the *via negativa* definition of Chomsky's program was most effective—the words *extended standard theory* into a synonym for "restrictiveness." Chomsky and the interpretivists regularly pointed to the whole, growing menagerie of generative semantics descriptive devices in horror, but conveniently ignored most of their own growing menagerie when making comparisons. The generative semanticists were naturally outraged, regarding the whole argument as smoke-and-mirrors logic, and complaining loudly about Chomsky's move. They still recall it with images of prestidigitation. As Postal puts it,

> Chomsky had these—what did George call them?—these wild cards that he could pull out of his hat whenever he wanted, and somehow they didn't count when it came to talking about restrictiveness.
>
> Whenever he was doing something descriptive, where he needed to describe facts that generative semantics would talk about in terms of transformations—linking meanings to deep structures, or to other kinds of structures, by way of global rules—Chomsky would appeal to semantic interpretation rules.
>
> He would never define them. He has never, to this day, given any content to that notion. He's never said what they were. But he could have as many as he wanted. Whenever he needed one, he could pull one out of his hat and use it. Now, when it came time to compare generative semantics to his framework [in terms of restrictiveness], those were never included. He never felt he had to say anything about them.
>
> It seems, *a priori,* implausible that he could get away with that. But he did.[26]

The generative semanticists noticed what Chomsky was up to, but almost everyone else took his restrictiveness arguments at face value. Even many of his most fervent detractors frequently take time out from attacking him to bash his generative semantics scapegoat for its descriptive licentiousness (for instance, Hagège, 1981:83).

In fact, his victory on this issue was so complete, that it is now difficult to appreciate its dimensions, especially for anyone unfamiliar with the rhetorical history of transformational grammar. We have only looked very casually at the appeals which constituted the rhetorical aresenal of transformational-generative grammar in its rapid climb to the top of linguistics. One appeal in particular was slighted, the great dependence on the notion of simplicity—as a goal of linguistic research, as the central criterion in theory comparison, and as a methodological principle. The evaluation metric with which Chomsky thumped the Bloomfieldians, Halle's case against the phoneme, the daily warrants for specific analyses, instruments, and hypotheses—all leaned heavily, in some cases exclusively, on the value of simplicity.

With the central, virtually defining role of simplicity in Chomskyan linguistics, one would have thought (Postal surely thought) that Postal's "Best Theory" case would be enormously appealing. It is a straightforward minimalist argument that the grammar with the fewest theoretical devices is the simplest, and therefore, the most highly prized. On these grounds, generative semantics wins, hands down. In particular, the model Postal dubs *Homogeneous I* is "the best grammatical theory *a priori* possible" (Postal, 1972a [1969]:136). It has semantic representations at one end, surface representations at the other, and a relatively uniform component of transformational rules explaining their correspondences; the *Aspects* theory had three levels, and two sets of rules.

Unfortunately, Postal concedes, there is a rub. His best of all possible theories just can't accommodate the facts, and it has reluctantly to embrace Lakoff's new rule-type, global derivational constraints. But the result, he claims, Homogeneous II, is still much simpler than *Aspects*—not to mention the vague post-*Aspects* interpretivist theory that Chomsky was marketing at the same conference.

Newmeyer observes, correctly, that despite what should have been an extremely winning case to formal linguists, "probably no metatheoretic statement by a generative semanticist did more to undermine confidence in that model than Postal's paper, 'The Best Theory'" (Newmeyer, 1980a:169; 1986a:135). The reason goes far beyond the arrogance many found in Postal's title, and far beyond the culprit that Newmeyer himself cites, the character (or, in Newmeyer's view, lack of character) of the new rule-type that moved generative semantics from Homogeneous I to Homogeneous II. The reason, simply put, is that Chomsky accomplished a remarkable change of agenda in the debate.

He raised the alarm—"the gravest defect of the theory of transformational grammar is its enormous latitude and descriptive power" (1972b [1969]:125)—and as grave as that defect is, he said, generative semantics makes it worse by adding global rules to the transformational arsenal. A grammar organized solely around transformations (that is, Homogeneous I) "is a rather uninteresting theory," because of its immense descriptive power; "it can be made still more uninteresting by permitting still further latitude, for example, by allowing rules other than transformations that can be used to constrain derivations [that is, by adding global rules and becoming Homogeneous II]" (1972b [1969]:126). Thus far, the argument does not add much weight to Chomsky's position; indeed, it works against him. It says that transformations are bad, and that adding more rule-types makes any grammar that incorporates them even worse. *Aspects,* then, looks worse than generative semantics, with its semantic interpretation rules, and post-*Aspects* interpretivism looks worst of all, with distinct types of semantic interpretation rules coming in almost daily. But Chomsky isn't through: "Notice that it is often a step forward . . . when linguistic theory becomes more complex" (1972b [1969]:126). The grounds of theory comparison changed dramatically: simplicity was shuffled off into the wings and restrictiveness took over at center stage, complexity at its side.[27] Chomsky, however, did not manage the feat on his own.

Although Chomsky frequently dismisses their influence, the single most important reason for the success of his restrictiveness case was a series of papers by Stanley Peters and Robert Ritchie (1969, 1971, 1973a, 1973b), which provided mathematical results demonstrating transformational grammar's virtually complete lack of discrimination; it is in a passage on the Peters-Ritchie results that Bach made his lightbulbs-and-raccoons comment about transformational grammar. Specifically, the Peters-Ritchie results show that the class of grammars described by *Aspects* is so all-encompassing that it can't distinguish between any indiscriminate list of strings of symbols (say, all the decimal places of π, divided into arbitrary sequences and enumerated by value of the products of their digits) and a list of actual strings that people use to communicate (say, English). These results formalize notions that had been present in transformational theory for some time, but a mathematical proof brought them home very powerfully.[28] The concern with restricting transformational grammar that led Chomsky and Ross and Postal to work on constraints had been one of many concerns about the *Aspects* theory, but Chomsky's urgings at the Goals conference, coinciding with the publication of the Peters-Ritchie results, brought it to the fore. Lack of restrictiveness precipitated something of a crisis in linguistic theory in the seventies, and generative semantics became a very convenient donkey on which to pin the tail of descriptive profligacy.

The most fascinating aspect of the restrictiveness counter-argument, however, is that, in 1969, generative semantics showed very few signs of the descriptive wantonness for which Chomsky indicted it. Indeed, since *Aspects* generative semanticists had done a great deal more to constrain transformational grammar than anyone in the interpretivist camp. Ross and Postal had both done extremely important work on constraints (as had Perlmutter, in a more local way). McCawley had reanalyzed phrase structure rules in a way that made them serve as filters, and argued to extend Ross's movement constraints to the lexicon. Ross and Lakoff had done crucial work on the cycle. Ross and Ronald Langacker had made parallel proposals for restricting the application of pronominal transformations. And the focus of Chomsky's attack, Lakoff's proposal of global rules, was an attempt at further restriction; the expanded name is global derivational *constraint*.[29] In effect, Lakoff argued that virtually all of the serious work in transformational grammar had involved ways of constraining derivational relations. Transformations, he said, constrain two contiguous trees in a derivation. Their application is local. They apply only to two trees standing side by side. Other principles and rules—in particular the Ross-Perlmutter-Postal line of research, but also such transformational traditions as rule ordering and cyclicity—constrain noncontiguous trees in a derivation. Their application is global. They apply to any pair of trees ("or perhaps sometimes triples"—Lakoff, 1970b:638) in a derivational arbor, irrespective of the distance between them.

Global rules, in short, are just derivational constraints with a wider province than transformations, and offer the important nominal advantage of putting a label to the seemingly disparate research into the restriction of transformations. Lakoff's main argument is that linguists should recognize the necessity for all the extra paraphernalia beyond transformations and begin exploring them as a class of rules, rather than as a mixed bag of ad hoc devices. Lakoff calls for the development of a "theory of global grammar" (1970b:638), although—and this is probably the other main factor in the success of Chomsky's restrictiveness argument—Lakoff never took up the mission himself.[30]

In contrast, the interpretivists had done very little work on derivational constraints (except, of course, the ones generative semanticists began calling *local derivational constraints*). Chomsky (1964b [1962]) had inaugurated this area of investigation, with his A-over-A principle, and Joseph Emonds's important (Chomsky-supervised) 1970 dissertation on a new type of filtering was just about to hit the market. But the bulk of the work on constraining the transformational component—in 1969, when Chomsky called the lack of restrictiveness "the gravest defect" in transformational theory, and 1970, when Lakoff urged a theory of global grammar to correct that defect—had been done by generative semanticists. Moreover, as Postal pointed out, the interpretivists were guiltier than their whipping boys of the "illegitimate appeal to overly powerful devices" (1972c:215)—vague or completely unspecified rules of performance, partially sketched semantic interpretation rules, and syntactic features.

The situation changed dramatically in the early seventies, however. Postal and Perlmutter moved on—proposing new and interesting constraints, but in another framework altogether, relational grammar—Ross and Lakoff lost interest in constraints, and Chomsky's camp took up the job of grammatical restriction with a vengeance. And—in a fit of suicidal strangeness—many generative semanticists warmed up to the role of bogey man. In a climate where the most urgent problem in transformational grammar seemed to be restricting descriptive power, and global rules were painted as the most serious offender, Lakoff said that "the real problem with global rules is not that they are too powerful, but that they are too *weak*" (Parret, 1974 [1972]:176; Lakoff's emphasis), and accordingly proposed more powerful devices—in particular transderivational constraints, which relate not two noncontiguous trees in a derivation, but two trees in different derivations (and, with the introduction of these devices, Lakoff became explicit that it no longer made sense to maintain the Katz-Postal hypothesis—1975:283–84).[31] Sadock even proposed meta-transderivational constraints (which involve "two derivations and an aspect of the real world"—1974b:604). The liabilities of these descriptively powerful devices were compounded by terminological confusions. McCawley, for instance, introduced the term *panderivational constraint* (1982b [1973]:54), and adopted *extraderivational constraint* as a generic for both Lakoff's *transderivational constraint* and Sadock's *meta-transderivational constraint*—attempts at clarification which probably did little more than contribute to the generative semanticists' growing reputation for theoretical extravagance. Such additional descriptive devices as meaning postulates, conversational postulates, and syntactic amalgams all entered the generative semantics picture, with very little clarity as to how or if they related to transderivational constraints, or even to derivations. All the whole, both Lakoffs, Ross, and a good many second-generation generative semanticists were spending a good deal of their time and effort mucking around in data that appeared to call for more powerful devices yet; George Lakoff even entertained "such madness as ordering of transderivational constraints, cyclical transderivational constraints, exceptions to transderivational constraints, and perhaps the elimination of transformations altogether" (1973a [1970]:452).

In the other camp, the interpretive semanticists took up Lakoff's call for a theory of global rules (all the while attacking Lakoff and shunning the word *global*). Emonds (1970) worked out an elegant way for the phrase structure rules to exercise direct control over every tree in a derivation (expanding on some of McCawley's ideas), and Chomsky (1973a [1971]) developed Ross's constraints in such a compelling way that they became the focus of interpretivist work for the rest of the decade, and continue to be crucial elements of Chomsky's program. The most obvious interpretivist excursion into globality was Chomsky's (1973a [1971]) introduction of the trace convention which gives transformations the power to mark sentences at one stage so that other transformations, arbitrarily later in the derivation, can tell they have applied.[32] This convention says that a movement leaves behind a phonologically null, but syntactically and semantically important, "trace," as in the examples of 11—11a representing the deep structure, 11b the surface structure, and *t* marking the place where the noun phrase, *Tyler* was before it moved transformationally to the front of the sentence.

11　a　it seems to each of the girls Tyler to like the other
　　　b　Tyler seems to each of the girls *t* to like the other

This convention looks to be the height of absurdity to nonlinguists, and it means, among other things, that surface structures were becoming increasingly abstract in the interpretive camp. But traces are extremely attractive little devices. Their principal use is to maintain aspects of deep structure (like the original location of *Tyler* as the subject of *to like*) for later transformations or semantic interpretation rules—and this is their clear global aspect, since they allow noncontiguous trees to communicate—but they also have some very interesting side effects. The most celebrated of these side effects is the account they offer of the lack of ambiguity in 12b, in contrast to 12a.

12　a　Bleeding Gums Murphy is the man I want to succeed.
　　　b　Bleeding Gums Murphy is the man I wanna succeed.

Sentence 12a could mean that I want Bleeding Gums to succeed (be successful), or that I want to succeed Bleeding Gums (follow him in some way, maybe as a saxophone legend); 12b can only mean that I want to succeed Bleeding Gums. The *Aspects* model explains the two different meanings of 12a by saying they spring from two different deep structures (12c when I want Bleeding Gums to succeed, 12d when I want to succeed Bleeding Gums).

12　c　Bleeding Gums Murphy is the man [I want the man to succeed]
　　　d　Bleeding Gums Murphy is the man [I want to succeed the man]

The *Aspects* model, that is, handles the ambiguity quite well, but it says nothing about the fact that 12b, similar in almost every respect, is univocal, necessarily deriving from only one deep structure, 12d. Trace theory to the rescue: with Chomsky's convention, the trace-enriched surface structures for the two meanings are, respectively, 12e and 12f.

12　e　Bleeding Gums Murphy is the man [I want *t* to succeed]
　　　f　Bleeding Gums Murphy is the man [I want to succeed *t*]

With the reasonable extension that traces block contraction, then, the explanation is clear: either 12e or 12f can be a surface structure for 12a (hence, its ambiguity), but only 12f can be the surface structure for 12b (hence, its univocality).[33]

And there was one more extremely attractive feature of traces. Remember, Jackendoff's ugly interpretive model (figure 7.2)? Traces streamlined it considerably. All the access Jackendoff's semantic component needed to the derivation between deep structure and surface structure was necessary to keep track of when and where movement transformations scooted constituents around. With traces (since every movement leaves a little bit of itself behind) all that information could now be represented in the surface structure.

Trace theory became a cottage industry for interpretivists, part of the dedicated effort to work on constraints, but the few generative semantic attempts to propose and explore specific global rules or conditions (such as Lakoff, 1971b:238ff, 1974; Ross, 1972a), were restless and abortive, betraying little conviction; indeed, Ross discards his global rule in the last few pages of the paper, calling for a transderivational solution. Most generative semantic invocations of globality were hand-waving affairs: "here are some phenomena, and it looks like we're going to need a global rule to handle them;" no specific rules offered. (Postal was an honorable exception—1972b [1970].) That is, the arguments ran exactly and ironically parallel Chomsky's early arguments for surface-structure semantic interpretation rules.

The situation with transderivational constraints was worse yet. Lakoff introduced them in a very informal paper, "Some Thoughts on Transderivational Constraints," which doesn't so much as offer an example of this new rule-type (1973a [1970]). Yet further discussion of transderivational constraints was largely of only two, equally unproductive sorts: (1) uncategorical denunciation, from the interpretivists; and (2) unwarranted invocation, by the generative semanticists.

Chomsky's camp simply wrote off transderivational constraints as the complete abandonment of formal grammar, and, indeed, their introduction coincided with a very sharp decline in such interests within generative semantics.[34] The generative semanticists simply appealed to transderivational constraints when other theoretical mechanisms broke down (an extremely frequent occurrence, given the data they were exploring), with very little justification, and without specifying the constraint formally, or even examining its application very carefully. The only explicitly proposed transderivational rule came very late in the schism, and its author quickly repudiated it (Gazdar, 1977; 1979).

Notice, however, that we have moved a long way from Chomsky's initial charges of descriptive profligacy. For one thing—and, from their perspective, it is by far the most important—the generative semanticists had expanded their data concerns substantially, involving a great many pragmatic ("performance") phenomena well outside the bounds of Chomsky's theory. But they had also reconceived or rejected the defining notions of transformational grammar, like competence, performance, and grammaticality. And many had abandoned formal theory construction altogether—most notably, Ross and Robin Lakoff—while others were barely hanging on by their fingernails. Interpretive and generative semantics were no longer comparable on metrics like restrictiveness; indeed, generative semantics$_{1973}$ was no longer comparable on such metrics to generative semantics$_{1970}$, which was markedly different from generative semantics$_{1967}$.

Grammaticality

When a linguist presents a list of sentences with and without stigmata (*, ?, etc.), he is normally reporting not "grammaticality judgements" (even though he may well say that they are grammaticality judgements) but rather judgements as to the normalness of that sentence as an expression of a given meaning under given contextual conditions. The ability to make "grammaticality judgements" as popularly understood, that is judgements about the "goodness" of a string of words independently of its syntactic structure or meaning in context, is extremely rare and appears to be a concomitant of virtuosity in constructing puns.

James McCawley

While Chomsky and his kith were lecturing piously against the descriptive promiscuity of generative semantics, the return charge was of a hermetic and unwholesome descriptive asceticism. Much of this counterattack was unfocused, but some generative semanticists—most notably, McCawley—brought it directly to bear on the concept of grammaticality. Remember what he told *The New York Times:* "Chomsky assumes that there are sentences which belong to the language and other sequences of words which don't—and the grammarian's task is to write rules [that] determine which belong and which don't. Postal and Lakoff and I say this isn't a coherent notion" (Shenker, 1972).

Although he hasn't been entirely consistent in application, Chomsky has always said that *grammaticality* is a technical term, relative to some specified grammar.[35] A sentence is grammatical if and only if there is a grammar, a body of rules, which generates it. A sentence is grammatical or not relative to, for instance, the *Aspects* grammar or the *Syntactic Structures* grammar. *Acceptability,* on the other hand, is relative to a specified speaker in a specified context. A sentence is acceptable if a speaker says it is okay.

Since grammars are ideally models of linguistic knowledge, *grammatical* pertains to the theory of competence, *acceptable* to the theory of performance.

There are obvious overlaps between *grammatical* and *acceptable,* of course, but they are very distinct notions; grammatical sentences, for instance, can be very, very long, so long as to be unacceptable to some speakers, or they can lead speakers down a garden path so winding that they refuse to accept them, and acceptable sentences can be, as in poetry, deliberately ungrammatical. The position that grammaticality is relative to an abstract grammar ("model," "theory"), rather than to a speaker's judgment, has led to some confused ridicule of Chomsky, mostly from humanists and Bloomfieldian holdovers, about such theoretical phenomena as the shifting status of "Colorless green ideas sleep furiously"—sometimes grammatical in his framework, sometimes not. But this sort of drift is very common in science: as the grammar changed, so did its grammaticality implications. The atom is sometimes the smallest piece of matter in physics, sometimes not, depending on the theory; light sometimes corpuscles, sometimes waves, sometimes both; space sometimes full of ether, sometimes not, sometimes peaceful, sometimes turbulent.

The generative semanticists came increasingly to view the competence-performance distinction, on which the grammatical-acceptable distinction rests, as artificial—worse, incoherent—and the rejection of Chomskyan grammaticality came along for the ride. Ross and the Lakoffs attacked the concept somewhat obliquely, by cataloging phenomena clearly required by language (or, perhaps more accurately, clearly required by speakers), but wholly indifferent to grammaticality—

phenomena like *please* and *thank-you* and the relative appropriateness of 13a and 13b for a lecture presented to an anthropological society.

13 a   Defecation is generally expedited by the use of large banana leaves, or old copies of the New York *Daily News*.
    b   *Making number 2 is generally expedited by the use of large banana leaves, or old copies of the New York *Daily News*.

(R. Lakoff, 1973a:301)

Lakoff's use of the asterisk here is telling, since she makes it a signal for inappropriateness in a given context, not for ungrammaticality, and her opposition to Chomskyan grammaticality had much to do with her increasing interest in ordinary language philosophy, for which context is extremely important. Ordinary language philosophy grew more important for generative semantics in the seventies. Beginning with Ross's fairly direct importation of Austin's insights about performatives, which Sadock and Davison took up at Chicago under McCawley, it gained considerable momentum under the influence that philosopher H. P. Grice's conversational implicature work had on both Lakoffs, on their students, and on McCawley.[36] It is from this period that linguists began to develop a sense of something they called *pragmatics,* as distinct from what it called *semantics.* For generative semanticists in particular, and linguists in general, the latter was virtually a synonym for 'meaning' until the early-to-mid-seventies, and the former belonged among the exoterica in philosophy. Now, *semantics* began hardening into a term for the truth conditions of sentences (of the Cormorant-Island sentences type) and a hatful of related notions (principally, paraphrase and entailment).

Pragmatics, . . . well, *pragmatics* never hardened into anything particularly, but it pretty much stands for 'everything else about meaning' in linguistics now—in particular, everything related to the influence of context. To repeat Gerald Gazdar's crude equation (1979:2),

PRAGMATICS = MEANING − TRUTH CONDITIONS

Non-truth-conditional meaning began to get serious attention for the first time in the history of linguistics when generative semanticists began snooping around in the data and implications of ordinary language philosophy, a great deal of which played havoc with Chomsky's notion of grammaticality.[37]

The main, and most vocal, opponent of Chomskyan grammaticality wasn't Ross or one of the Lakoffs, though. It was McCawley, and his disaffection followed a somewhat different route. In 1970, James Heringer did an unusual study of quantifier-negative idiolects—unusual for transformational grammar in that it gathered data empirically rather than introspectively. The study was very modest, but among its results were that a number of his informants found 14a, a sentence most transformationalists would brand *ungrammatical,* to be acceptable given the context supplied in 14b:

14 a   All the applicants didn't fail the test we so carefully rigged, didn't they?
    b   Two industrial psychologists, Mitt and Matt, have a grudge against

their supervisor and are attempting to retaliate by allowing incompetent people to get hired by their company. To do this most quickly, they have designed a screening test which doesn't screen out anyone and have tried it out. Matt has just gotten the scores back and Mitt, anxious to determine if their plan is working, asks this question. (Heringer, 1970:295)

Although this study is explicitly about acceptability, McCawley took it as a confirmation of his "long-held suspicion . . . that (contrary to claims by Chomsky, Katz, and others) native speakers of a language are not capable of giving reliable judgments as to whether a given string of morphemes or words is possible in that language" (1979 [1972]:218); in brief, that Chomskyan grammaticality has no psychological basis, or, reverting to earlier terms, that two Chomskyan goals are in irremediable conflict. Grammaticality runs afoul of mentalism. Indeed, McCawley came to regard the pursuit of grammaticality as completely wrongheaded: "[it is something which] I would not label as linguistics" (1979 [1972]:217–18); worse, it is unethical, belonging to his collection of "ideas not to live by" which "I hold to be pernicious in that they have retarded our development of an understanding of how language functions" (1979 [1976]:234); worse, it is a personal embarrassment, for which "I hang my head in shame at seeing how many times I have spoken of sentences as being 'grammatical' or 'ungrammatical'" (1982b:8).

As one might expect, McCawley's use of *grammaticality* and its related terms in the debates is frequently confused (hence, the head-hanging). For instance, he retroactively qualifies his use of such terms in his 1973 review of Chomsky's *Studies on Semantics* by denuding them of technical significance and asking the reader to "take 'ungrammatical' as simply an informal English equivalent for . . . the kind of anomaly that I happen to be talking about at the time" (1982b:8). On another occasion, he says that grammaticality is something he applies not to strings of words, but to complexes of semantic structures, surface structures, all intermediate structures, contextual information, and the speaker's intentions, adding a telltale generative semantic *etc.* (Parret 1974 [1972]:250). But such confusions were extremely common in the dispute, and, in any case, the force of his objection is extremely clear: grammaticality, as a property of sentences relative to a grammar isolated from context (a Chomskyan grammar), is not a feature which corresponds to anything in the head of a language user, and is therefore detrimental to the pursuit of psychologically real models of language. The argument, in fact, is a permutation of a typical Chomskyan argument, and McCawley still insists he is pursuing competence—though a much different variety than the one that concerns his former teacher (1979 [1972]:220).

Chomsky's response to this line of argument had been on record for a long time. Some Bloomfieldians objected to his concept of grammaticality when he first proposed it (see, especially, Hill, 1961), as did his old mentalist mentor, Roman Jakobson; and Chomsky had some words for them:

Linguists, when presented with examples of semi-grammatical, deviant utterances, often respond by contriving possible interpretations in constructed contexts, concluding that the examples do not illustrate departure from grammatical regularities. This line of argument completely misses the point. It blurs an important distinction between a class of utterances that need no analogic or imposed interpretation and others that can receive an interpretation by virtue of their relations to properly selected members of this class. Thus, e.g., when Jakobson observes that "golf plays John" can be a perfectly perspicuous utterance [see Jakobson, 1959:144], he is quite correct. But when he concludes that it is therefore as fully in accord with the grammatical rules of English as "John plays golf," he is insisting on much too narrow an interpretation of the notion "grammatical rule"—an interpretation that makes it possible to mark the fundamental distinction between the two phrases. (1964a [1961]:385)

Chomsky is up to familiar tricks here—in particular, calling the much broader Jakobson "much too narrow"—but his principal point is inescapably clear: no matter what one can do with *All the applicants didn't fail the test we so carefully rigged, didn't they?* and **Making number 2 is generally expedited by the use of large banana leaves, or old copies of the New York Daily News*, and *Spiro conjectures Ex-Lax*, one still needs to distinguish them somehow from more canonical sequences.

Finally, however, the argument on this front comes down, like so many scientific disputes concerning goals and methods, to a matter of faith. Chomsky acknowledges that the limits one puts on the study of grammar by roping off certain sections from others are necessarily arbitrary (1964a [1961]:385n5), but he is willing to live with the arbitrariness because it buys him a more manageable data set. As Holton (1988:39) puts the familiar position, "we are always surrounded by far more 'phenomena' than we can use and which we decide—and must decide—to discard at any particular stage of science." This, of course, is exactly what Bloomfield was up to when he consigned some linguistic phenomena to sociology and psychology, and banished other phenomena altogether, by invoking mentalism. And Chomsky is up to the same thing—his performance comes extremely close on a number of criteria to Bloomfield's mentalism. The trouble is that one can discard too much data, and end up with an unproductively narrow view of one's business. A frequently invoked criticism of the Bloomfieldians is that they discarded syntax, to concentrate on phonemes and morphemes, and missed a great deal of what is going on in language. Generative semanticists saw Chomsky committing the same sort of error by discarding pragmatics. They saw his ropes as excluding valuable avenues of research, perniciously directing linguistics down a blind alley.

Rug Bulging

> Chomsky's shifting definitions of performance provide him with a rug big enough to cover the Himalayas.
>
> George Lakoff

The generative semanticists also saw Chomsky's strategic roping-off of data to be evil in another way, reflected in Postal's complaint a few pages back. Beyond the methodological error generative semanticists believed Chomsky to be committing, they also saw a deliberate and disingenuous attempt to cloud the discussion of everything about descriptive power, a beclouding carried out by virtually every interpretivist who raised the issue. This last phrase is redundant: to be an interpretivist in the seventies was to raise the issue. Whenever the interpretivists attacked generative semantics as too powerful, the entire generative semantic panoply of devices was brought into the tally—in particular, transformations, global rules, and transderivational constraints. On their side of the ledger, however, the interpretivists listed only transformations and constraints (which, of course, were not called *global rules,* and were held to reduce power, not increase it). Conspicuously absent were the increasingly powerful semantic interpretation rules. And since the generative semanticists felt responsible for a much wider class of phenomena than the interpretivists—phenomena of competence and of performance—there was another loophole in the accounting procedure. Many of the facts that generative semantics addressed were simply disregarded, postponed, left out of the comparison.

Interpretive semantics, that is, was in an extremely good rhetorical position on restrictiveness. Its architecture was more complex, which Chomsky claimed as an advantage. It could handle many phenomena essentially for free, by leaving them to semantic interpretation rules. And it could relegate the most troublesome phenomena to performance—at best waving in the direction of a solution, such as the analogic rules of "Remarks." As we have seen, the first of these maneuvers, complexity, caught the generative semanticists rather slack-jawed. They couldn't believe Chomsky would pull it, or that anyone would fall for it, and offered no counter-arguments. The other two, shuffling data into the "free" semantic component or into the netherworld of performance, they found equally blatant, but here at least they drew attention to what they regarded as very suspect argumentation.

In fact, even before Chomsky had launched the restrictiveness assault, Lakoff was complaining about his use of wild cards—beginning with "Remarks." Recall Chomsky's trick with analogy from the last chapter. If the grammaticality judgments go the right way, Chomsky said in "Remarks", well and good. If they go the other way, then we can attribute them to a rule of performance, and that's not my concern. Heads, I win; tails, you lose. Given Chomsky's position on grammaticality and acceptability, the *criticism/criticizing*-analogy-argument is a fairly standard example of phenomena saving. But the generative semanticists saw it more as phenomena flushing, Chomsky using the possibility of analogic rules as a porcelain bowl down which to wash recalcitrant data. All of the interpretivists continued this tack throughout the debate, shuffling troublesome data out of their competence-centered purview and exhibiting the attitude "that theoretical innovations need no particular justification if they can be relegated to 'performance'" (McCawley, 1982b [1973]:29). It continued to drive the generative semanticists to distraction.

Performance wasn't the only culprit. Semantic interpretation rules were also brought in, not just to handle semantic phenomena of the sort generative semanticists were handling with transformations or global rules, but for a surprising amount of (formerly) syntactic phenomena. Postal proposed his Crossover principle to explain straightforward grammaticality facts, and grammaticality was a wholly syntactic beast for Chomskyan linguists.[38] But Jackendoff (1972:145–59) said that the data ought to be handled by semantic interpretation rules, an account that requires a sequence of words like "Himself was shaved by Jeff" to be grammatical ("syntactically well formed") but semantically anomalous. The most notorious example of this application of semantic interpretation rules occurred at the 1969 Texas conference, where Lakoff presented an argument he had from Perlmutter concerning *respectively* sentences in Spanish. In some dialects, goes the argument, 15b is a legitimate sentence, indicating that it derives from 15a (Spanish requires adjectives to agree in gender with the nouns they modify):

15 a Mi madre es alta y mi padre es bajo.
    b Mis padres son respectivamente alta y bajo.
       [My parents are respectively tall_feminine and short_masculine]

The data clearly supports lexical decomposition (*padres* from *madre y padre*), and clearly suggests that the gender-agreement transformation is conditioned semantically (since the adjectives do not agree with *padres,* which is masculine, but with its implied constituents)—seeming to offer solid evidence for generative semantics. Chomsky is reputed to have declared gender agreement to be a wholly semantic phenomenon, something handled by semantic interpretation rules, rather than transformational rules, a position which implies that 16 is perfectly grammatical, but semantically anomalous:

16 Mis padres son alta.

Since gender agreement had traditionally been treated the same as other "syntactic" properties, like person, number, and case, a corollary claim in English would be that 17a and 17b are both grammatical, but 17b is semantically troublesome:

17 a My parents bought themselves a motor home called *Moby.*
    b My parents bought herself a motor home called *Moby.*

We obviously can't reconstruct the conference argument with any reliability, but Chomsky did comment on the exchange in "Some Empirical Issues," observing that "if [the facts in 15a and 15b] are accurate, it would appear that gender agreement may be a matter of surface interpretation, perhaps similar to determination of coreference. This would seem not unnatural" (1972b [1969]:155n26). As the prophylactic *would,* the cautious copula, and the double negative all indicate, such a conclusion is in fact quite unnatural, but if it allows him to keep syntax in a petri dish, away from semantic contamination, Chomsky is willing to embrace it.[39]

The generative semanticists were incredulous. No matter what arguments they came up with, Chomsky just calmly moved the goalposts back on them. McCawley uses this incident to demonstrate the violence Chomsky is prepared to commit on the "commonly held conceptions of 'grammaticality'" and the power he is prepared to give over to semantic interpretation rules (1982b [1973]:89–90) in order to keep semantics away from syntax; Lakoff uses it as an illustration of how counter-evidence can push the interpretivists "to ever crazier positions" (Parret, 1974 [1972]:169).[40] One day a given sentence is grammatical, the next day it is ungrammatical but acceptable because of rules of performance; one day a sentence is ungrammatical, the next it is grammatical but semantically anomalous.

For their part, the interpretivists just regarded these moves as responsibly arranging their data into the piles that would make it most amenable. To them, the generative semanticists, with their ill-specified global rules and completely ungovernable data, were the ones who looked crazy. Worse: they were slipping back into the alchemical dark ages from which Chomsky had rescued the field.

The Bloomfieldian Backslide

Is generative linguistics infiltrated by a counter-revolutionary underground?

Ray Dougherty

Chomsky's routing of the Bloomfieldians had been so complete that by the late sixties any of the synonyms for that school (*taxonomic, descriptive*—even *structuralist,* which described Chomsky as well as anyone, better than some) were also synonyms for *misguided, unscientific,* and *blockheaded.* So it was inevitable that these synonyms would be sprinkled on the newest unscientific blockheads, generative semanticists, and the final offensive against generative semantics was that it represented a backslide into the prescientific era from which Chomsky had raised linguistics; or, in Dougherty's quasi-Kuhnian terminology, a Bloomfieldian counterrevolution. Chomsky was not active in the final campaign, though he may well have been its sponsor; much of the argumentation that flows out of MIT begins with him, and he certainly endorsed the generative-semantics-as-Bloomfieldian-backslide case.[41] The case is in fact very strong, though it is difficult to see the same amount of evil in a return to some Bloomfieldian tenets that Dougherty (1974, 1975, 1976a, 1976b, 1976c), Brame (1976), Katz and Bever (1976 [1974]), and Ronat (1972 [1970]) all see. The evil of empiricism was so self-evident to them that none apparently felt any burden to establish why the return is so calamitous. Dougherty and Brame are content with name-calling; Katz and Bever show carefully how generative semantics opens the door for the return of Bloomfieldian epistemology, and then end their paper with an ominous *"Caveat lector."*

The backslide arguments have two primary components, one methodological, the other philosophical. The methodological part of the case was silly and vitriolic, and belongs mostly to Dougherty, who asked such memorable questions as

Whatever became of those linguists who were thoroughly trained in taxonomic methodology? Where are those old students who brought joy to the taxonomic hearts of their old masters? Where are the old students who, while suffering through a sequence of field methods, relentlessly pursued the phoneme from teepee to teepee? (1974:278)

After jumping all over McCawley and Ross and Lakoff for many haranguing pages, and dropping such broad hints as "having cut their eyeteeth on Bloch, etc.," and diagnosing Postal's work as symptomatic of the dread disease "Generative Breakdown-Taxonomic Relapse," Dougherty leaves his readers to find the answers on their own.[42] And we can leave Dougherty at this point, too, and leave the methodological issues with him. Chomsky *added* some methodology to linguistics—in particular, an emphasis on introspective evidence and on the value of deviant data—but he didn't subtract any. And, in any case, he argued that data-collecting techniques and tools of analysis were far less important than the sense one could make out of the data once it was collected and analyzed.

The philosophical component, represented most cogently by Katz and Bever's

(1976 [1974]) "The Fall and Rise of Empiricism," requires a little closer examination, however. As the title suggests, the case concerned empiricism, and therefore the terminological yin to its yang, and calls for a rehearsal of those terms:

Empiricism: most knowledge is acquired through the senses.
Rationalism: most knowledge is not acquired through the senses.

The case proceeds, then, that Bloomfieldian linguistics falls into the former category, empiricism, and Chomskyan linguistics falls into the latter category, rationalism—points that very few people would dispute. To the extent that Bloomfield nodded in the direction of the mental aspects of language, he subscribed largely to behaviorism, a thoroughly empiricist psychology; Chomsky, who is preoccupied with the mental aspects of language, holds a thoroughly rationalist psychology. Bloomfield and Chomsky are explicit about these allegiances. The generative semanticists, however, are not, which is where the argument gets interesting. In fact, Katz and Bever never argue either that generative semantics entails an empiricist epistemology, or that generative semanticists have deliberately adopted empiricism:

> We do not claim that the linguists who are bringing it back are necessarily empiricists or are aware that their work has this thrust, but only that their work clears the way for the return of empiricism. (Katz and Bever, 1976 [1974]:30)

Generative semanticists, it seems, are unwittingly providing a medium for the reemergence of empiricism. Katz and Bever sound the alarm about the direction in which generative semantics will lead linguistics, not about its current (mid-seventies) stance. But, they believe, this is a very dangerous direction. Trotting out the rhetoric of cold-war American foreign policy to strike fear in the hearts of linguistic consumers, they propose a domino theory of encroaching empiricism in linguistics, built around the notion of grammaticality: first it is relaxed; then it is modified; eventually it must be discarded. "Once it fell," they argue, "so would each other domino: conversational bizarreness; next cultural deviance, then, perceptual complexity, and so on" (1976 [1974]:59).

Their case is more detailed than we need explore here, and there are complications—relative grammaticality is not quite the terror they paint it to be, for instance (Chomsky, for one, finds it compatible with his own work—1972b [1969]:121), and Katz's notational variants position clouds the discussion—but generative semantics unquestionably represents a return to empiricism. Look, for instance, at how McCawley criticizes a diagram Chomsky had made famous in defining the central focus of his program (given as figure 7.3). "The flaw in this account," McCawley says, is that it makes the child look like "a linguist who elicits ten notebooks full of data from his informants in New Guinea and doesn't start writing his grammar until he is on the boat back to the United States" (1976b [1968]:171); that is, the model doesn't provide for hypothesis-testing, experimentation, and game-playing—in a word, it doesn't provide for *feedback*—all of which are defined not by exposure to data (the rationalist position) but interaction with data (the empiricist position). McCawley suggests the diagram be revised along the lines of figure 7.4.

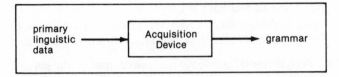

Figure 7.3. Chomsky's "hypothetical language-acquisition device" (adapted from McCawley, 1976b [1968]:171, and Chomsky, 1966b:10).

McCawley's commentary is a willful misreading of Chomsky, who would not deny that "primary linguistic data" includes feedback, and that the acquisition device involves the formulation and modification of successive grammars before full competence is achieved; indeed, his notion of an evaluation metric is predicated on selective modification.[43] But the two diagrams are nonetheless very revealing. Chomsky is not especially concerned with the process of language acquisition in any detailed way; he rarely cites empirical acquisition studies in his work, for instance, and the diagram McCawley modifies here (that is, figure 7.3) shows up recurrently in his work, essentially in the same form. He is interested far more in the properties of the cognitive mechanism, Acquisition Device, than in its specific employment—a definitively rationalist concern. McCawley is interested at least as much in the way the mechanism is put to work, and in the way it interacts with general-purpose learning strategies, and in the character of the acquisition data— empiricist concerns all. He doesn't reject Chomsky's rationalist arguments about language acquisition. He just has somewhat broader interests:

> Chomsky's well-known arguments that language acquisition cannot be accomplished purely by general purpose learning faculties should not lead to the non sequitur of concluding that general purpose learning mechanisms play no role in language acquisition: General purpose learning faculties clearly exist . . . and it is absurd to suppose that they shut off while language is being acquired. (1980b:183)

Returning to the main critique, however, we find that Katz and Bever barely consider McCawley. They focus almost exclusively on Lakoff, which, among other things, illustrates how fully generative semantics had come to be associated with Lakoff by the mid-seventies, especially in interpretive eyes,[44] but they don't look at anything he said about the issue either. What he said was fairly consistent by the

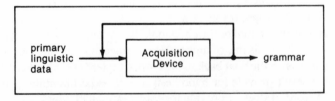

Figure 7.4. McCawley's modification of Chomsky's Acquisition Device (adapted from McCawley, 1976b [1968]:171).

early seventies. There was a little I-know-you-are-but-what-am-I? taunting about empiricism, as in his remarks about (who else?) Chomsky:

> I would say that, of contemporary linguists, Chomsky is among the more empiricist linguists . . . in the sense that he is still interested in accounting for distributions of formatives in surface structure without regard to meaning. (Parret, 1974 [1972]:172; see also McCawley's comments, p. 251)

But, for the most part, Lakoff was happy to admit his interests—he calls himself, in fact, a "Good Guy Empiricist"—and happy being associated with Chomsky's immediate predecessors, celebrating the Bloomfieldians for their creation of "a broad, diverse, and interesting field, which happened not to be very good at dealing with the syntactical problems raised by Chomsky, and which showed little interest in formalized theories" (1973c). Further, he claimed that "when transformational grammar eclipsed structural linguistics, it also eclipsed many of these concerns, much to the detriment of the field" (Parret, 1974 [1972]:172). More explicitly, even his interpretation of Chomsky's famous nativist arguments leaves a good deal of room for empiricism, going considerably beyond McCawley's live-and-let-live, don't-forget-general-purpose-learning-faculties interpretation. Lakoff puts the implications of the argument in binary terms, and lobbies heavily against the rationalist side of the coin:

> What Chomsky has shown is that *either* there is a specifically linguistic innate faculty *or* there is a general learning theory (not yet formulated) from which the acquisition of language universals follows. The former may well turn out to be true, but in my opinion the latter would be a much more interesting conclusion [though see Lakoff, 1968b:1–4, when he found the former more interesting]. If I were a psychologist, I would be much more interested in seeing if there were connections between linguistic mechanisms and other cognitive mechanisms, than in simply making the assumption with the least possible interest, namely, that there are none. (Lakoff, 1973c; his italics).

Or, a few years later, take his depiction of the workaday grammarian's approach to linguistic theory:

> In practice, you try to set up your linguistic theory in such a way that linguistic abilities will ultimately turn out to be special cases of nonlinguistic abilities as much as possible. There is good reason for going about linguistics in this way. It seems highly implausible that linguistic ability has nothing whatsoever to do with any other aspect of our being human. For me the most interesting results in linguistics would be those showing how language is related to other aspects of our being human. (Lakoff, 1977a:238)

Meanwhile, Ross began to emphasize Zellig Harris's influence more and more—citing him, for instance, as marking the major conceptual break that led to modern linguistics, in a paper remarkable for its odd sense of history. The paper is addressed to cognitive psychologists, outlining the sorts of contributions linguists can make to their field, and Harris, as thoroughgoing an anti-mentalist as they come, gets the lion's share of the credit for making these contributions possible; Chomsky, incredibly, is not even mentioned until well into the article, when he is conspicuously introduced as Harris's student (1974b:64, 68).[45] Pursuing the same theme, Lakoff takes the implications of Chomsky's apprenticeship to even further absurdities, suggesting that his work is mindlessly derivative:

> Chomsky was extraordinarily dependent on his teachers for his intellectual development. Most of his early linguistic analyses are taken directly from Harris, as is the idea of the transformation. The idea of evaluation metrics was taken over directly from Nelson Goodman. (Parret, 1974 [1972]:172–73).

There was a schizophrenia about the links generative semantics were reforging with Bloomfieldian linguistics. On the one hand, there was something of the standard reaction to reach back historically to embrace the enemy of your enemy, as Chomsky had reached back to Humboldt and Descartes. On the confused other hand, Chomsky was just a derivative Bloomfieldian anyway.

Parting Company

> When two opponents have been arguing, though the initial difference in their positions may have been slight, they tend under the 'dialectical pressure' of their drama to become eventually at odds in everything. No matter what one of them happens to assert, the other (responding to the genius of the contest) takes violent exception to it—and vice versa.
>
> Kenneth Burke

Shenker describes Postal's main occupation in the early seventies as "proliferating exceptions to Professor Chomsky's theories" (1972), which is actually a pretty good phrase for what both sides were up to for a goodly portion of the dispute. A great many exceptions to Professor Chomsky's work, of an unfocused sort, percolated out of abstract syntax—Postal's verb-adjective conflation, Ross's auxiliary analysis, McCawley's reanalysis of selectional restrictions as semantic rather than syntactic. When the threshold to generative semantics was crossed, the exceptions became far more specific, zeroing in on deep structure—McCawley's *respectively* argument, Postal's *remind* argument, the Predicate-raising, CAUSE-TO-DIE argument. On Professor Chomsky's part, he too started out with somewhat loose and unfocused exceptions. The "Remarks" proposals gnawed away at the abstract-syntax foundations of generative semantics, but the exceptions it spins out are somewhat scattered, going after abstract verbs here, category-reduction there, "the transformationalist hypothesis" somewhere else.

For the next two years, however, the whole MIT program, Chomsky at the helm, did little more than proliferate a class of very specific exceptions, directed at the Katz-Postal hypothesis. Meanwhile, the generative semanticists switched from their assault on deep structure and began proliferating pragmatic and grammaticality exceptions to Professor Chomsky's work, left, right, and center.

At this point a rather clear difference surfaced. The generative semanticists loved data. And the more problems it caused for various bits of theoretical machinery, the more they seemed to love it. They kept their exception-proliferating noses to the grindstone throughout the seventies. The interpretivists loved theory. And the more problematic the data was, the more eagerly they shunted it off to grammatical provinces for which they felt little or no responsibility. And, since exception-proliferation is a data-heavy activity, they quickly tired of it. The brunt of their attack on generative semantics became conceptual. In fairly rapid succession, the arguments came: generative semantics is just a new name for the same old grammar (a notational variant); it is licentious in its use of theoretical mechanisms (unrestrictive); and it has the wrong philosophy (backsliding into Bloomfieldian empiricism). At times, the combination of these arguments strains credulity to the limits; for instance, in one paragraph Chomsky says that global rules "are quite similar, if not identical, to the interpretive rules proposed by Jackendoff and others," but that they add "immense descriptive potential" to grammatical theory, and thus their introduction "constitutes a highly undesirable move" (1979 [1976]:152): global rules are notational variants of a good rule-type, but they are a bad rule-type.

The generative semanticists kept proliferating their exceptions, and were stunned at the interpretivist reaction; that is, at Chomsky's reaction. They accused him of an absurdly narrow conception of grammaticality, a conception invoked only to buffer his theory from the cruel world of linguistic facts. They accused him of sweeping those excluded facts under the performance or semantic interpretation carpet, or waving half-baked, inexplicit solutions at them. Then, suddenly, they found that he just didn't care. After the barrage of counter-arguments in his 1969 Texas paper, "Some Empirical Issues," he just turned his back on them. He was still happy to bash generative semantics in class, in interviews, and other informal settings, and remains happy to do so, but it rarely earned even a contemptuous footnote in his formal work after the Texas paper. In 1971, his trace-proposing paper, "Conditions on Transformations," began circulating underground (published 1973a; 1977:81–162), and it marks his official withdrawal from the debate. Before that paper, Sadock says, Chomsky was "still talking the same lingo." Afterwards, "there was a new philosophy," a philosophy of restrictiveness, and a complete inattention to any of the issues that were driving generative semantics. His students continued to press the attack, becoming more and more savage, but Chomsky had barely a sneering allusion left for his old enemies. He gave up his rhetoric of dissent to pursue positive work.

What Sadock fails to notice, however, is that generative semantics wasn't talking the same lingo it started with either. Generative semanticists had also gone their own way. By the mid-seventies, they were no longer "engaged in 'generative grammar'" (McCawley, 1982b [1973]:11).[46] The most succinct turning point—or, perhaps, *burning point,* as they torched their bridges behind them—was over the notion of grammaticality. But there were many, many divergences.

The rhetorical breakdown was complete. Witness Geoffrey Pullum's recollection of the reception for a paper about auxiliaries he and Deirdre Wilson wrote in the mid-seventies (Pullum and Wilson, 1977):

> It tried desperately to separate the issue of whether auxiliaries are main verbs (they are) from the issue of whether generative semantics was right. Hardly anyone was listening. We might as well have suggested the Israelis and the Palestinians sit down together and talk.

Each side went its own way. To the generative semanticists, it looked pretty much like they were going off in victory. As Chomsky recalls, in the mid-seventies, "the overwhelming mass of linguists interested in transformational grammar were doing some kind of generative semantics" (1982a [1979–80]:43). Not only was the first generation of MIT linguists enamored of it, so were their students. Influential generative semantics work was being done at most of the major institutions—at the University of Texas-Austin, at the University of Massachusetts-Amherst, at the University of California-Los Angeles, at University of Illinois-Urbana, at the University of Michigan (where the Lakoffs taught briefly), and, of course, at the University of Chicago. Chomsky really had only MIT, and even there Postal, Perlmutter, and Ross had toeholds in the period. Transformational textbooks from the seventies virtually all include a sort of cutting-edge section that declares or suggests generative semantics to be the linguistics of the future. The most dramatic illustration is John Kimball's *The Formal Theory of Grammar*. Kimball was a 1971 MIT graduate, and the final section of his book doesn't so much as hint that Chomsky has a contemporary position on semantic questions. Generative semantics shows up as the natural successor to the "standard theory," just as natural a successor to that theory as the standard theory was to the one presented in *Syntactic Structures* (Kimball, 1973:116–26); it's almost as if Chomsky retired after *Aspects*.[47]

Nor was it just the younger linguists who were signing up to generative semantics. Many linguists from earlier generations also found the model very attractive. Emmon Bach was influential early on, and Lees endorsed it with enthusiasm. Wallace Chafe had proposed a semantically-based transformational grammar himself, independently (1967a, 1970a, 1970b), but aligned himself with generative semantics (especially, 1970b:56, 68). Even some Bloomfieldians, from the linguistics boneyard, rattled in approval. Hill's LSA memoir captures the general mood of the old guard (or maybe "the very old guard," since Chomsky now looked like the old guard) in the seventies—Hill recalls receiving Ross's abstract syntax work fondly, delighting at a Robin Lakoff paper for blowing the lid off Chomsky's Cartesian claims, and even praises the other Lakoff for his humorous sentences. Eugene Nida painted his mid-seventies work as having "a dependence on the generative semantics approach" (1975:8), and Householder commented that "the views of Ross, Lakoff, McCawley, [and] Mrs. Lakoff on the nature of the base . . . are the ones which I find most congenial" (1970:35). Dwight Bolinger went a step further, becoming an active ally in the movement, and shows up widely in generative semantics footnotes for perceptive or inspiring assistance; Ross singles him out for "a special kind of thanks" in one article, because he "has been saying the kind of things I say in this paper for a lot longer than I have been able to hear them" (1973c:234). Even the transformational granddaddy, Zellig Harris, was being touted by his supporters as developing a theory "similar to a generative semantics theory" (Muntz, 1972:270).[48]

Psychologists found generative semantics appealing, for obvious reasons—"the attraction for psychologists of generative semantics is the greater plausibility of supposing that a speaker begins by generating the basic semantic content of 'what he wants to say', only then going on to cast it in an appropriate syntactic form" (Greene, 1972:85)—and the leading generative semanticists had no trouble gaining a hearing with them. McCawley and Ross, for instance, were featured presenters at the 1972 Conference on Cognition and Symbolic Processes (McCawley, 1974b;

Ross, 1974b). Philosophers, too, showed considerable interest, welcoming in particular the explorations of logical form by Lakoff and McCawley. Donald Davidson and Gilbert Harman, for instance, invited them to "a cozy cross-cultural colloquium at Stanford" in 1969 (Quine, 1985:357), and then included a paper by Ross in the collection of work stemming out of that colloquium (Lakoff, 1972b; McCawley, 1972; Ross, 1972d. See also Harman, 1972).[49]

Europeans, most of whom only became interested in transformational work in the late sixties, also signed on to generative semantics in large numbers. Pieter Seuren, Rudolf de Rijk, and Werner Abraham all became influential generative semanticists. The model had notable followings in Sweden, France, and Germany, where Herbert Brekle's *Generative Satzsemantik* went through two editions (1970 [1968], 1978), and where the important American papers all saw translations (Abraham and Binnick, 1972; Lakoff, 1971d; Vater, 1971). In Czechoslovakia, where Petr Sgall had independently proposed a semantically based transformational grammar, the term gained rapid acceptance. There was even a growing following in Japan, and in Australia, Frans Liefrink proposed his *Semantico-Syntax* (1973).

Generative semantics, it seemed, was the future of linguistics; Chomsky and his interpretivists, the past. Guy Carden even remembers his old teacher expressing puzzlement over one of his mid-seventies papers that attempted to keep a dialogue open with the interpretivists: "Guy, why are you still talking to those people?" Lakoff asked him. "Haven't you noticed the war is over? We've won." Getting in what he may have taken to be the last word, Lakoff remarked at about the same time that "one of the joys" of debating is that "the winner gets to say 'Nyaah, nyaah!' to the loser," a few pages later, of course, ending the paper with "Nyaah, nyaah!" (1973b:286, 290). The comment tells us something about the dynamics of the dispute, a little about the style of generative semantics argumentation, and a lot about Lakoff.

Hold the phone, George. The story's not over yet. But the important point here is the triumphant mood it conveys.

Generative Semantics 3:
The Ethos

Is it not, then, better to be ridiculous and friendly than clever and hostile?

Socrates

Euphemistic genital deletion.
(A generative semantics rule name)

Paul Postal

Not Being Chomsky

All through the war, the generative semanticists were having a lot of fun, fun that spilled into and all over their work, which had everything to do with the times. We are, don't forget, in the late sixties and early seventies, when the cultivated alienation of the hippies had a pervasive influence on all things academic and most things non-academic, and the generative semanticists—"a bunch of people who got together at conferences to make puns and play Fictionary and smoke funny cigarettes" (R. Lakoff, 1989:972)—are as representative of that period as beads or bomb threats or bongs.[1] Their work teems with the themes of drugs, music, casual sex, and, of course, politics.

The stylistic proclivities of generative semantics, like those of the period generally, were not just a veneer of cultural references and allusions glued onto an otherwise conventional form. They run deep. The hippies (forcing this word to carry much more freight than it did when it was, briefly and narrowly, the preferred label for the group I am gathering under it here) were as seriously at odds with the preceding generation as any group could be, with perhaps the exception of the Khmer Rouge. The early term for their ethos was *youth culture*, which makes the dividing line clear. It was replaced by *counterculture*, which makes the scope of the rejection equally clear. The final, even more transient, label was *Woodstock Nation*, which was coined in a formal declaration of independence from, and state of war with, "the Pig Empire" (Hoffman, 1971). At the major risk of trivialization and other

198

distortions, at both ends of the analogy, it is impossible to miss the parallels with generative semantics, which ended, for many, in a total rejection of the transformational home that turned them out; George Lakoff, the chief spokesman of this rejection, sounds like Jerry Rubin or Abbie Hoffman when he decries Chomsky's work as "the epitome of emptiness" (1977a:284).

Just as the hippies defined themselves in pained relief against "the establishment," a central tenet of generative semantics for Lakoff, was "the idea of not being like Chomsky." The generative semanticists saw Chomsky as dishonest in his handling of data—reworking it to serve his temporary purposes, discarding what he couldn't rework, ignoring vast regions altogether—and, in very sharp contrast, they jubilantly celebrated all kinds of data, not only data that gave the interpretivists trouble, but data that gave their own theory the fits as well.

A similar pattern holds for theoretical machinery: where Chomsky's style is to redefine and retain modifications of his flawed proposals, usually in terms which point to their new and improved abilities, not their former inadequacies, generative semanticists tended to renounce theoretical innovations very publicly, to proclaim their errors from the rooftops. And a related pattern, at least in the eyes of the generative semanticists, holds for the social extensions of the two models: they found Chomsky exclusionary in his disciplinary politics, defining other linguists in a concentric Dantesque vision—"There were the inner circle," as Robin Lakoff characterizes it, "the various outer circles, Limbo, and Bad Guys" (1989:972)—but they wanted everyone to join their party. Well, almost everyone. They, too, had a Limbo and a mob of Bad Guys, Chomsky riding at the head of the latter. But where Chomsky turns his back rapidly on people from his program who head off in directions he does not personally find promising, generative semanticists encouraged diversity, welcoming any and all forays into uncharted data; their Limbo wasn't populated with hopelessly misguided researchers (the way Chomsky views, say, sociolinguists), but with people working on things generative semanticists just hadn't got around to yet. Where Chomsky works on a rather narrow, almost individual, set of linguistic problems—always using images of isolation to describe his own relation to the rest of the field—generative semanticists eagerly looked for problems in other areas of linguistics, in psychology, in philosophy, in logic, in literature, even in the medicine cabinet. The Oedipal reflex has rarely been carried to such extremes in science; because Chomsky was regarded with almost religious fervor when the schism began, he was promptly demonized: "Once Chomsky was seen not to be an idol," as Robin Lakoff told us in chapter 4, "he was recast as satanic, the Enemy" (1989:970).

The political elements of the generative semantics style, however, throw something of a monkeywrench into the simple anti-Chomskyan account of their ethos. The generative semanticists opposed the war in Vietnam, and Chomsky was one of its most outspoken critics. A simple analysis of this twist in the Oedipal story might put the political explicitness of many generative semantics into the trying-to-out-Chomsky-Chomsky file, much as a simple analysis of the generative semantics genesis has Lakoff and Ross going deeper into transformational grammar than its leader dared, having more courage in their convictions than he did. But simple analyses don't work in either case.

It is, first, extremely difficult to out-Chomsky Chomsky on political issues. He does not use sample sentences like "In a real sense, Nixon is a murderer" in his articles, as the generative semanticists did (this one is from Lakoff, 1972a:210), but he is on record, repeatedly and forcefully, as calling Johnson, Nixon, Kissinger, and their ilk, "genuine war criminals," on a level with the thugs of the Third Reich and Stalinist Russia—more specific, more caustic and far more thoroughly prosecuted sentiments. He was (and remains) a more tireless, dedicated political rhetor than any of the generative semanticists; in fact, it is only mildly hyperbolic to say he is more tireless and dedicated to political concerns than all the generative semanticists combined. Even the comparison between the respective political efforts of Chomsky and the generative semanticists is otiose and misleading, like saying Milton was a better writer than most poets, Pavarotti a better singer than most jingle-artists, Gretzky a better hockey player than most bush leaguers. The difference in scale is several orders of magnitude. The generative semanticists were more explicit than Chomsky, in linguistic articles, about their political concerns, and certainly more irreverent. But they were not, by even the wildest stretch of the term, more political (the way they might easily be called, for instance, "more semantic").

It is, second, not coincidental that the shared political outlook of Chomsky and most generative semanticists (and, indeed, most interpretive semanticists) was shared dissent. They were united in opposition to the war, and the political figures behind it. On positive issues (like what should be done to correct the mistakes of Vietnam, who should do it, how it should be carried out), there would have been far less agreement, maybe none at all.

And, third, the war was unspeakably obscene. It was, fortunately, impossible for many people to remain silent in the teeth of both America's vicious program of terror in Indochina and the parade of ugly, blatant lies coming from the government about that program. What distinguished the generative semanticists on this count was their willingness to let their outrage into their linguistics.

It may seem digressive, even diversionary, to identify a generative semantics style at all, let alone to spend a chapter examining it. One prominent opponent of the movement, for instance, is fond of writing it off as "an irrational cult" and "a fad," just another symptom of the early seventies collapse of values, and the kind of attention we're giving it here can play into such dismissals. But, facts are facts: all the major figures partook in an identifiable style, in very identifiable ways. Everyone who talks of the period both remembers the style and has an opinion on it. For better or for worse (in fact, for both), a loose-jointed, absurdist, politically direct style suffused much of generative semantics.[2]

That style—to take the dry, reductionist, analytic approach that is death to any style, but at least gives us some postmortem results—had three principal manifestations: humor, politics, and data-worship. We begin the autopsy with humor.

Humor

> Fuck my sister tomorrow afternoon.
> *Fuck those irregular verbs tomorrow afternoon.
> Fuck my sister on the sofa.
> *Fuck communism on the sofa.

Fuck my sister carefully.
*Fuck complex symbols carefully.

Quang Phuc Dong

The age of the hippy was not a subtle age, and the defining style of generative semantics was not a subtle style. The sample sentences implicate all three of the most visible elements of the counterculture, elements which (like much of the spirit of the sixties) rapidly degenerated in the following decade, in this case, into one of the most mindless and hedonistic slogans of all time:

"Sex,
1 a   The fact that Max plorbed Betty did not convince Pete to caress her on the lips. (Postal, 1988a [1969]:74)
  b   The M.C. introduced Mick Jagger's penis as being large enough to amaze the most jaded of groupies. (Borkin, 1984 [1974]:18)
  c   Let's fuck. (R. Lakoff, 1977:82)
and Drugs,
2 a   Hey, if John went to Chicago, that means we'll soon have a big supply of dope. (Schmerling, 1971:249)
  b   My cache of marijuana got found by Fido, the police dog. (R. Lakoff, 1971b:154)
  c   Fred does nothing but smoke hashish and play the sarod; John is similar. (McCawley, 1976b [1972]:304)
and Rock 'n' Roll!"
3 a   *Sam snarped 10 Beatle records for a nude photo of Tricia Nixon. (McCawley, 1982b [1973]:80)
  b   Paul is dead and I do not believe he is dead. (Lakoff, 1975)[3]
  c   She left one too many a boy *behind.* He committed *suicide.* (Bob Dylan, cited in Zwicky, 1976:683)

There was also plenty of room for other prominent counterculture characteristics—politics, scatology, and general absurdity:

4 a   Amerika's claim that it was difficult to control Vietnamese aggression in Vietnam surprised no one. (Grinder, 1970:300).
  b   *The shit that John took weighed 600 grams. (Quang, 1988 [1971]:96)
  c   *I don't want to kiss no gorillas. (Postal, 1974:236)

Humor was a point of pride with many generative semanticists, in large part because it was a very clear break from Chomsky, a declaration of the spiritual, in addition to the intellectual, schism with the master. Chomsky's prose has all the humor of Aristotle. (One of the late sixties joke templates was title-of-the-world's-shortest-book, like *Theories of Racial Harmony,* by George Wallace, and *Problems of the Obese,* by Twiggy, and a prime candidate for this title circulating linguistics at the time was *The Bawdy Humor of Noam Chomsky.*) "We felt Chomsky took himself too seriously," Lakoff remembers. "We thought it was extremely important that people be able to laugh at themselves and what they were doing, as well as have fun."

It didn't hurt that the youthful academic audiences around the turn of the decade were very well disposed to appreciate absurdity, boundary-pushing, and celebrations of their own cultural artifacts.

Like other stylistic proclivities of generative semantics, these built slowly, and have their roots in the first transformational forays into semantics. Katz and Fodor include the occasional stab at humor (such as sentences like *Occulists eye blondes*), as do Katz and Postal (who introduced the technical term *G-string*—1964:57). But it was in the abstract syntax period that these traits began to show up with regularity, the first symptom being a penchant for flamboyantly named protagonists in sample sentences, an appropriately contra-Chomskyan symptom. Chomsky's sentences are notoriously dull (from any but a strictly linguistic perspective). His favorite protagonist is *John,* and even outsiders have remarked on the flat, characterless quality of such sentences as *John is easy to please* (a lineal descendant of such flights of imagination as Bloomfield's *Poor John ran away*).[4] The abstract syntacticians, on the other hand, went for *Max* and *Seymour* and the ever-popular *Floyd;* generative semanticists, for *Knucks McGonagle, Figmeister,* and *Norbert the Nark.* Under the genius of abstract syntax—namely, Postal—the sentences also began to populate with gorillas, wombats, penguins, toads. With generative semantics the situations got weirder, the cast of characters expanding to include cultural icons like Willie Mays and Yoko Ono. Patterns and themes recurred. One leitmotif had Richard Nixon loitering in men's rooms.

Sample sentences were the thin edge of the weird-name wedge. Postal called one transformation, *Flip.* Ross named another one *Slifting;* then, following an alliterative theme, proposed *Sluicing* and *Stuffing.* Carden proposed *Q-magic.* These names at least had some reference to the actions performed by the rule (Slifting raised a Sentence node, an S node, to a higher clause, so its etymology is *S-lifting;* Flip exchanged noun phrases; Q-magic concerned quantifiers). Later in the game, as respect for the descriptive machinery of transformational grammar declined, the names became deliberately arbitrary; in particular, there was a fad of adopting proper names for rules, like *Irving, Ludwig,* and *Richard.* As Lakoff explained it, "one way to [remember where you're fudging] is to use obviously arbitrary names like CLYDE instead of arbitrary names that sound profound but aren't, like Determiner" (1971a:iii; see also Ross's justification for the name *Do-Gobbling*—1973a:70). Other names went in the opposite extreme, becoming absurdly specific, like Grinder's *Apparel Pronoun Deletion* (for the syntactic behavior of certain sentences concerning disrobing) and—the hands-down winner in this category—Postal's *Euphemistic Genital Deletion* (for sentences where certain graphic nouns are demurely spirited away, as in *John is too big for Mary,* or *Max is playing with himself again*).[5]

For many generative semanticists, this style leaked out of the sample sentences and rule names to permeate the prose. Anecdotes, inside allusions, and schoolyard expressions like "one swell foop" showed up regularly, most notably in the second-generation papers. Jerry Morgan, for instance, offers this account of some missing arguments:

> [My earlier paper] contained the ultimate solution to the problems of pronominalization, reference, identity, as well as an item of overwhelming and irrefutable empirical

evidence against the lexicalist position. Unfortunately, it was handwritten on a package of Puritan Hog Chow, and was eaten by a hungry Chicago policeman who tore it from me during a tear-gas attack on three jaywalkers, thereby being lost to mankind. (1976 [1968 or 1969]:340n1)

But, lest we get carried away, it is important to note that most of the humor, while far from subtle, was not exactly center stage. There are few belly laughs in generative semantic papers. Aside from some of the wilder underground papers (many collected in *Studies out in Left Field: Defamatory Essays Presented to James D. McCawley*—Zwicky and others, 1970), the papers don't exist to tell jokes; serious work goes on in and around the humor. Even the Quang papers have important linguistic points. The jokes participate in the papers—sometimes propelling them along, sometimes offering a commentary on the author's confidence in the analysis—but mostly just contributing a general tone of goofiness to the work, in deliberate counterpoint to Chomsky. To use a slightly effete word for the practice, it is more whimsical than humorous. It depends on odd situations, insider references to other linguists or to cultural figures, and a general attitude of surrealism, not on punchlines or strenuously sustained metaphors. There is also a very clear gradation of the amount of whimsy as a function of audience. The rabid-hog quotation above, for instance, comes from an underground paper circulated amongst core generative semanticists, of a piece with "Camelot, 1968." To stay with the same author, Morgan's CLS contributions are still loose and whimsical, with subheads like "Pickles and Strawberries" and the extended participation of Ernie Banks, but they are markedly less informal. His more mainstream publications, as in *Language* (1977), are positively tame (though certainly not Chomskyan).[6]

The presence of second-generation generative semanticists here raises another factor in the levity of generative semantics papers, a factor Hagège (1981 [1976]:21) contemptuously dismisses as the "whiff of amateurism": many of the more outrageous generative semanticist disquisitions did come from amateurs, graduate students. The chief generative semantics schools (Chicago, Illinois-Urbana, and Michigan) all followed the practice that Chomsky and Halle had begun in MIT of urging students out into the fray as soon as possible. But where inexperience and exuberance tended to manifest themselves among MIT students as bloodthirsty polemics, in nascent generative semanticists the trend was toward looney humor, political asides, and declarations of awe at the complexity of the data.

And, of course, style is a personal trait, like speech pattern; there were several very distinctive people at the helm of generative semantics. Newmeyer's (1986a:137) *rambunctious* captures much of their collective spirit, but there was individual spirit at play as well, and not all the humor had the same motivations. With Lakoff, humor may well have been in some part the "concealed intellectual aggressiveness" that Arnold Zwicky recalls of the movement generally. Lakoff is an aggressive guy, given to some rather grating rhetorical tactics, and it would be difficult to deny that this aggressiveness had vent in his publications, one of which, remember, ends with "Nyaah, nyaah!" (1973b:290).[7]

For McCawley, who has a very widespread reputation for gentleness, even when directly challenging someone's argument, the humor was probably more celebratory. Ross falls somewhere in the middle; Robin Lakoff a little closer to McCawley;

and Postal's humor, a juxtaposition of deadpan scholarly prose and sample sentences teeming with wombats and gorillas and anaphorically orphaned blondes, is more difficult to pin down. Many of the second-generation generative semanticists also had their own cluster of stylistic character traits—some leaning more toward politics, others toward absurdity—each one indicating a slightly different configuration of motives, and a different hierarchy of role models.

Politics

> This History is humbly dedicated to those Valiant young Knights who, in quest of the Holy Grail, shed their Blood under the onslaught of Savage wild boars in the Forests of Lincoln and Grant, and in the Stone Valley of Michigan, in the Duchy of Czechago, Summer, 1968, Richard the Leatherbuttocked, Lord Mayor.
>
> Sir Lancelot of Benwick, Morgan le Fay, and The Green Knight

Humor and absurdity were principal devices of counterculture rhetoric. There were certainly earnest dissidents at the time—Dave Dellinger, Cesar Chavez, Noam Chomsky—but, for better or worse (actually, for both), the temper of the times was caught more fully by less focused, more boisterous people, like Ken Kesey, Jerry Rubin, and Abbie Hoffman. The dominant social spirit was the loose mélange of anger, liberation, and confusion that I have collected under the umbrella, *hippy,* and, even in my broad usage, that umbrella does not cover Dellinger or Chavez or Chomsky. Nor do all the young, shaggy radicals, no matter how angry and confused (Tom Hayden and Mark Rudd spring to mind) fit comfortably under the umbrella. The counterculture is Kesey seducing the young and appalling the old with his portable acid tests, Rubin and Hoffman leading phalanxes of outrageously clad longhairs on a mission to levitate the Pentagon. It is the Women's International Terrorist Conspiracy from Hell (WITCH) casting hexes on the administration during sit-ins at Chicago; the slogan of young Parisian radicals, *"Je suis marxiste, tendance Groucho"*; a young linguist in a tie-dyed T-shirt lecturing the staid professorate at Yale about the theoretical ramifications of words like *Kalamafuckingzoo.* Much of the energy was political, but much of it was also just the sheer, loopy celebration of life.

Politics and absurdity, in any case, are not very easy to disentangle at the turn of the decade. The political situation was a marriage of Carroll and Kafka. King, the Kennedys, and Malcolm X had all been shot; George Wallace, who was about to be, stumped around the country denouncing "left-wing theoreticians, briefcase-totin' bureaucrats, ivory-tower guideline writers, bearded anarchists, smart-aleck editorial writers and pointy-headed professors" (O'Neill, 1971:389), while his running mate advocated a nuclear solution in Vietnam; little bags of hair from the Chicago 7, shorn after their arrest, were auctioned off at Republican fundraisers; the government's response to increased protests over the Vietnam War was to attack Laos and Cambodia; Ronald Reagan ordered over two thousand National Guardsmen to "quell" a demonstration against the University of California-Berkeley's expansion into what became People's Park, resulting in one death, 128 wounded;

the Weatherman staged ridiculous occupations of working-class schools to wake up the proletariat, usually being driven off by hippy-hating toughs; enraged at the violence, repression, and deafness of the power structure, or possibly just bored, various collectives and individuals bombed over three thousand public structures in 1970 alone, and made over fifty thousand bomb threats; protesters were killed at Kent State, Jackson State, and Berkeley, maced, clubbed, and arrested elsewhere; Chicago police mugged proudly for photographers after killing Fred Hampton; other black dissidents, and some whites, were jailed on weak drug charges or driven into exile; the reduction of American casualties by turning over much of the ground-fighting to the South Vietnamese in order to concentrate more fully on bombing, defoliation, and napalm attacks, actually reduced public outcry; the temper of President Nixon, or, more accurately, his distemper, paled even Garson's notorious off-Broadway satire of his predecessor in *MacBird!:*

CRONY: Peace paraders marching.

MACBIRD: Stop 'em!

CRONY: Beatniks burning draft cards.

MACBIRD: Jail 'em!

CRONY: Negroes starting sit-ins.

MACBIRD: Gas 'em!

CRONY: Latin rebels rising.

MACBIRD: Shoot 'em!

CRONY: Asian peasants are arming.

MACBIRD: Bomb 'em!

CRONY: Congressmen complaining.

MACBIRD: Fuck 'em!
             Flush out this filthy scum; destroy dissent.
             It's treason to defy your president. (Garson, 1966:73–74)

Linguists were among the most concerned protesters. A very Beat looking Háj Ross, for instance, can be seen in the foreground of a picture in *Time* accompanying the story of Dow Chemical's lab director being barricaded in a Harvard conference room for seven hours, in a protest against Dow's napalm production (3 November 1967:57). McCawley was a vocal participant in the 1966 University of Chicago protests against allowing the Selective Service access to class rankings; at a particularly vulnerable time in his career, he snuck into a meeting for tenured-only faculty on the question, took notes, and reported back to a student meeting. Jerry Morgan was very active in the Chicago convention protests. One of the more important collections of linguistic papers from the period is dedicated "To the Children of Vietnam" (Jacobs and Rosenbaum, 1970). And, of course, there was Chomsky.

Much of the political activity of linguists may have been indirectly related to the tough, impassioned stand Chomsky took against American imperialism in Indochina. Certainly a partial motivation for many students who joined the MIT linguistics program at the time (and at the present) was a political affinity with Chomsky, just as Chomsky's principal motivation for studying at the University of Pennsylvania was his political affinity with Zellig Harris. More generally, a partial motivation for many linguists going into the field, and particularly into transformational grammar, was Chomsky's political reputation. Although he kept his politics and linguistics fairly distinct at the time, his political feelings were extremely well known, and widely applauded, in the linguistics community. McCawley, for instance, ends one of his long letters to Chomsky about the *respectively* controversy with the postscript, "There is no truth to the nasty rumor going around that the CIA is subsidizing my research in hopes of thereby diverting your energies from the war" (18 January 1968).

The generative semanticists, while no more deeply concerned about the social and political pathologies of the day than the interpretive semanticists, were far more overt about those concerns in their linguistics. This difference was undoubtedly a function of the leading figures in each camp. Chomsky did not engage much in blends of politics and linguistics during the linguistic wars. The closest he came was to mention that one of his sample sentences came from an activist associate (1972b [1969]:67), or to relate discourse situations concerning his refusal to pay taxes and his anti-war speeches (1975b:61–62)—fairly simple illustrations taken from his life, with no overt political meaning in the context, though they surely lent some of his political reputation to his linguistic work and helped contribute to the sense that linguistics was a field for people with social consciences.[8] Jackendoff was only marginally more political in his linguistic work, occasionally incorporating sentences such as *The president is insane, one suspects, beyond all hope* (1972:97). On the other side of the debate, the picture initially doesn't look much different. Postal never blended politics and linguistics, Ross and Lakoff only occasionally and peripherally, taking a few swipes at Republicans, for instance. But McCawley was a different story. He was more politically active than the others—getting involved in the Selective Service protests, demonstrating for civil rights in Jackson, Mississippi, burning his draft card, refusing to pay his taxes—but two other factors are considerably more important than his political activism. First, there was his complete lack of reserve about involving his political views in his linguistics—under his famous *nom de guerre* founding what Pullum (1991 [1987]:101) calls "the *Fuck Lyndon Johnson* school of example construction," and, as that designation indicates, it was a distinctively counterculture sort of involvement. Second, McCawley was also the most influential generative semantics teacher, and it is with the second generation that the FLJ examples really began to flower.

Too, there are the matters of space and time. The official home of generative semantics, the site of its yearly CLS festivals, was Chicago—which witnessed the Democratic convention riots, by the police; the subsequent conspiracy trial and its attendant circus; the killing of Fred Hampson; the Weatherman's Days of Rage; assorted acts of autocratic weirdness by Lord Mayor Richard Daley, and assorted acts of anarchist weirdness by groups like WITCH. As for timing, generative semantics and the counterculture became popular in exactly the same period. There had to be cross-pollination. Interpretive semantics didn't take off fully until the popularity of the counterculture began to wane.

Among the more interesting sidelights to the political aspect of generative semantics is the fate of Georgia Green's McCawley-supervised thesis, which was held up for a year because of Cambridge University Press's fear of libel suits (entertainingly documented by Geoffrey Pullum's "Trench-mouth Comes to Trumpington Street"—1991 [1987]:100–111). The process bordered on the ludicrous, as legal advisors cautioned against including sentences like 5, as defamatory of Mary:

5   Mary gave John a hickey.

But it brings to light another element of the generative semanticist style, a subtle step beyond the FLJ examples—the sentence *à clef,* where the names refer, more or less obliquely, to living people. Mary and John in 5 are very likely arbitrary, but it doesn't take much imagination to find referents for 6.

6   Martha gave John trench-mouth, and he gave it to Ted.

As Pullum projects the publisher's concerns, "Could the *John* be the indicted Attorney General John Mitchell? Could *Martha* be his wife Martha Mitchell? Was *Ted* perhaps Senator Edward Kennedy?" (1991 [1987]:107). Probably. The generative semanticists had a track record in sample sentences of this sort, with a wide cast of characters, including, in addition to the notorious Mitchells and other political figures, cultural icons, and one another:

7   a   Martha knows that it will be necessary to behave herself until the election is over. (Neubauer, 1972:287)
     b   Who does John think the police will say Nixon ordered to kidnap Martha? (Morgan, 1973:739)
     c   No one but the bastard himself pities Lyndon. (McCawley, 1976b [1970]:302)
     d   Only Muriel voted for Hubert. (Horn, 1988 [1969]:164)
     e   John will not tickle Yoko when she belches. (Yamanashi, 1972:388)
     f   I saw Che alive last week. (Green, 1973:264)
     g   Jim resembles Quang in accent. (Ross, 1972a:162)
     h   George likes Peking Duck, but all linguists are fond of Chinese food. (R. Lakoff, 1971a [1969]:138)

Even members of the interpretivist camp dabbled in the sport. Given the animosity between Lakoff and Jackendoff, it is hard to imagine that this sentence is entirely innocent:

7   i   Although the bum tried to hit me, I can't really get too mad at George. (Jackendoff, 1968:13).

Beyond the more overt political content of much generative semantic work, there were two other ways in which the approach touched more effectively the social mood of the times. First, there was the growing influence of sociolinguistics. While generative semanticists were not particularly engaged in sociolinguistics, they were unflaggingly sympathetic toward the field. Lakoff, for instance, included some of Labov's work on black American English in his most famous policy statement, "On Generative Semantics" (1971b:280f), and Robin Lakoff began to carve out her own unique blend of sociolinguistics and ordinary language philosophy. More generally, generative semantics' concern for language use and social context ensured that they benefited from the growth of social conscience at the time:

> The political atmosphere of the late 1960s and early 1970s contributed to this feeling of discontent [with Chomsky's approach] and was a major factor in drawing many serious students of language into adopting the generative semantics program, which, by combining work on formal grammar with concern for the use of language in the real world, promised to satisfy their intellectual interests as well as the demands of their social conscience. George Lakoff [in Parret, 1974 [1972]:172] was undoubtedly correct when he [said] "Nowadays students are interested in generative semantics because it is a way for them to investigate the nature of human thought and social interaction." (Newmeyer, 1986a:228)

The second way that generative semantics more effectively fit the political temper at the turn of the decade was its rhetorical emphasis on dissent. Dissent was a very marketable property in the early-to-mid-seventies. The hippies took generational discord to extremes (Roszak, 1969:1–41), making any assault on authority attractive almost by definition. "Bring the revolution home: kill your parents," went the Weatherman credo. "The Oedipal Conflict has replaced Marxian Dialectics," said Abbie Hoffman (1971). Jerry Rubin, going upper case, screamed "KILL YOUR PARENTS!" (1971:194).

Generative semanticists were not as extreme as the Weatherman or the yippies. No one was demanding Chomsky's head. But Lakoff was not above describing one assault on interpretivism as "an exercise in anti-establishment thinking," or calling transformational grammar "as much a part of the intellectual establishment as General Motors is a part of the military-industrial establishment" (1971a:ii). In the more muted terms of an interview, he discussed the broader implications of dissensual data:

> Teaching linguistics these days is not without some indirect—very indirect—political consequences. In linguistics, as in politics, much of the relevant data to support or refute many claims are available to the average person. In linguistics, it is in your mind and all you have to do is train yourself to recognize it. In politics, it is all around you, in the newspapers and on TV. Again you just have to be trained to recognize it. Just about any beginning linguistics student, with some careful thought, can in an afternoon think up enough crucial examples to show the inadequacy of our most sophisticated current theories. Similarly any citizen of average intelligence can pick out many of the lies that his government tells him. The thought processes are not all that different, though the subject matter is. Any beginning linguistics student will discover with a little thought that men of great stature in the academic establishment, even very bright ones like Chomsky, can be wrong on just about every issue. It makes one wonder about the 'experts' who are running our governments. (Parret, 1974 [1972]:170)[9]

Most data-gathering papers, which came to characterize generative semantics, in fact followed precisely the prescription Lakoff offers here, except that they also often nod favorably in the direction of generative semantics: here are some facts which any theory of language should address; no current theory comes anywhere near treating these facts naturally; generative semantics, however, comes the closest; interpretive semantics is completely out to lunch.

Data-Worship

> I can't resist giving one more example. This one is from Geoff Nunberg (God
> bless him).
>
> George Lakoff

Generative semantics developed out of a concerted attempt to make the *Aspects* model work on a growing spectrum of data—including issues like category membership, lexical composition, pronominal relations, conditions on transformations, performative sentences, quantifier scope, the implications of symbolic logic, and considerably more. It was a research project to extend the power of transformational description. But one of its principal side effects was increased suspicion of the descriptive powers of transformational grammar. For every generalization, there were pockets of facts that slipped out of grasp. For every rule there were lists of exceptions ("the blood of a wounded theory"—Green, 1974:4). For every theoretical principle, there were areas in which it fell drastically short.

Inevitably this suspicion of transformational machinery came hand in glove with an increased reverence for data, particularly when ordinary language work and Dwight Bolinger's research came to exert more influence. Chomsky (although many of his detractors deny it) has a healthy respect for linguistic data, but his principal allegiance is to the formal properties of grammar. His is the attitude of theoretical physics. The generative semanticists rapidly developed an attitude more like that of an observational science, like astronomy, and began to display the same sense of awe characteristic of that field. The stylistic offsprings of these twin notions—theory-suspicion and data-reverence—were also twins: incessant discussion of the inadequacies of theoretical proposals, and humility in the face of vastness of natural language.[10]

One of the principal motives for close attention to the data was educational, a motive best illustrated by a small, peculiar, barely-above-the-ground pamphlet— perhaps the single most illuminating document for the attitudes that came to be pervasive in late generative semantics—called *Where the Rules Fail: A Student's Guide. An Unauthorized Appendix to M. K. Burt's* From Deep to Surface Structure (Borkin and others, 1968). Even its genesis is offbeat enough to fit the generativist narrative like a tie-dyed T-shirt. It started life as a homework assignment that Lakoff gave a first-year graduate syntax class at Michigan: take a rather typical explication of a number of the conventional arguments for the *Aspects* model and tear it to shreds. Lakoff chose a workbook by an MIT student, Marina Burt, who was briefly persuaded to appendicize this attack to the second edition of her *From Deep to Surface Structure* "as a kind of antitext within a text." However, when she saw the virulence of the assault, and showed it to "several persons for whom I have great respect" (read *Noam Chomsky* and *Morris Halle;* possibly some others), she thought better of it. In her declining letter, she defended her decision with the very reasonable argument that "although you present excellent counterexamples, there is no analysis of the facts which replaces what exists even though we both know it's not adequate. The feelings are that it is not enough to pick apart an existing theory without offering some kind of alternative." There is little value in shouting no! when you don't also offer something worth saying yes to.[11]

The students were irate at being denied the opportunity to thumb their noses at Burt with her sanction, and at the person for whom she was, to some extent, a synecdoche, Chomsky. They had the antitext circulated, with a foreword by Lakoff, through the Indiana University Linguistic Club, along with a petulant preface that complained about being treated like children who "are too immature to be confronted with the realities of science."[12] Lakoff very likely had a hand in IULC's decision to publish, and, in any case, his foreword contributes much to the pamphlet's fascination. It is pitched directly at beginning students of the sort who would use Burt's workbook, and gives a very good picture of his exhortative teaching style in the late sixties and early seventies. It is explicit in its appeal to youth, though it adopts the curious language of pessimistic optimism:

> By pointing out some of the failures of the rules in Burt's book, we hope to give students a feel of linguistics as a living discipline, where most analyses are hopelessly far away from their goals, and where the old goal of actually writing a complete grammar for a language has become at best a hope for the future centuries and at worst a joke. (i–ii)
>
> Finding out that very little works the way most introductory textbook writers would lead you to believe can be a frustrating experience. On the other hand, it can and should be an exhilarating one. After all, the less that is known, the more there is for you to find out. If you want to do something interesting with your life and are contemplating doing work in linguistics, it should be anything but frustrating to find that there is a lot for you to do. (v)

The point, clearly, is that the old-timers were mistaken about how language works. Burt's workbook, he says, "dates from the days when many linguists thought that transformational grammar in its classical form was basically correct." The old-timers were not just wrong in this, they were arrogant in their wrongness, and Burt is guilty of adopting "the sanctimonious pose of presenting solid results." Still, Lakoff generously adds, Burt's book does present a very valuable opportunity for beginning students, since, if used in the right pedagogical spirit, "it is a good vehicle for teaching students that transformational grammar is not all that it is cracked up to be." The important point is not to blame Burt, though she makes a convenient scapegoat, but to blame the entire transformational enterprise prior to the new generative semantics enlightenment.[13]

Among the many ironies in this approach to linguistics is the distance it places Lakoff from his dissertation, only a year after its publication. His dissertation was Lakoff's first major contribution to linguistics, and it took the most decisive early steps towards generative semantics, and it was data-happy. Lakoff's thesis project was essentially to snoop out apparent counter-examples to the *Aspects* theory and tame them, to save irregular phenomena with transformations. As the generative semantics program grew, Lakoff continued to work on taming the most difficult data he could uncover, but soon gave up using transformations to do so. And he taught his students to approach linguistics by engaging in anti-establishment thinking and trafficking in counter-evidence. Somewhere, however, the imperative to save the phenomena was misplaced, and much generative semantics energy was increasingly devoted simply to celebrating anomalies.

Running Out of Patience

> Not all generative semanticists would agree with one ex-partisan that "we went out of our way to be funny in our papers so that once our ideas were refuted we could get ourselves off the hook by saying, 'Oh, did you take us seriously? Couldn't you see that we were just fooling around?' [personal communication]." But most, I suspect, would acknowledge a kernel of truth in it.
>
> Frederick Newmeyer

Newmeyer's comment about generative semantics humor is a bone sticking in the craw of almost every ex-partisan.[14] For good reason. It is far too bald to be true.

There are certainly harsh words for generative semantics humor; the few published comments about it at the time rebuked the generative semanticists for the "extreme of cultivating the bizarre or the risqué for its own sake" and for their "juvenile political and sexual references" (Percival, 1971:184; Sampson, 1976:179), and Lees, with typical bluntness, just says that "It was smartass humor." The interpretivists were generally unimpressed—Lasnik says "You have to take your field seriously. You can't convey to the world that it's like a standup routine in a nightclub"—though Chomsky himself is somewhat neutral:

> Science is like any other human activity. You don't have to put a straightjacket on it. If people like to give papers with jokes, that's fine. It's neither good nor bad. Maybe it's more fun to listen to their papers, I don't know. But it shouldn't [affect one's evaluation of] their results.

Nobody, however, seems to share Newmeyer's feeling that there was bet-hedging going on, that the proposers wanted to be able to excuse themselves later with the claim that it was all a joke. The central generative semanticists regard his observation as flatly stupid, if not malicious; Lakoff (demonstrating a characteristic he shares with Lees) terms it "an utter falsehood," and calls the informant "a jerk."

There is, though, some muted truth behind Newmeyer's observation.

The truth behind it has more to do with the general informality of presentation, of which humor was an ingredient, than with humor per se. Informality was part of the working-paper mode of much generative semantics; its literature almost never made any attempt to hide the tentativeness of the proposals. Many generative semantics papers, particularly the ones in the CLS volumes, are rife with expressions of hesitancy about their proposals (though they almost always claim superiority over interpretivist treatments of the same phenomena), and stylistic looseness reflected this hesitancy, perhaps tacitly requesting some lenience from the audience. It expresses, in effect, what Lakoff put more explicitly very late in the debates:

> I do not have anything even close to an adequate theory. What I do have, by way of formal treatment, are some grossly inadequate but suggestive ideas. . . . So far as I can tell they are only slightly better than what is now available in various versions of gen-

erative grammar—and you know how bad they are! I hope you will approach [the proposals] with a charitable heart. The ideas are young and need to be cared for. (1977a:259)

Lakoff's statement is the fullest one on the tentative nature of many proposals, but it is very far from unique. Nor were generative semanticists shy about recanting their proposals when clear proof of something better came along—although no one, as far as I can determine, ever used the excuse "Couldn't you see that I was just fooling around?" in such a recantation.

The most obvious example of this circuitous connection between humor and recantation is from Postal, who proposed (1970 [1969]) the feature [DOOM] for marking noun phrases which faced elimination later in a derivation, and shortly thereafter renounced it (1972a [1969]:140), when global rules entered the picture and seemed to offer a superior treatment of the phenomena. The name *DOOM* was a stylistic recognition that his solution was artificial, that it was something of an epicycle on the theory of transformations—it was doing exactly what Lakoff advised students to do with names like *CLYDE*—but it was a perfectly serious proposal. Later, when he found a mechanism to do the same job less artificially, Postal was happy to let the feature go.

In short, while there was clear hedging going on, and humor contributed to that hedging, it is unfair to say that humor was a deliberate escape hatch. The proposals were seriously advanced, but they came with a warning label against unqualified acceptance. This truth-in-advertising was rarely appreciated. Transformational-generative grammar had been in flux almost from the outset—certainly from as soon as it grew beyond Chomsky's private toy. Lees's 1962 preface to the second printing of his *English Nominalizations,* just two years after the first, apologizes that it "has long since become antiquated" (1968, xxvii). But such confessions were rare in the early years, and have always been rare in the work most closely associated with Chomsky, which projects confidence and certainty.

Consumers—of theories just as of washing machines—want confidence and certainty. They don't want to invest their intellectual capital in something that may be out of date in a few years, or a few months, or a few days. Listen to this complaint about the immediate dissemination of partially digested notions:

> What is at issue is whether every new speculation, no matter how ephemeral, should be broadcast in the literature when the development of ideas is so continuous and rapid that the latest speculation is already out-of-date by the time it is published. (O'Donnell (1974:79n1)

More acerbic, at least in its adopted persona, is Wall's suggestion to linguists who have trouble keeping up with syntax:

> Look, maybe the thing to do would be to go into historical linguistics and let syntax go for a while—at least until things settle down a little and these assholes stop printing every hare-brained hallucination that afflicts their heads. (Wall, 1970:167n15)

O'Donnell and Wall come from very different perspectives in these critiques. O'Donnell is confused and hostile, but aligned to neither the generative nor the interpretive side of the debate, and is therefore a pretty good barometer of the overall linguistic community in the mid-seventies. Wall, as should be clear from the tone, is conducting a satire. But he was something of an insider, so the satire is deadly accurate, a terrifically biting view of the movement (though Wall certainly does not spare Chomsky); *hallucination,* for instance, is not a randomly chosen term.

Things were beginning to look bad for generative semantics, and ethos was one of the reasons for its failing health.

Generative Semantics 4:
The Collapse

The transcendent truth of my position has been buttressed time and again, most recently by the splendid work being accomplished in progressive (as opposed to regressive, although of course no directionality is implied) semantics. I refer here not so much to the writings of McQuarrelly, whose thought is not always sound dogmatically, but rather to the output of Coughlake, that prolific exponent of generative power (see, inter alia, Coughlake to appear a, to appear b, to appear c, to appear N_0). Coughlake's irrefutable, nay absolutely crushing, indications of the necessity for wholly novel forms of grammatical apparatus—approximately one revolution in theory each week, beginning with the Ann Arbor Film Festival winter, Durational Constraints—quite boggle the mind. Nothing could prove the correctness of the ESP [Erector Set (British Meccano) Proposal] more convincingly than these repeated demonstrations that the required pulleys must be larger and stronger than we were inclined to believe.

Ebbing Craft
(a.k.a. Arnold Zwicky, in a multipronged parody)

Unsurprisingly, I have no explanations to offer here.

Háj Ross

The Cessation of Hostilities

Judith Levi set to work on *The Syntax and Semantics of Complex Nominals* in 1974, an extensive reworking of her doctoral dissertation under McCawley that does for generative semantics what Lees' *The Grammar of English Nominalizations*, a reworking of his doctoral dissertation under Chomsky, does for early transformational grammar. Levi's book proves—within all the variations, uncertainties, and controversies that characterize any scientific attempt to model reality, especially a reality as mucky as language—that generative semantics works.[1] But when she finished, and the book was in press (published 1978), Levi "looked around and found that nobody was interested. There was no community."

On the other side of the battlefield, an emerging interpretivist scholar, Howard Lasnik, noticed that the burning issues of the last five years had disappeared overnight from the MIT hallways: "People just stopped talking about them. I have this distinct impression. It was like someone put his head out the window and it wasn't raining anymore, and that was that."

The debate was over.

Chomsky, it very shortly became clear, had won.

The victory was crushing, and Chomsky deserves a good deal of praise or blame, depending on your perspective, for the death of generative semantics. But not all. Some of his counter-argument bullets were very well placed, but generative semantics also shot itself in the foot, often, very publicly, just as Chomsky's model began making impressive headway.

It was Chomsky's positive proposals—providing an alternative to a model undergoing a crisis of confidence—far more than his negative attacks on generative semantics, that pulled his interpretive bacon from the fire, albeit badly singed.

The Chomskyan Ebb

> Let's face it. The publication of *Reflections on Language* [Chomsky, 1975b] leaves little doubt that transformational-generative grammar has become an intellectual fraud.
>
> Bennison Gray

In the early-to-mid-seventies, Chomsky looked to be, in Zwicky's biting parodic pseudonym, Ebbing Craft. He had been outflanked by his progeny, and had retrenched to a Bloomfieldian anti-meaning position rather than embracing their advances. His counteroffensives and attacks—the misunderstood lexicalist and x̄-syntax proposals, the vaguely adumbrated post–deep structure semantic proposals, the peculiar notational variants position—all looked feeble. And that was just the start of his troubles.

His reputation in psychology was taking a beating. In the early sixties, he was hailed as a founding father of the cognitive revolution; by the early seventies, even as his previous accomplishments were making their way into textbooks and popularizations, working psychologists were rapidly losing faith in his program. For one thing, it wasn't panning out experimentally. Psychologists had come up with a very pretty cognitive theory of transformational grammar that had clear empirical consequences: The more transformations involved in a sentence, the longer it should take for someone to understand it. A passive sentence should take longer than its active counterpart. A negative should take longer than a positive. A negative-passive should take proportionally longer than a positive-active. You get the picture. At first, this model—known as the *derivational theory of complexity*—seemed spectacularly successful, giving a psychological boost to transformational grammar, a grammatical boost to cognitive psychology, and an empirical boost to the hybrid fledgling, psycholinguistics. It *did* take people longer to understand sentences with more transformations in their derivation. But all too soon, when sentence length and meaning were factored in (passives are longer than actives, for instance, and have subtle differences in meaning; negatives are slightly longer than positives, and very different in meaning), transformations receded in importance; at best, they now seemed untestable. Worse, transformations which had no differences in length or meaning (relating sentences like *Debbie called up Jeff* and *Debbie called Jeff up*) also had no appreciable impact on comprehension time.[2] A similar story unfolded for experimental attempts to confirm the psychological reality of deep structure: initial success, followed by reinterpretations of that success considering other factors, and then outright failure.[3]

Even more problematically, the consequences of Chomsky's theoretical and methodological positions in *Aspects* were beginning to make psychologists nervous. In particular, the confluence of three factors—his competence-performance distinction, his insistence that transformational-generative grammar directly concerned only competence, and his focus on ideal constructs—seemed like a conspiracy of sneaky maneuvers to relieve his work of any empirical responsibility. "The adventure into transformational grammar," recalls Eric Wanner, looked to many psychologists to have "reached a dead end" (1988:150).

And the generative-interpretive squabbling didn't help. Some psychologists, the ones who managed to maintain their interest, found in generative semantics the same attractive and inevitable outgrowth of Chomsky's early work that most linguists initially found in it, and preferred the more elegant, more intuitive picture of language it offered. But it proved no more amenable to experimentation, so there was little for them to choose between, and most just found the dispute arcane and ill-mannered. Chomsky's reputation suffered the most, in exactly the way it suffered in the media accounts of the dispute. A savior only four or five years back, Chomsky now didn't have the courage of his own convictions to follow through with the more explicitly semantic and mentalist program that grew out of *Aspects,* let alone the strength of character to keep his former devotees at his side.

Chomsky's notational-variants argument was especially destructive with outsiders. In the version that reached the broadest public, Katz's *New York Times* comments, the argument looked like this:

> You can measure things by yards and inches, or you can use the metric system, and both are correct. Here we have a parallel in linguistics. One theory starts with the meaning of words and combines them to arrive at the meaning of the sentence. The other starts with the meaning of the sentence and breaks it up to obtain the meanings of the words. Both methods employ the same rules. (Shenker, 1972)

A dispute involving as much heat as this one, but which reduces to something as trivial as a confusion over measuring rods could not possibly be taken seriously by its spectators. It was also easy for process-minded psychologists to view the two models as completely inverse—*interpretive* suggests a grammar for the hearer, *generative* suggests a grammar for the speaker. Why, then, couldn't the linguists get their acts together? Moreover, for any innocent bystander who had read Chomsky's popular books (especially *Cartesian Linguistics* or *Language and Mind*), it did look as though Chomsky was saying exactly what the generativists were saying. Deep structure had seemed to people on the outside to be a semantic representation all along; why so much strife now?

The more linguists I see, psychologists were beginning to feel, the more I like my rats; someone finally published a response to Chomsky's *Verbal Behavior* review (MacCorquodale, 1970), and reports of the death of behaviorism began to look somewhat exaggerated. Worst of all, the damn theories refused to keep still. The most dedicated, hardworking psychologist couldn't keep up with the literature on even one side of the debate, let alone both.[4]

Psychologists were not the only disaffected observers. Things were just as bad for Chomsky in English studies, where the hoped-for panaceas in composition and poetics failed to cure any ills, and complicated many; from where the semantics wars looked arcane and ill-mannered; and from where they could see that the damn theories wouldn't keep still. English folk went from delight to disillusion even more rapidly than psychologists, and by the mid-seventies articles like "Why Transformational Grammar Fails in the Classroom" were common—hysterically drawing attention to the negative correlation between national ACT scores and "the rise of the linguistic revolution," and using analogies to the New Math, once a recommendation, now with derision (Luthy, 1977). While nobody seems to have articulated it in these terms, one of the principal reasons for disaffection was the increasing forcefulness with which Chomsky made it clear he cared only about the mental aspects of language; English folk are almost exclusively concerned with the social aspects, with communication.

One English professor was so aghast that he published a wonderfully malicious satire, *Oh's Profit,* which attacked the species-specific module of Chomsky's nativist hypothesis. The hero of the novel is a gorilla, the eponymous Oh, who speaks perfectly good American Sign Language, and hence raises the ire of the Great Leonard Sandground, director of the Institute of Cortextual Commitment, inventor and chief proponent of Genesis Grammar, and a prominent anti-war activist in the bargain. Even Chomsky's by-now-notorious talent for shifting the grounds of his theory at the drop of a theoretical or empirical hat comes under biting attack. For Sandground, running through an internal monologue, the main problem with Oh is not his linguistic abilities, but his timing:

> When the next talking ape came down the road the Institute would have been theoretically prepared. The proper articles could be disseminated to the appropriate journals. Yes, the solution was, ironically, at hand. Genesis G. *could* handle talking apes, *could* make room for them. With a loss of elegance, of course . . . (Goulet, 1975:154; Goulet's italics, abbreviation, and trailing dots)

Unfortunately, no short-term solution presents itself to Sandground, and he sets out to murder the ape.[5]

Even the moral implications of Chomsky's work—or, rather, the lack of them— were under attack, many people noting the irony that one of America's premier social critics avoided the social aspects of language like a case of the hives. American linguistics prior to Chomsky had long been a field with a social conscience—in a real sense, it was born out of a social conscience—most notably in its approach to indigenous languages. In the late Bloomfieldian period, this conscience had atrophied somewhat. With Chomsky, it withered away to nothing: If linguistics is a branch of cognitive psychology, its relevance for social issues is negligible at best. But the sixties and seventies saw a renewed interest in the sociopolitical dimensions of language, though now it tended to focus more on the urban poor, and with that focus came what Newmeyer calls "the moral critique" of Chomskyan linguistics.[6] The criticism had a number of reflexes, but revolved around the program's emphasis on English data, on introspection, and on the resultant ivory tower into which it locked the field:

> The almost exclusive study of their own language, English, by so large a proportion of the world's linguists, has seemed to the participants a source of deepening insight into the underlying structure of all languages. Leaving aside the methodological difficulties that have become increasingly apparent, we must consider that to many other communities, including those of American Indians, such a concentration may seem an expression of ethnocentrism at best, a hostile turning of the back at worst. . . . Many participants in formal linguistics are liberal or radical in social views, and yet their methodological commitments prevent them from dealing with part of the problems of the communities of concern to them. (Hymes, 1974b:21–22; Hymes, not coincidentally, is an ethnological linguist with some pre-Chomskyan allegiances).

The social climate of the sixties and seventies was extremely receptive to this style of attack on Chomsky, and many people found it very telling, seeing additional support for it in the overwhelmingly predominant funding source for transformational grammar, the U.S. military.

There were more attacks. Aarsleff jumped all over Chomsky's claims to Cartesian roots, and his scholarship, and his honesty (1970; 1971).[7] Gray denounced the "Alice in Wonderland state of affairs" in Chomsky's idealizations and formalisms (1974:5), and his "inversion of priorities" (1976:38), and the "intellectual fraud" of his work (1977:70). Maher railed against Chomsky's school—MITniks, MITnik myths, MITnik methods, general MITology, and sundry MITnikia (1973a:passim). Derwing launched a book-length assault on Chomsky's rhetoric, puzzling in particular over his success "despite key arguments which include the fully specious, the mainly irrelevant, and even the out-and-out false" (1973:222).[8] Other books joined in: Koerner (1975), Robinson (1975), Hagège (1981 [1976]), and Anttila, in an all-purpose, anti-Chomskyan harangue, celebrated these proliferating assaults:

> So much of the current criticism is directed against transformational-generative grammar and its various offshoots. It does not mean that an innocent is brutally drawn under gang attack, but that they are indeed real offenders who have brought scholarly discourse to an all-time low. (Antilla, 1975:172)

But there was little in these attacks to comfort generative semanticists. As Anttila's joy indicates, it was not only Chomsky, but all things Chomskyan, coming under attack in the seventies, especially the squabble that had broken out among his formerly close-ranked followers. Maher complained that "every MITnik is a revisionist" (1973b:30), and Talmy Givón muttered from the wings about a discipline going to hell in a handcart, "rife with fads, factionalism, and fratricide" (1979:xiv).

A few observers were surely pleased by the dispute. Searle noticed some of the Bloomfieldians "rubbing their hands in glee at the sight of their adversaries fighting each other" (1972:20), and they weren't alone. A fair number of psychologists and philosophers were pleased at Chomsky's fall from the graces of his erstwhile disciples. Chomsky had denounced Skinner, but had not been able to replace his research program with anything as expedient for framing problems (he had said no to behaviorism, but hadn't given psychologists something they could easily say yes to), and disaffected behaviorists were happy to see the revolution founder. Chomsky had also alienated many philosophers by rejecting empiricism, and some of them were happy to see him in a less angelic light. The Chomskyan revolution was far-reaching in its implications, and could not help but stir up cognitive dissonance in the fields it impinges upon. Add to this natural dissonance the arrogance with which many of the revolutionaries plied their trade, and it was inevitable that there were pockets of scholars in the wings waiting for any sign of internal discord, and that they would be plenty happy to crow about it.

But most bystanders, the neutral and hopeful majority, were just bummed out by the dispute. They wanted Chomsky's work to make their lives easier, or, at the least, they had not ruled out the possibility that his research would have value for them. Language scholars outside of linguistics, who had been told (usually by false prophets in their own discipline who had only vague notions of Chomsky's work) that Chomsky's framework would revolutionize the way they taught Spanish, or studied the novel, or examined arguments, were disappointed at the sight of a bitter feud among the people who were supposed to lead them into the light.

The bickering, that is, did not help the discipline. Fights are usually healthy for a field, and the interpretive-generative dispute illustrates this point very well—it churned up a great deal of knowledge and promoted a diverse spectrum of perspectives—but bickering, a decidedly pettier pursuit, is not so healthy, and again the interpretive-generative dispute illustrates the point very well. Bickering has two general effects: (1) It maintains a state of bitterness, the squabblers accumulating more and more personal baggage, which virtually guarantees they will never reconcile. And (2) it annoys bystanders. They move away.

Generative Semantics Gets Fuzzy

> [Generative semanticists] have said a great many potentially interesting and illuminating things. . . . [But] even the most illuminating suggestions are bound to lose a great deal of their light if put forward in a theoretical vacuum, and, furthermore, . . . without the constraining influence of a coherent theory there is bound to be a mixture of proposals, going from the brilliantly illuminating to the downright silly.
>
> W. R. O'Donnell

Back in linguistics, the wheels were falling off generative semantics. First it lost direction, then almost immediately it fractured into a number of loosely aligned interests. Just as quickly, there was a gushing hemorrhage of confidence, and generative semantics expired.

Once it became clear that transformational grammar could not handle meaning as elegantly as Homogeneous I had promised—that the strong form of the Katz-Postal principle could not stand—the movement went in several directions at once, and by the mid-1970s it was inescapably clear, to friend and foe alike, that there was no longer a single beast called *generative semantics*. It was legion. Initially, there were promises for definitive texts. First, Ross and Lakoff were to write a text, *Abstract Syntax*. Then Lakoff was to write another text on his own, *Generative Semantics*. But new data and new theoretical devices kept derailing their plans. In the end, the closest thing to a definitive text was a quasi-publication from the Indiana University Linguistic Club by one of McCawley's students, Donald Frantz, and his *Generative Semantics: An Introduction* carries the qualification in its preface that it doesn't quite meet the demands of truth-in-advertising, that *Generative Semantics According to Frantz* would be a more appropriate way to label its contents (1974:2).

The diversity of generative semantics in the early-to-mid-seventies is most concisely illustrated by a quick look at the paths taken by the four horsemen.

Postal dismounted quite early (around 1972–73, though some later published work still participates in generative semantics), to work on an alternate formal model with Perlmutter, relational grammar.[9] This grammatical theory was in partial competition, partial alliance, with generative semantics, but its effects on the fortunes of that model were uniformly detrimental (which, please note, is a long way from saying its effects on *linguistics* were detrimental; it has had a powerfully beneficial impact on the field). It looked like Postal was deserting a sinking ship.

Ross drifted into murkier and murkier realms of data, and began to argue that the distinctions among grammatical categories and constructions were not, as virtually all grammatical theory had hitherto treated them, distinctions of kind, but rather only of degree.[10] So, for instance, he proposed that the following continuum (which he calls a *squish*) more realistically captured the differences between its categories than did the discrete approaches typical heretofore of transformational grammar, and of virtually all other forms of linguistic analysis (Ross, 1972c:316):[11]

Verb > Participle > Adjective > Preposition > Noun

So, some words were sort-of, kind-of verbs; others were sort-of, kind-of nouns, and in between them were verby nouns, usually called *prepositions,* and nouny verbs, usually called *participles,* and a bunch of hermaphrodites called *adjectives.* These new sort-of classifications all came embedded in a mesh of very subtle arguments, depending heavily on personal intuition and extremely informal surveys, and involving the interaction of controversial or poorly explored processes, and leaning on metaphors:

> To pass from left to right [on the verb-to-noun continuum] is to move in the direction of syntactic inertness, and to move away from syntactic freedom and volatility. To wax metaphorical, proceeding along the hierarchy is like descending into lower and lower temperatures, where the cold freezes up the productivity of syntactic rules, until at last nouns, the absolute zero of this space, are reached. (1972c:317)

We don't have time to look at any of these reason-meshes of observation and metaphor, fascinating as they are, but squishiness was not a hit. The biggest problem was a familiar one: Ross argued repeatedly that a discrete grammar (like all contemporary transformational work, interpretive and generative semantics alike) faces intractable problems which a nondiscrete grammar would not face, but he never offered a nondiscrete grammar. Ross never, in fact, went beyond the most general suggestions about what such a grammar might look like. There is talk of a "radical departure from the previous transformational [work]" (1973c:138), and intimations of freezes and "funnel directions" (1972c:325), but no solid proposals; Gazdar

and Klein (1978:666) invoke Pirandello to describe one of his squishy papers, calling it "little more than a collection of data in search of a theory." Ross's good buddy
Lakoff went even further than Gazdar and Klein, saying "no current theory of
grammar can even begin to accommodate the facts that Ross has observed"
(1973b:271). Apparently, Lakoff meant this as a recommendation, but it helped
breed a certain lack of confidence that Ross knew where he was going. No one is
eager to follow someone who is groping around in the dark.

Actually, one person was, and he was not so much willing to follow Ross, as to
lead the way himself, Lakoff. He launched a campaign to retool generative semantics into something capable of embracing squishes. The critical notion in Ross's
arguments is clearly the notion of degree, and Lakoff proceeded by way of splicing
into his program ideas from parallel disciplines involving degrees of variability; in
particular, he was strongly influenced by Eleanor Rosch's work on the psychology
of nondiscrete representation, and Lofti Zadeh's work in fuzzy set theory, from
which Lakoff took the label for his new approach, *fuzzy grammar*. What this
means, in the context of this discussion, is that we need a brief, necessarily inadequate, glance at an external research program—two, in fact—of the sort that says
much about generative semantic research strategies, especially those associated
with Lakoff.

Rosch's work in mental representation hinges on two related observations, the
primacy of categorization for cognitive processes, and the centrality of prototypes
in the definition of categories. Briefly, the category "bird" has many members,
which standard biological taxonomies distinguish among. But such taxonomies do
not distinguish among those members in terms of "birdiness." In a wide range of
experiments, prototype theorists have demonstrated what is clear to all of us with
a moment's thought, that people regard some birds (like robins and sparrows) as
more central to the category, more birdlike, than other members of the category
(such as ostriches and penguins). As members of the category move away from the
prototype on such dimensions as size, capacity for flight, and nesting habits, they
are perceived as less and less birdlike (though still birds). Prototype effects are why,
for instance, most of us hesitate when asked to categorize an olive, but not an apple
or a carrot.[12]

Zadeh's work concerns formal set membership, rather than mental categories
directly, but the same notion of degree is clearly at play. Consider a collection like
the set of tall people. Tallness clearly is a relative notion. Some people are obviously
tall (say, those over seven feet); some people are obviously not tall (say, those under
five feet); many people fall in between. Moreover, estimations of tallness vary by
gender, age, race, occupation, and perhaps several other variables; the height of
someone who is tall for a jockey would be much lower than the height of someone
who is tall for a basketball center. Zadeh proposes that set membership, accordingly, not be a binary notion, but be assigned by degrees; say, a real number between
0 and 1. In this way, the proposition that *Leila is tall* would not be true or false in
an absolute sense, but true to some quantitative degree (say, 0.38). All of the *says*
in this discussion make Zadeh's work look very informal; in fact, it has achieved a
good measure of precision, and represents a very fruitful research program, particularly in artificial knowledge representation.[13]

Lakoff published a number of investigations blending Rosch and Zadeh's work—and, in fact, such investigations have now become a focal point of his research (see Lakoff, 1987)—but the principal paper on this approach is an analysis of linguistic hedges, like *sort of* and *pretty much* (1973d [1972]). He offered analyses of the truth of such statements as those in 1, suggesting that their truth values move from 1a, which is absolutely true, to 1e, which is absolutely false, through several intermediate degrees:

1   a   A robin is a bird.
    b   A chicken is a bird.
    c   A penguin is a bird.
    d   A bat is a bird.
    e   A cow is a bird.
      (Lakoff, 1973d [1972]:223)

Then, he charted the interaction of certain hedges and intensifiers (sometimes called *degree words*) on the truth values of such sentences:

2   a   A robin is sort of a bird. (false—it is a bird, no question about it)
    b   A chicken is sort of a bird. (true, or very close to true)
    c   A penguin is sort of a bird. (true, or close to true)
    d   A bat is sort of a bird. (pretty close to false)
    e   A cow is sort of a bird. (false)
      (Lakoff, 1973d [1972]:234; the truth judgments are his)

The whole point of Lakoff's importing exercise is to investigate ways of putting such observations to work in grammatical theory, in part to help make sense out of Ross's squishy material. But he doesn't get very far. He proposes a rough sketch of logical apparatus for treating the semantics of hedges, and, appropriately, he hedges it substantially (briefly, he treats hedges as predicate modifiers which affect the degree of truth assigned to the proposition in principled ways).[14]

And fuzziness was not all Lakoff was up to in the mid-seventies. He was also exploring ways to incorporate H. P. Grice's informal work on conversations into formal grammar, looking at syntactic amalgams, and linguistic gestalts, and generally investigating pragmatics, and expanding his investigations of logic, under the slogan of "natural logic" to the point where he was calling for "hundreds, perhaps thousands more" concepts (Parret, 1974 [1972]:162).[15] Lakoff was very busy.

McCawley was busy, too, but in a much more low-key and clearheaded way. Early in the debates, he advanced some of the strongest theoretical arguments, largely around the theme of reducing complexity by treating syntax and semantics as a unitary phenomenon. But McCawley dropped this meta-theoretical approach in the early seventies to pursue more specific work, a shift in approach reflecting his conscious decision to be more responsible scientifically (and he was, with Jackendoff a close second, the most responsible scientist of the dispute): to concentrate on trenchant critiques of the interpretivists and on the application of generative semantics principles to specific phenomena. His work in the seventies simply assumed generative semantics, rather than championing it (as Postal had before he dismounted) or celebrating it (as Lakoff always did), though he didn't make his motives clear until the end of the decade:

> I do not mean to suggest by [assuming, rather than explicitly marketing, generative semantics] that I am so arrogant as to regard all the controversies over claims I and other 'generative semanticists' have made about the unity of syntax and semantics as having been settled in our favor. Rather, I simply think that for further discussion of these questions to be productive, the disputants need a much broader and deeper understanding of the relevant factual areas than any of them (myself included) had around 1970. (1979:viii)

He worked assiduously to replace "the sweeping and often rash generalizations in my earlier work about the relationship of grammar to logical structure and the lexicon by more detailed proposals whose backing is less anecdotal" (1979:viii).

McCawley, in short, spent most of the seventies buttering syntactico-semantic parsnips. In particular, he followed Postal's reductionist program to find the atomic categories; arguing, for instance, that tense (1976b [1971]:257–72) and *not* (1976b [1971]:277–84) were underlying verbs, and defending Ross's analysis of modals (1979 [1975]:96–100). Eventually, this line followed the path of the modern atom very closely, and McCawley finally argued that there were no syntactic categories at the deepest level after all, that they were composites of smaller particles yet (1982b [1979]:200). His most notorious claim from this period was that the underlying word order of English, was not, as had been assumed since Harris, a reflection of the canonical surface order, subject-verb-object, but more like Polish-notation symbolic logic, verb-subject-object (1976b [1970]:211–28). The argument, regarded as something of an unwitting *reductio* by many anti-abstractionists, is a clear, well-reasoned application of simplicity to the organization of transformational grammar:

> Of the 15 transformations of English that I can argue must be in the cycle, there are ten for which it makes no significant difference whether they apply to structures with predicate [verb] first or predicate second. . . . For the remaining five cyclic transformations, the underlying constituent order makes a significant difference in the complexity of the conditions under which the transformation applies, or in its effect. In each case, the version of the transformation that assumes predicate-first order is significantly simpler in the sense of either involving fewer elementary operations or applying under conditions which can be stated without the use of the more exotic notational devices that have figured in transformational rules. (1976b [1970]:217)

Adopting a deep verb-subject-object order complicates no transformations, and simplifies fully a third of them. The VSO hypothesis was adopted by most generativists (though not Newmeyer) and rejected by all interpretivists: the expected pattern.[16]

But, of course, McCawley's VSO hypothesis has nothing to do with generative semantics. While it complicates some of Chomsky's more cherished working assumptions (in particular, that there is an underlying verb phrase), it is fully compatible with an interpretive semantic component—along with the position that *not* is a verb, or that tense is a verb, or that syntactic categories are not primitive. Much of McCawley's analytical work in the seventies, although conducted under the flag of generative semantics, was, more properly, abstract syntax. And abstract syntax was out of favor. It had been rejected virtually wholesale by the interpretivists, since it was incompatible with the lexicalist hypothesis, but other generative semanticists appeared to have little use for it either. Lakoff was working on everything but. Ross was wading in squishy data. Robin Lakoff, who published the important early book, *Abstract Syntax and Latin Complementation* (1968), had pretty much evolved into an ordinary language philosopher. Aside from McCawley and a few of his students (most notably, Judith Levi and Georgia Green), abstract transformational analyses had gone the way of the kernel.

By the mid-seventies: Postal was gone; Ross was hip-deep in murky data; Lakoff was as inventive and daring as ever, but wouldn't sit still long enough (figuratively or literally) for other linguists to get a fix on his work; and McCawley was out of step with the times.

The Ethos Backfires

> I have never made but one prayer to God, and a very short one: 'O Lord, make my enemies ridiculous.' And God granted it.
>
> Voltaire

Without a defining center, and with the times changing (protests were out, discotheques were in), many of generative semantics' identifying traits became liabilities, and Chomsky's reputational ebb did not benefit them. Most commentators found them to be even worse than their former master. The generative-interpretive squabble certainly hurt both sides, but the boisterous generative semanticists were a much easier target. For Gray, "Chomsky sounds quite Trotskyish compared to the logico-linguistic Stalinists of the younger generation [specifically citing McCawley]" (1974:4), and Hagège complained that "the promoters of generative semantics have only prolonged, extending them to the point of caricature, [the] already existing procedures" he found repugnant in Chomsky's work (1981 [1976]:83). Maher, giving full vent to his fetish for capital letters, identifies the worst MITniks as the authors of "innumerable facetious pieces of juvenilia presented as scholarly papers at LSA, CLS, and other TG club meetings," particularly "QPhD's [Quang Phuc Dong's] pupils Binnick, Morgan, and Green" (1973b:30).

The *juvenilia* in Maher's complaint had mostly to do with the loony humor infusing generative semantic work, which was leading many linguists to the conclusion that lack of seriousness ran deeper than style. O'Donnell (1974:75), for instance, sniffs "serious grammatical studies may, as [Lakoff] claims, be in their infancy: serious grammarians, however, are not." But there was another problem with the humor. It tended to be directed, rather narrowly, toward other generative semanticists. It was not invitational; Robin Lakoff calls the characteristic style of her community "a kind of secret handshake" (1989:977). The most disastrous example of this invitational failing is in a major article by Lakoff (1971b), written as a reply to Chomsky's "Deep Structure, Surface Structure, and Semantic Interpretation." The publication circumstances were very tight, and Lakoff decided to parody Chomsky:

> Chomsky wrote this paper ["Deep Structure"] attacking generative semantics, for a collection by Steinberg and Jakobovits [1971]. Danny Steinberg called me up and asked me if I wanted to write a reply. I said "Sure." And he said "Well, you have a month."
> I looked at this thing. It was this huge paper. It was full of quotes taken out of context, and characterizations of our position that were really wrong. I had a month to reply to it. I didn't know how to deal with it, so I decided to write a parody. I took his own style, and tried to turn the style against him. Apparently, he never realized that it was a parody.[17]

Satire always misses some people—if literary critics had such things, this would be one of the Ironic Laws—and there is no question but that Chomsky, for this satire at least, was one of those people. He completely failed to recognize the parody, welcoming Lakoff's paper warmly for having adopted, "with only a few changes, the general framework and terminology of ["Deep Structure"], so that the differences between [interpretive] and generative semantics, as so conceived, can be identified with relative ease" (1972b [1969]:134). Chomsky's missing the joke, in fact, became the source of chuckles for some generative semanticists, showing just how irredeemably humorless their opponent was.[18] But it was an easy joke to miss. Once the paper is identified as a parody of "Deep Structure," it is extremely effective.[19] Lakoff, however, completely fails to flag his paper as a parody. The title—"On Generative Semantics"—is serious, and, in fact, promises a rather definitive treatment of the movement Lakoff was advocating. Nothing in the text is so outrageous as to provide an unequivocal clue. Lakoff apparently regarded the following definition of "the semantic representation SR of a sentence" as patently absurd (attacking the sort of work Jackendoff published in *Semantic Interpretation*):

> SR = $(P_1, PR, Top, F, . . .)$, where [P_1 is the first phrase marker in a derivation,] PR is a conjunction of presuppositions, Top is an indication of the topic of the sentence, and F is the indication of the focus of the sentence. We leave open the question of whether there are other elements or semantic representations that need to be accounted for [hence, the ellipses at the end of the formula] (1971b:234–35)

But without knowing Lakoff's intention, the definition just makes him look like a wild-eyed semantic imperialist, a conception most interpretive semanticists already held of him, and one that many spectators held of transformational grammarians in general. Nor is this proposal much stranger than the one that Lakoff was seriously advocating soon afterward.[20] Much of the paper is also written utterly deadpan, particularly the opening pages, and it is not until late in the article, when Lakoff starts proposing rule features like [$\pm$ PEDRO] that there is a solid hint the author is not simply drab by constitution, but drab by choice; until that point, it is more reasonable to regard the author stylistically as a clone of Chomsky, whatever their differences, than a parodist. Even the sample sentences are very restrained, prominently featuring Chomsky's *John*.

More problematically, the paper contains no direct indication that it is a specific reaction to "Deep Structure," which would at least have alerted readers to the passages and terms in Chomsky's paper that were under ridicule. And "On Generative

Semantics" contains the one ingredient that muffles irony better than anything, a good deal of serious discussion, which is frequently impossible to disentangle from the smirking. The passage just quoted, for instance, includes a footnote discussing the possibility of reducing the formula to (P_1, PR), precisely what one would desire. Similarly, consider this admixture in Lakoff's introduction of the term, *basic theory:*

> I will refer to the above theory of grammar as a 'basic theory', simply for convenience and with no intention of suggesting there is anything ontologically, psychologically, or conceptually basic about this theory. Most of the work in generative semantics has assumed the framework of the basic theory. It should be noted that the basic theory permits a variety of options that were assumed to be unavailable to previous theories. For example, it is not assumed that lexical insertion transformations apply in a block, with no intervening nonlexical transformations. The option that lexical and nonlexical transformations may be interspersed is left open. (1971b:236)

In some ways, the passage is very deft. It mocks Chomsky's use of labels to disparage or elevate concepts and theories *(taxonomic linguistics, deep structure, standard theory, extended standard theory),* all the while protesting that the labels are mere expository conveniences, wholly without evaluative implications.[21] It makes a barbed comment on Chomsky's "Deep Structure" definition of *the standard theory* in such a way as to allow quite radical changes, like the lexicalist hypothesis, to fall naturally within its scope. Perhaps there is even a specific sneer at Chomsky's still broader definition in "Some Empirical Issues" (1972b [1969]:130–36), where he sneaks output conditions into the standard theory through the back door. But it also includes some important information about generative semantics; namely, that lexical and nonlexical transformations are interspersed, and consequently that there is no level of analysis parallel to *Aspects'* deep structure. The paper is also full of arguments for legitimate generative semantic positions: that semantics and syntax are inextricably intertwined, that global rules are necessary theoretical tools, that lexical insertion has to follow some transformations, that elements of the Chomskyan tradition are arbitrary and should be scrapped. Nor does one generally attack the object of the parody explicitly from within the parody—parody has to do its work indirectly, by exposing latent absurdities, not by direct attack—and Lakoff's paper contains many direct assaults on Chomsky.

In any case, almost no one beyond the inner generative semantics circle appears to have gotten the joke, and its effect was uniformly deleterious, contributing as much as any individual paper could to the communication breakdown that characterized the dispute in the early seventies. Howard Maclay's introductory discussion of Chomsky's and Lakoff's papers, for instance, takes Lakoff's at face value (pausing to remark that "the extent to which the structure of argumentation in [Lakoff's paper] is modeled on Chomsky is quite striking"—1971:178), and then attempts to straighten Lakoff out for misrepresenting Chomsky in a passage where Lakoff is in fact obliquely illustrating what he takes to be Chomsky's rhetorical sneakiness.[22] Spectators to the dispute generally took "On Generative Semantics" as a principal sign of the departure of coherence—illustrated most strikingly by Dubois-Charlier's (1972:43) comments on "the paradoxical aspect of the matter . . .

that both Chomsky and Lakoff end up by saying at the same time that the two theories are identical [notational variants], but that the other's is wrong!"[23]

As Lakoff should have expected, "On Generative Semantics" was soon taken to be "an important manifesto of the approach" (Raskin, 1975:462). It was the only publication from any of the leading figures in the movement to include *generative semantics* in the title and it appeared in a very prestigious anthology. An earlier version of the paper (1969b) was presented at a CLS session, where it very likely went over splendidly. Most of the audience would have been sympathetic to both its aims and its style. Everyone would have been familiar with Lakoff's personality and Chomsky's tone, as well as with the issues surrounding the definition of *the standard theory*. Even the published version of that talk, in the preceedings of a conference becoming famous for its sense of goofiness, may not have caused much confusion. But, in a thick, staid volume, alongside papers by Grice, Strawson, Searle, Quine, and Chomsky, among many others, almost all of them very serious in tone, it is extremely difficult to take the parody as anything but a singleminded promotion of generative semantics, and a befuddled one at that. McCawley's rebuke is rather mild in these circumstances:

> Lakoff is guilty of . . . [a] failure to distinguish adequately between what he would seriously propose as a correct theory of grammar and what he offered (partly in jest) as a general framework for the discussion of theories both correct and incorrect. (1982b [1973]:75)

There is other guilt on his head, too. Radzetsky is reported to have told his troops, retaking Sardinia for the Hapsburgs in 1848, "Spare the enemy generals—they are too useful to our side" (Robertson, 1952:354). Lakoff was such a general.

By the time Katz and Bever put together what Newmeyer (1980a:169; 1986a:134) calls "the consummate critique of the philosophical implications of late generative semantics," a few years later, they zeroed in on "Generative Semantics, Lakoff Style" (1976 [1974]:30), and O'Donnell describes the leaders of generative semantics as "Lakoff and those associated with him" (1974:54).[24] In some degree—with Postal gone, Ross on the fringe, and McCawley avoiding strong theoretical claims—the role was his by default. But Lakoff also courted it. As early as 1972 he was defining generative semantics very clearly in terms of his own (broad) personal interests, making little reference to the work of Postal, Ross, or McCawley, and citing his uninfluential 1963 paper as the starting point (see especially Parret, 1974 [1972]:151–78). (His sentiments have only strengthened in this regard. See his recent discussion of logical form—in a festschrift for McCawley!—where McCawley is mentioned in only the most cursory way, where his own work is considered definitive, and where the logical enlightenment begins in 1963–1992.)

The late stages of the schism, accordingly, are often characterized as Chomsky vs. Lakoff (or Lakoff vs. Chomsky; most linguists, then as now, have a preferred home team). The contrast could not have been stronger. Chomsky defined his mid-seventies program in terms diametric to Lakoff's most notorious theoretical proposal, global rules, and by 1975, whatever else Lakoff called his work, it could no longer be called *Chomskyan* by any stretch of the term.

With Lakoff at the helm, promoting his very wide conception of linguistic theory, and a growing concerted interpretivist attack based on the virtues of restrictiveness, the general perception came to be that generative semantics was theoretically promiscuous, incapable of saying no. It opened its doors to any and all phenomena impinging upon language. It adopted increasingly powerful formal mechanisms: global derivational constraints, transderivational constraints, even meta-transderivational constraints. It was given to frighteningly naive and sweeping claims ("What we have done is to largely, if not entirely, eliminate pragmatics, reducing it to garden variety semantics"—Lakoff, 1972b:655). It was out, not to borrow from logic, but to redesign it from the ground up. It was going to accommodate

> not just syntax-semantics, phonetics-phonology, historical linguistics, anthropological linguistics, etc., which form the core of most academic programs in this country, but also the role of language in social interaction, in literature, in ritual, and in propaganda, and as well the study of the relationship between language and thought, speech production and perception, linguistic disorders, etc. (Lakoff, in Parrett, 1974 [1972]:151)

One gets the impression that Lakoff stopped the list more because he ran out of breath than because he ran out of vision, carefully remembering to throw in that *etc.* before gulping some air. Generative semantics under Lakoff tried to do too much, this reasoning goes, and it burst at the seams. There is much to recommend this version of events. Certainly, generative semantics tried to do a great deal, and certainly Lakoff was one of the prime forces pushing at the seams, and, certainly, it burst. It burst very publicly: as a matter of course, the great majority of generative semantics papers included several comments like:

> We are forced to conclude that, awkward though it may seem, the similar properties of *both* and *each* cannot be accounted for by the same formal mechanisms in our existing theory. An explanation of whatever underlying regularity there may be will have to wait for a cleverer linguist. (Carden, 1970:189n10)

And:

> This paper was undertaken as an attempt to shed light on some very mysterious problems. I fear I have done little more than show which lamps have cords too short to reach the outlets, but hopefully this information will be helpful eventually in finding explanations for these mysterious distributions. (Green, 1972:93)

And, most tellingly:

> It is not a very satisfactory experience to write an entire paper without being able to offer any decent analyses or explanations for the phenomena I have discovered. It is, however, an enlightening one, and I believe, a necessary one. (Lawler, 1972:255)

These last two quotations belong to papers which participate in a genre that pushed this data-mongering tendency to the limit—a genre that Gene Gragg calls "creature features" and defines as a type of article that "is intended to point out some oddities which a theory of . . . speech will eventually have to come to grips with" (1972:75), but which nothing on the horizon appears capable of treating. In Thomas Kuhn's terms, the word is not *oddities* but *anomalies,* his word for data that strains the current paradigm, potentially to its breaking point, and the seminal document in this genre—Postal's highly corrosive underground classic, "Linguistic Anarchy

Notes"—makes it very clear that straining the paradigm is exactly what these papers are about. The first note begins:

This is the first in a random, possibly nonfinite series of communications designed to show beyond any doubt that there exists no linguistic theory whatever. There are apparently endless numbers of fact types not incorporable within any known or imaginable framework. In particular, what has been called the theory of transformational grammar, seems to have only the most partial relation to linguistic reality. (1976 [1967–70]:203)

The notes turned out to be finite, but they sparked an increasing number of similarly dissensual efforts, papers whose sole aim was bringing to light data that would give any and all pretenders to theoryhood the heebee-jeebees. The high point of the cycle came in the first few years of the seventies, with creature features like "Semi-indirect Discourse and Related Nightmares" (Gragg, 1972), and "Read at Your Own Risk: Syntactic and Semantic Horrors You Can Find in Your Medicine Chest" (Sadock, 1974b).[25] These papers were only just the most overt symptoms of a mood pervading generative semantics of the period, one which shows up in the nooks and crannies of the overwhelming majority of papers. Even Postal's public persona, which, whatever he said in private or circulated underground, was normally cocksure and authoritative, cracked in one straightforward *Linguistic Inquiry* paper, which ends "There is, of course, an explanation of these [very strange and mysterious properties] but, believe it or not, the present writer does not know what it is" (1972d:400).

His Anarchy Notes raised something of a stink. In particular, it is said that Halle thought them very ill-advised, probably fearing that they would spark exactly the kind of attitude that they did spark—though, perhaps, *fan* is more appropriate—and Kuroda chided him gently with his "Linguistic Harmony Notes" (the first number being "Charms of Identity"—1976 [1967]). But it is important to notice that the motive behind this traffic with embarrassing data is not to deal a nihilistic blow to the heart of the entire transformational enterprise. One of Postal's avowed short-term aims in the Notes, true, is to demonstrate in the most graphic way possible that the enterprise is "not just slightly in error and rather incomplete"—the attitude that characterized much of the interpretivist camp—"but in deep ways hopelessly far from linguistic reality" (1976 [1967–70]:215); his ultimate aims, though, are much higher. His goals, he says, "are entirely positive." He is trying to save grammatical theory from itself:

Many people today are engaged in the attempt to construct linguistic theories. My view is that an important difficulty with all such attempts is that there is not a good *a priori* statement of the full range of known facts which a theory must handle. To the extent that theories are formulated in the absence of explicit awareness of this range of facts, they are dreamlike. (1976 [1967–70]:205).

It is safe to say that this desire to save grammatical theory from its own shortsightedness was the strongest single motive fueling the generative-semanticists' preoccupation with problematic data.

Whatever the motives behind this endemic fussing with inadequate theory and celebration of challenging data, however, it is clear that it had powerfully negative effects, many resulting from the existence of the *CLS Proceedings*. The youthful exuberance of Chomsky and Halle's students, the vicious polemics, remained primarily oral, printed versions circulating, if at all, in mimeograph, and only among the faithful. The youthful exuberance of McCawley's, Zwicky's, and the Lakoffs' students went to press. The impulse was admirable: to get the key ideas out rapidly for criticism and development, and give graduate students a high-profile forum. But it had serious drawbacks.

In particular two traits—its embrace of a wide range of interests and its self-definition primarily in the rhetoric of dissent, in saying no to Chomsky—were perhaps the principal reasons the movement fell apart. The most frequently repeated (posthumous) diagnosis of the decline and fall of generative semantics is "it tried to swallow the world, and choked on it." It tried to incorporate too many diverse phenomena within a narrow Chomskyan framework, and they wouldn't fit. It stretched that framework by adding a number of powerful devices, and finally stopped trying to save the phenomena and rejected the framework. This assessment certainly captures some aspects of the demise. But many generative semanticists kept an essentially Chomskyan framework, rejecting or modifying only peripheral assumptions. A strong case can be made, in fact, that Chomsky has deviated further from *Syntactic Structures* than Postal or McCawley has. The more explanatory component of the collapse does not involve the clash of data and framework, so much as the absence of a center of gravity.

Generative semanticists dispersed like the crowd after Woodstock, everybody wandering off, individually or in clumps, to pursue other interests. Some straggled off under the force of Chomsky's restrictiveness arguments, and the concomitant unwillingness of any prominent generative semanticist—in particular, the unwillingness of Lakoff, who had issued the call for global grammar—to meet the challenge of those arguments. Generative semanticists didn't up and cross the ring to raise Chomsky's arm in victory—or, very few did—but other approaches were developing (most notably, Montague grammar and relational grammar) that disaffected generative semanticists found attractive. In part the attraction was because of a more focused opposition to Chomsky, in part because of more tractable semantics, in part because of solutions to problematic data.[26] And there were new phenomena to worry about that no model handled very well—Sadock, Green, Davison, and Morgan, for instance, began exploring pragmatics; Robin Lakoff started looking at social control and power issues in language. Some, principally Lakoff and Ross, never really left, but their work mutated in directions that bore less and less resemblance to their original starting point. They also adopted new labels for their work. First Lakoff began talking about *global grammar,* which was clear a continuation of generative semantics, but the name implied a qualitative leap to something new. Then Ross became associated with *squishy grammar,* Lakoff with *fuzzy grammar,* approaches that had shared perspectives, but which were different enough to have separate names. Lakoff also embraced labels like *cognitive grammar* and *experiential linguistics,* which identified different elements of one evolving program, but which made it look as if he was hopping from theory to theory depending on what sample sentences he heard on the way into work that morning.[27] Ross eventually rounded to an "almost terminal distrust" for labels, noting of his more recent work that "I have already tried on modifiers for *linguistics* like *human, holistic, ecological*—and each time they tended to sound like some increasingly more horrific concoction of the trendy, the buzzword, the big deal" (1991:2).

McCawley kept doing what he always did, producing perceptive and challenging syntactic, semantic, lexical, and phonological analyses, but even he abandoned the label, *generative semantics*. By a 1979 conference on syntactic approaches, he gave only *unsyntax* as the name of the work he was doing (Kac, 1980; Moravcsik and Wirth, 1980). No one was using the label *generative semantics* at all anymore, except in time-delayed surveys written for nonlinguists, which were still hailing the Homogeneous I model as the "latest development." Everyone close to the scene knew that it was finished. Lawler (at the same conference where McCawley was discussing unsyntax), lamented being a fluent speaker of a "dead metalanguage" (1980:54, 59n12), and Elgin's second edition of her *What Is Linguistics?* (1979), excised the original cutting-edge invocation of generative semantics (1973:34–35). Gazdar, reworking his 1976 thesis for publication (1979), put all the present-tense references to generative semantics into past tense (and, apropos of an earlier topic, also changed the bathroom-haunting *Nixon* of his sample sentences to *Dixon*). The tag was on its toe, the drawer closed.

The Chomskyan Flow

Where then does the expression "generative semantics" come from? It is from a general attitude or point of view which was expressed, for example, by Lakoff in an article entitled "[On] Generative Semantics," or by Postal in his 1969 article "The Best Theory." But nobody—at least not to my knowledge—has actually accepted this theory, which in the form presented was virtually empty. What the theory asserted was that there exist representations of meaning, representations of form, and relations between the two. Furthermore, these relations between the two representations were virtually arbitrary. . . .

A theory that permits global rules has immense descriptive potential. . . . To approach an "explanatory" linguistic theory, or—which is the same thing—to account for the possibility of language acquisition, it is necessary to reduce severely the class of accessible grammars. Postulating global rules has just the opposite effect, and therefore constitutes a highly undesirable move. . . .

[My own work has led to hypotheses which] restrict very severely the expressive power of transformational rules, thereby limiting the class of possible transformational grammars.

Noam Chomsky

Interpretive semantics began to thrive, though it too was not the same old model. Chomsky's architecture changed several times. Most dramatically, the campaign against the specificity of transformations was carried about as far as it could go: All specific transformations were eliminated. Some were axed outright by various technical moves. Deletions, for instance, were tossed out in favor of base-generated, phonologically null pronouns. So, where the *Aspects* model derived sentence like 3a from deep structures like 3b by deleting *someone* and the second occurrence of

Tania, the new model just plugs empty pronouns like PRO and *e* directly into the deep structure (as in 3c), with surface structure semantic rules making sure that *e* is interpreted as coreferential with *Tania.*[28]

3 a Tania is easy to please.
    b Tania is easy for someone to please Tania
    c Tania is easy for PRO to please *e*

But, in the most interesting (and, for awhile, the most controversial) policy change, all the remaining movement rules were collapsed under one very general transformation. Passive was out, Topicalization was out, Wh-fronting was out, and Move-α was in, where α is a variable that stands for any constituent. Now, Move-α on its own just produces word salad for the surface structure, any conceivable scramble of words. But a rich array of filters and constraints ensured that (in theory) only the grammatical sequences made it to the surface.

His current model—"sometimes called 'the Extended Standard Theory (EST)'" (Chomsky, 1986:67), though it has only a passing resemblance to the 1969 model with that name—looks roughly like figure 9.1. This picture is deceptively simple, since the Roman numerals all represent distinct rule systems: *I* represents phrase structure rules; *II* represents transformations; *III* represents phonological rules; and *IV* represents "rules of the LF component," essentially semantic interpretation rules concerned with quantifier scope and anaphoric relations. The other elements of the model are, as follows: *PF* is "phonetic form"; *LF* is "logical form"; *D-structure* and *S-structure* are abstract levels of syntactic representation related by way of transformations, which, however coy Chomsky is in treating *D* and *S* as arbitrary letters, inescapably evoke *deep structure* and *surface structure,* respectively.[29] Lexical insertion is nowhere to be seen.

Chomsky no longer concerns himself very directly with semantic argumentation. In fact, he never did. Semantics was a carrot, attracting a good deal of attention to his model in the late fifties and early sixties. He endorsed the work of Katz, Fodor, and Postal, but he did not engage in it. When semantics became a stick with which

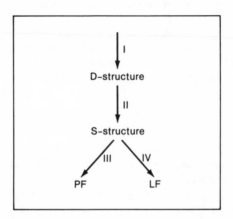

Figure 9.1. An outline of Chomsky's current grammatical model (From Chomsky, 1986c:68).

to beat the generative semanticists in the late sixties and early seventies, the same was true. He raised problems for the Katz-Postal principle—by attending to focus and presupposition phenomena, and by discussing quantifier-scope difficulties— and he endorsed solutions to these problems when it suited him, but he never proposed any solutions on his own. Once generative semantics was pretty much dispatched, he stopped discussing these issues. His interest in explicit semantics has always been quite modest.

He has been called, rightly, a *syntactic animal* (Passmore, 1985:39), and Jackendoff characterizes the attitude around MIT in the mid-seventies as "Let's do some syntax again, now that this whole schmeer is over." The whole schmeer wasn't quite over at this point—in fact, the interpretivist assault was building up a good head of steam—but the threat was gone, and the theoretical questions in Chomsky's framework took a decided turn back toward syntax.

Chomsky turned his back on generative semantics, and started work on another strand of his program, alternatively known by a couple of different labels—the *revised extended standard theory,* signalling increased modification, and *trace theory,* taking the name of its single most prominent modification. One of them was used mostly by linguists tired of Chomsky fiddling with his model, and often took the form of a derisive acronym, as in "Give it a REST." The other was used more neutrally. But neither is particularly useful. In fact, it became clear at about this point that putting labels on Chomsky's work at all is an iffy business at best, even the ones he himself endorses (*Government-Binding Theory* for awhile, giving way to *Principles-and-Parameters,* possibly on the way to being displaced by *The Minimalist Program*). A more productive approach might be to treat him like a great and restless artist—like Picasso, with his postimpressionist period, his blue period, his rose period, his cubist period, . . .

We have been looking, in increasing detail, at Chomsky's Harrisian period, his anti-Bloomfield period, his *Aspects* period, and his anti–generative semantics period; as with Picasso, one merges into the next, positive moods alternate with negative ones, elements and themes connect through them all. The next period, when he broke rather dramatically from his anti–generative semantics focus to concentrate on the more positive work, is marked by "Conditions on Transformations" (1973a [1971]), the paper in which, Sadock says, Chomsky began speaking a new lingo. The majority of Chomsky's work in formal linguistics between 1966 and 1971, that is, was directed *against* generative semantics, rather than *for* interpretive semantics. Even his positive suggestions—x̄-syntax, for instance, and the lexicalist hypothesis—seemed to interest him primarily for the obstacles they placed in front of generative semantics. With the conditions period, he began again to articulate a promising, workable linguistic model.

The proposals contained in that model rapidly bore fruit: at exactly the time that it became almost impossible for linguists to find a coherent program under the label of *generative semantics,* interpretive semantics was getting its act very much together. With Jackendoff (1972) and Chomsky (1973a [1971]), offering a genuine alternative, rather than a few sketchy and barbed suggestions, increasing numbers of students, junior professors, and neutrals found the new package persuasive. The package included not only increased attention to lexical items, an elegant new treatment of phrase structure, and an enriched transformational component, but the forceful rejection of generative semantics. In particular, Chomsky and (especially) his former students defined his more "restrictive" theory largely against the bogey man of generative semantics.

The one common denominator among the scattering generative semanticists is that, whatever else they were up to, they continued, until the very last minute (well, some of them are still at it), rejecting Chomsky. Even the gracious, gentle McCawley couldn't have thumbed his nose at the syntaxicentric Chomskyan vision any more explicitly than by choosing to call his work *unsyntax.*

This man, Chomsky, obviously, is the key. He gave interpretive semantics an irresistible center of gravity. Anyone in his immediate framework who begins working on a strand that is uncongenial to him, or even just uninteresting, rapidly becomes, by definition, out of the program—not necessarily for reasons of pique or malice, perhaps not even by design; simply because Chomsky's concerns automatically set the agenda for the community. If he supports research which involves his framework, as he supported Katz and Postal's pre-*Aspects* proposals about kernel semantics, it is in the program. If he stops talking about someone's work, as he largely stopped talking about Katz's semantics in the post-*Aspects* period, or if he rejects work, as he rejected Postal's abstract syntax in the post-*Aspects* period, it is out of the program. Katz certainly promoted an interpretive brand of semantics, and expended a good deal of energy doing battle against Chomsky's principal enemies, but he was not, in the usage that best suits the pattern of the debate, an interpretivist. Linguists soon realized that Katz no more spoke for Chomsky than did Lakoff or Ross. With Postal, the case is even clearer. He was (and remains) a Chomskyan linguist in the most pristine, *Syntactic Structures,* sense of the term, interested primarily in building "precisely constructed models for linguistic structure" (Chomsky, 1957a:5), and he was the driving force behind research that everyone thought would have been very dear to Chomsky's heart, seeking out deep lexical regularities and explaining surface diversity with transformational mechanisms. But Chomsky rejected the work and Postal rapidly moved into the role of his opponent.

There was some of this in-the-movement, out-of-the-movement categorization to generative semantics as well; Newmeyer, in particular, felt ostracized when he rejected McCawley's VSO proposal, and Lakoff still sullenly insists that "Newmeyer was never a generative semanticist." But it was much less prevalent, and much less obvious. The major reason for this difference is probably that generative semantics was not organized around a single individual. All four of the horsemen had their own interests, and collectively their range was much broader than Chomsky's. But the movement was also inherently more open and anarchic than interpretivism (or, for that matter, any of Chomsky's other frameworks). If someone with generative semantics sympathies began working on a particular data set or formal mechanism, it was usually assumed that they took the theory with them. Sadock examined the syntax, semantics, and pragmatics of medicine bottle labels (1974b). Green looked at proverbs (1975). Ross looked at his squishes (1972d). Lakoff looked at Rosch's category work, Zadeh's fuzzy set theory, Grice's conversational maxims. And, automatically, the phenomena or principles were within the province of generative semantics.

Problems and Mysteries

> I think the argument will be most fairly tested, if we take the 'if' out of it.
>
> Socrates

One of the more implacable foes of generative semantics analyzes its death, in large part, as the implosion of ethos: "They destroyed themselves by insisting on taking on the mannerisms, the style, of one of the crazier sixties cults." There is truth here, and it is difficult to avoid analogies with the youth-culture-cum-counterculture-cum-Woodstock-Nation, which changed its shirt almost daily. The issues of hippydom were real, and many people were committed, and worked very hard to bring about social and political changes of huge magnitude. But the majority of people were committed, if at all, for only a short period. The issues were real, but they were also faddish, falling out of favor with long hair and beads. This book is certainly not the place to explore why they fell out of favor, although much of their transience surely had to do with the complete lack of focus in (and, in many cases, outright lack of) positive alternatives to the institutions and practices under attack; it was, says Roszak (1969:34) in the thick of the flight, "much more a flight *from* than *toward.*" Indeed, even what was being rejected grew rapidly, from a fairly tight set of concerns about civil rights inequities and the invasion of Vietnam, into a great amorphous stew of beliefs, objects, people, culture. The same trends are apparent in generative semantics: the flight from Chomsky; the rapid expansion of dissent from a few specific technical questions; the ever-widening range of phenomena to be embraced. But these are very deep reasons for failure, utterly distinct from a few epithets, some absurdist humor, and a political jab or two. These traits contributed in some measure to the decline of generative semantics, but it was the generative semanticists' remarkable frankness about the shortcomings of the theory, and their tendency to celebrate rather than marginalize anomalies, both reflections of deep methodological and philosophical positions, that truly marked the theory [+DOOM].

Much of the generative semantics style is, unfortunately, exclusionary. Certainly, it alienates people with more conservative politics, and people more squeamish about epithets. Hagège says, with an adjective that is clearly meant to cut two ways, "that more serious scholars, who already consider that amateurism does not stop with form, risk finding elsewhere a justification for their reservations" (1981 [1976]:22).[30] That unsympathetic scholars would find more reasons for hostility is not, or should not be, much of a concern. Their predispositions would be satisfied no matter what style they encountered. More worrisome is that neutral scholars might simply be discouraged from reading material they found offensive for reasons other than its linguistic claims—because of general irreverence, or allusions to drugs, or aspersions on public figures. The Venerable Quine, for instance, seems only to recall that Lakoff and McCawley "were over-eager to amuse . . . with whim-sical examples" at a 1969 conference on logic and language, complaining of their "chatter and clowning" (1985:358). And the university publisher for Cambridge University Press sent these comments to Green when it had become clear that she would not expunge references to Nixon, Agnew, and crew from her manuscript:

> The sentences, even if in the event they were not to be libelous, are certainly offensive, and quite gratuitously so: a textbook of linguistics is not an appropriate stalking-horse from which to issue innuendoes against public figures. To use the book for this purpose would be likely to cheapen the reputations of author and publisher alike. (R. W. Davis, quoted in Pullum, 1987:141)

It is very difficult, at least for me, to fault someone familiar with the stunning brutality of Vietnam for expressing contempt of the men with the most guilt on their hands, and once the door was open on people like Johnson and Nixon, the sad-sack Humphrey and the buffoonish Mitchells were obvious (and easy) targets. But whether one faults this trait or not, it is undeniable fact of scholarship that neutral prose is the safest course. Scholars come in all persuasions, but a relatively common trait among them all is pricklishness, and annoyance is not the road to adherence.

Even these traits, however—humor and political aspersion—might not have done much damage in isolation. In fact, they contributed to the movement's attractiveness for enough young scholars that the aggravation of others might have been counterbalanced. The real damage was done by the closely related traits that Lakoff calls *honesty* and *data-love.*

Although the term *honesty* is obviously self-serving, and perhaps should be replaced with one like *frankness,* the generativists were clearly explicit about their shortcomings in a way very few scientific collectives are.[31] In particular, they were at pains to avoid what they took to be Chomsky's rhetorical underhandedness. As one additional example of these pains, consider attribution. Many generative semanticists regarded Chomsky as extremely selective in the way he acknowledged his intellectual debts, and they tried scrupulously to avoid falling into similar patterns. Since we are dealing with generative semantics, this scrupulosity led to some rather odd acknowledgments, like "Sentence (1) was brought to my attention by Haj Ross (who in turn had heard it from Avery Andrews)" (Lakoff, 1988 [1974] :25), and "I am grateful to J. L. Morgan for producing this sentence [*Since Nixon was elected, I've come to miss LBJ*], thereby setting me off on a productive train of thought" (Davison, 1970:199n3), and "No thanks is due to John Lawler for calling my attention to this ugly class of facts" (Ross, 1974b:100n30). The winner in this category, as in most categories of excess from the period, is George Lakoff and his "Linguistic Gestalts," whose main attributive note begins "This work has grown out of conversations and correspondence with more people than I can possibly remember, let alone list," but goes on to mention twenty-five people anyway, and two restaurants (1977a:236).[32]

This frankness had much more serious manifestations; in particular, in the frequent reminders that "the author is fully aware of the fact that there are many uncertainties, unclarities, and errors in the text as presented here" (Seuren, 1975:84). Uncontroversially, every transformational work of the period was full of uncertainties, unclarities, and errors, and most linguists acknowledged them in some way. Here is Chomsky at the end of "Some Empirical Issues":

> Very roughly, this seems to me a fair assessment of the state of the theory of transformational generative grammar—at the moment. Obviously, any such assessment must be quite tentative and imprecise at crucial points. I will be very surprised if in a similar review several years from now, or perhaps next week, I will not want to present a rather different picture—surprised, and not a little disappointed as well. (1972b [1969]:199)

But the incidence of such admissions, and the general tone of discussion, are vastly different in Chomsky and in the generative semanticists. Chomsky has the impressive rhetorical talent of offering ideas which are at once tentative and fully endorsed, of appearing to take the *if* out of his arguments while nevertheless keeping it safely around. Among the other values of this talent, it allows him to prune rather gracefully notions which wither on the vine, while retaining those that bear fruit. The most dramatic example of this talent is his treatment of the Katz-Postal principle in *Aspects*—subordinate clauses and footnotes express reservation, but the general impact of the discussion is that the hypothesis is the natural precipitate of a decade of research, and the centerpoint of a compelling competence model (which, of course, it was). Lexicalism is a case of the opposite type. The weight of "Remarks" is strongly behind lexicalism as a natural advance occasioned by the introduction of syntactic features. But the argument is conducted as an abstract case study in the possible effects of trading off the functionality of different components, and concludes hesitantly: "On the basis of the evidence surveyed here, it seems that . . . the lexicalist hypothesis [is correct] for the derived nominals and perhaps, though much less clearly so, for the mixed forms" (1972b [1967]:60). When the Katz-Postal principle failed, he had no trouble withdrawing his endorsement; when the lexicalist hypothesis proved fruitful, he looked visionary (as indeed he was).

Few of the generative semanticists had this talent (though both Lakoff and McCawley show flashes of it), or perhaps they did not have the inclination, and their hesitations, their *if*s, are incessant. The most extreme examples are in Ross's work, especially in the seventies. It contains the "embarrassing candor and intense emotional involvement" that Holton says has marginalized Kepler in the history of science (1988 [1973]:54). Ross's arguments are not as wild as Kepler's, but they are remarkable all the same—multilayered, tortuous, and subtle, very sensitive to fluctuations in the data. They contain dozens of threads, looping around one another in the main text and annotated with lengthy, contorted discussions, some of which offer counter-examples, some of which offer alternative analyses, virtually none of which offer any support for the annotated point. They consist of long catalogs of subarguments—some strong, some weak, the weak ones always painstakingly flagged as such—interspersed with declarations of mystification and awe. Ross is, in Parmenides' term, one of "the men with two heads," someone who has taken dialectic to virtually pathological levels; Langendoen is said to have summarized one of Ross's presentations, with "It's fifteen arguments for us and nine for them; so I guess we win."

If *honest* is the appropriate descriptor for Ross's style, and it does seem right, it is *painfully* honest. The effect is confessional, in a way that perhaps makes one empathetic to the difficulties of his program, but hardly kindles interest in joining it. The effect is also cacophonous, with argument countering argument until it is all but impossible to hear a clear line of thought rising above the noise. The following assessment is telling: "It must be seriously open to doubt whether there is a coherent point of view to communicate."

But Ross is symptomatic: this assessment is not of his work. He isn't even cited. The assessment is of generative semanticists as a group. The assayer continues:

> Add . . . that some of them appear to change their minds almost continuously and that they are addicted to somewhat tendentious publication of views they no longer hold at the time of publication and you have a recipe for an intellectual confusion which might daunt even the most committed seeker-after-truth. (O'Donnell, 1974:74)

Ross's papers are not only representatives of the frankness-run-amok style of many generativists, but microcosms of the entire movement.

O'Donnell's response to generative semantics, though hardly as neutral as he makes out (1974:53), was not uncommon; nor was his response to the data-celebration papers issuing from the generative semantics camp, silence. What could a working linguist do with a paper full of facts whose raison d'être was that there was no conceivable explanation for them in current theory?

Lakoff's term for this aspect of generative semantics, *data-love,* seems pretty close to the mark. Compare his interview in *Discussing Language* with Chomsky's (Parrett, 1974 [1972]:27–54, 151–78). It is full of examples, contexts, and compliments for linguists who have rooted out particularly troublesome linguistic facts; Chomsky's is almost entirely abstract, with only a few brief and scattered invocations of data. Or consider the title *Adverbs, Vowels, and Other Objects of Wonder.* There are not many interpretivists of whom there is much danger they would choose that title for a book, and it would certainly stand out like a sore thumb against the titles in Chomsky's canon—*Syntactic Structures, Aspects of the Theory of Syntax, Some Concepts and Consequences of the Theory of Government and Binding.* The title, in fact, is McCawley's (1979), but it would be equally natural coming from almost any of the generative semanticists, as would this praise for a particular data set:

> Her [Borkin's] examples are dazzling and remain deep mysteries to this day. Such mysteries are central to our vision of what problems should be addressed by the linguistics of the 1980s and beyond. We are grateful for the gift of mysteries so worthy of the attention of those who would understand how language works. (Lakoff and Ross, in Borkin, 1984:viii)

Again, the contrast with Chomsky is extremely sharp:

> I would like to distinguish roughly between two kinds of issues that arise in the study of language and mind: those that appear to be within the reach of approaches and concepts that are moderately well understood—what I will call "problems"; and others that remain as obscure to us today as when they were originally formulated—what I will call "mysteries." . . . [About mysteries:] although there is much that we can say as human beings with intuition and insight, there is little, I believe, that we can say as scientists. . . . Some would reject this evaluation of the state of our understanding. I do not propose

to argue the point here, but rather to turn to the problems that do seem to me amenable to [scientific] inquiry. (1975b:137, 138–39)[33]

The generative semanticists celebrated mysteries, Chomsky avoided them, and the implications of these two strategies for working linguists are equally antithetical. Lakoff and Ross's program offered them a chance to work on dazzling data that promises to remain a deep mystery, perhaps for centuries to come; Chomsky's offered them a chance to work on amenable problems. There is little mystery as to why the majority accepted Chomsky's offer.

Whence and Whither

Will [the victorious] group ever say that the result of its victory has been something less than progress?

Thomas Kuhn

It's one thing putting away the past, and quite another to tape its mouth shut.

Liam Lacey

Whence the Dispute

So, what happened?

Why was there a split in the first place? Chomsky was not moving quickly enough to the promised land. *Syntactic Structures* made modest promises about semantics, which helped attract a phalanx of eager recruits. *Aspects* compounded these promises. *Cartesian Linguistics* compounded them. *Language and Mind* compounded them. But, aside from his enrichment of deep structure by getting rid of generalized transformations, Chomsky did very little himself to satisfy those promises. The more direct work that had been done—by Fodor, Katz, and Postal—was suggestive, but dreadfully incomplete. There had been plenty of talk about meaning, but not enough action. Some form of program with a fuller, more explicit, semantics was inevitable, and it is clear from Chomsky's earliest writings that he was never going to give meaning the controlling vote in his model. It is even clearer, from every word Chomsky has ever penned, that Chomsky was not going to give up. So, there was a new program and there was resistance, a guarantee for a schism. These, at any rate, are the local reasons; the more general answer is Raup's observation that "new ideas breed disagreement" (1986:150).

Why was it so acrimonious? Again, there are local and general answers. Locally, the most obvious triggers are personal: Chomsky is biting, Lakoff is abrasive, Jackendoff is hot-tempered; stir in Postal's dogged commitment, Dougherty's scorn, McCawley's congenital inability to let flung gauntlets lie, and the wonder is that there were no early morning trips up the river with pistols for two and coffee for one. More generally, scientists are often opinionated, touchy characters, given to occasional bouts of ego besotment, who tend to set sparks flying quite regularly.

240

Science is a full-blooded activity; Hobbes calls it "a Lust of the mind" even more compelling than the lust of the body, paling the "short vehemence of any carnall Pleasure" 1991[1651]:42. It demands an extraordinary level of personal invest-ment. The personalities it both attracts and induces, especially among those who are most successful at it, exhibit intense, passionate commitment, and whenever passion is in the same room with disagreement or misunderstanding there will not be peace.[1]

Why did generative semantics die? The answer that most interpretivists and their sympathizers give is Fodor's straightforward: "It was disconfirmed." The answer is straightforward, but far too narrow. The Katz-Postal principle did not fit the data churned up in and around the early generative semantics framework, true—which is what most interpretivists mean when they say the program was disconfirmed— but the framework shifted to accommodate that data, and, more to the point, there were far more issues on the table than the vitality of a single pre-*Aspects* principle. The answer most generative semanticists give is more interesting and more com-plete: in Levi's terms, "it made promises it couldn't keep." Certainly it had trouble living up to many specific promises. Ross, for instance, never took his performative analysis very far, and Lakoff pretty much abandoned global grammar as soon as he suggested it. There was even a slate of promised publications (largely by Ross and Lakoff, though sometimes involving Postal) which were routinely cited to bolster various points, but which never saw the light of the press.[2] This inability to follow through on relatively minor promises was just symptomatic of an inability to come through on the larger ones, and it was here that the lack of delivery really hurt the movement. When Chomsky got himself into trouble in the mid-sixties by promis-ing more semantics than he could deliver, he simply retracted the promise; or, rather, he stopped talking about it. When the generative semanticists got into the same deep water in the early seventies, they made bigger promises. Not only seman-tics, but now pragmatics, was going to fall into line. Not only would it handle the cognitive aspects of language, but now the social aspects. Not only would it account for discrete phenomena, but now fuzzy phenomena. Not only would linguistics be rehabilitated, but now logic.[3] Its customers, however—the nascent, neophyte, and neutral linguists, and neighboring scholars—showed little willingness to accept these grander promissory notes.

You may have felt, when you first saw the deep, deep, growing deeper trees of early generative semantics a little like the fish in *The Cat in the Hat:*

> And this mess is so big
> And so deep and so tall
> We cannot pick it up
> There is no way at all!

But those trees came with a whole army of handles—performative handles, logical handles, transformational handles—that linguists could use to pick up such work. By late generative semantics, the situation was very different. The movement had grown so unwieldy that the handles were worthless. Even sympathetic linguists were left with nothing but a vast new view of the whole, messy, commingling, social-mental, signifying thingamajig, language.

Generative semantics, that is, failed for the reason most research programs fail. You can't keep scientists' attention for too long without giving them something to do. They need handles. They need results. The motive forces that gave rise to generative semantics never went away, but Montague grammar, and relational grammar, and nongenerative approaches now looked much more promising. Too, Chomsky engineered another motivation for linguists, restricting the grammar, and provided them with a new set of tools.

The Aftermath

1. Generative semantics is not sufficiently explicit. . . .
2. Generative semantics is a breeding ground for syntactic irregularity. . . .
3. Generative semantics is prescientific. . . .
4. Generative semantics is antiabstractionist. . . .
5. Generative semantics gives up the quest for universals. . . .
     In addition to these points, I think it can be accurately said that generative semantics fails on almost every single proposal or suggestion for the analysis of a fragment of English grammar that it has advanced.

								Michael Brame

The first generation of Chomsky's graduates had been indoctrinated with Bad Guys courses which focused on the Bloomfieldians; the next impressive crop went through Bad Guys courses which focused on the generative semanticists, and they came out of the gate with the same gusto that their predecessors had. The mid-seventies was not a good time to be a generative semanticist.

Generative-semantics-bashing articles were the order of the day. The themes were not "proposal x from linguist y is wrong" so much as "generative semantics is totally bankrupt, unscientific, and vacuous, which is illustrated by problems with proposal x, from the misguided y." Many of these articles were self-conscious proclamations of the death of generative semantics, and they didn't stop with counterarguments to one or two or three generative semantic positions. They reeled off long catalogs of failures—driving nails into the coffin, chucking it into a grave, shoveling dirt over it, and erecting tombstones with epitaphs like Brame's "Final Verdict" (1976:67).

Brame, in fact, is the epitome of the apogee of the trend, with his *Conjectures and Refutations in Syntax and Semantics,* whence the epitaph comes. It rarely matches the overt hostility of Dougherty's diatribes (which mostly date from this period), but it is irredeemably snarky. One section, for instance, begins with an epigram contrasting Einstein to an amoeba, "which cannot be critical *vis-à-vis* its expectations and hypotheses," followed by an epigraph from Lakoff which Brame clearly regards as demonstrating this amoeba-like behavior (1976:3).[4] In other places he resorts to academic priggeries, like slipping a *sic* snidely into a generative semanticist quotation ("if we adopt this explanation [*sic*]"—1976:58) and calling global rules a "theoretical prophylactic" (1976:45). Elsewhere, he simply insults the work directly, as when he calls one of Lakoff's analyses "no more than a tortuous description of uninteresting facts" (1976:15). Green's appraisal of the book—"so unreasoned and prejudicial a treatise (I refrain from calling it a tract)" (1981:704)—is positively tame in the face of its systematic, symptomatic vitriol.

The attacks and obituaries from the new interpretivist crop were augmented by several assaults from older loyalists, some with allegiances to Chomsky, others with allegiances to his 1965 position. Katz, of course, was at the head of this contingent (1970, 1971, 1972a, 1976), but he was joined by Bever (Katz and Bever, 1976 [1974]) and Stockwell (1977:131), among others. Ostensibly nonaligned linguists also began to chime in (Kuiper, 1975; O'Donnell, 1974; Sinha, 1977b). Chomsky did not address generative semantics in his theoretical work, except rather obliquely (for instance, in a footnote which does not so much as name one of its proponents— 1975b:238n2), but continued, and continues, to dismiss it in more informal settings, particularly in interviews (for instance, Parret, 1974 [1972]; Chomsky, 1979 [1976]; 1982a [1979–80]). Some disillusioned generative semanticists even threw a few handfuls of dirt, as in Sadock's "The Soft, Interpretive Underbelly of Generative Semantics" (1975).

The most interesting feature of these publications is the recurrent scapegoating of George Lakoff, which had two quite contrary manifestations. On the one hand, it was routinely implied that Lakoff was not responsible in any significant way for the beginnings of generative semantics. It was "really" Postal's theory, who had developed some pregnant suggestions from Chomsky. On the other hand, Lakoff was made to shoulder virtually all the blame for its excesses, largely on the basis of the sin of proposing global rules. That is, insofar as generative semantics had been a natural development, had been a promising hypothesis, had uncovered anything of value, and so on, this was due to Postal, and ultimately to Chomsky; insofar as it was wrong, unscientific, profligate, sloppy, and generally bonkers, Lakoff was to blame. The implication that Lakoff was little more than a loudmouthed opportunist, who perverted a potentially useful research program, has a number of underlying motives, but much of it surely stems from Chomsky's extreme disregard for Lakoff. Even after the shelling had pretty much stopped, for instance, Chomsky discussed some of Lakoff's criticisms and blasts him for a severely undernourished cranium (1980a:46):

[Lakoff's] remarks betray a very serious misunderstanding.

[He] shows no awareness [of important issues].

Lakoff seems totally unaware of the actual character of the technical work to which he refers.

[The semantic work of interpretive semantics is] a matter Lakoff has never understood.

Lakoff's misunderstanding of the technical work is so far-reaching that his comments on it are completely irrelevant.

Lakoff shows no awareness of these issues.

Chomsky's attitude must have permeated his discussions of generative semantics at MIT, and consequently permeated his students.[5]

For his part, Lakoff is less than fond of Chomsky. In conversation, this disaffection takes the form of concerns about his politics, his honesty, and his ego (three subjects that figure prominently in many unfavorable discussions about Chomsky). But in print he is considerably more circumspect than Chomsky. Still, Lakoff rarely foregoes the chance to attack a Chomskyan stance, with enthusiasm. For instance, the quotations from Chomsky in the last paragraph come from a journal, *The Behavioral and Brain Sciences,* which devotes issues to extensive criticism of individual thinkers and then gives them a chance to respond. This particular issue was devoted to Chomsky. Many very able critics, including McCawley, Lakoff, and Dennett, took aim at Chomsky's general program, some of them with very stinging blows; Chomsky took greatest umbrage at Lakoff's comments. For good reason. Chomsky takes offense quite easily, and Lakoff could likely have annoyed him in any number of ways, but he goes right for Chomsky's most sensitive area, calling him all talk and no science. Lakoff focuses on those aspects of Chomsky's work where "we are in the realm of rhetoric, not science." Much of Chomsky's work, Lakoff says, is vacuous for practical purposes, "but as rhetoric, it is effective—at least so far as academic politics is concerned." Chomsky is particularly guilty for having "artfully chosen" some of his terms, an accusation Chomsky finds deeply repugnant.

There is certainly no question—whatever Chomsky's distaste for the observation—that he is a tremendously skilled rhetor. He isn't an especially impressive prose stylist. His writing can be as dense, gnarled, and forbidding as a blackberry patch, full of fruit you can see but you just can't get to, though Chomsky can also reach moments of persuasive lucidity unmatched in linguistics. He is at his most powerful orally and in books originally meant for oral presentation *(Syntactic Structures, Reflections, Language and Problems of Knowledge).* All of which makes it no coincidence that Chomsky's version of the generative semantics story—that it was disconfirmed early, but its practitioners absurdly clung to it anyway by changing it into the worst possible theory, whereupon they were driven from the field for irrationality and error—is the received view. His opinions tend to stick fast in the minds of his audience.

Generative semantics is, first, usually ignored. Most current grammar texts, reflecting Kuhn's observation that sciences have "a persistent tendency to make [their] own history . . . look linear and cumulative" (1970:139), simply do not mention the schism at all. But two recent anomalies suggest the orthodox, driven-from-the-field-for-irrationality-and-error position informing that silence. Van Riemsdijk and Williams's *Introduction to the Theory of Grammar* mentions the dispute only to offer an excuse for ignoring it, since its "main historical role has been to serve as a catalyst in developing the current conception of semantics" (1986:88). Horrocks's *Generative Grammar,* similarly sweeping it aside, tells us that generative semantics

> was abandoned as it became increasingly clear that the rule systems necessary to carry out the mapping operation [from semantics to syntax] would have to be enormously complex and riddled with exceptions. This was a sure sign that things were being looked at in the wrong way. (1987:14–15)

Popularizations of Chomsky's work follow the same dismissive pattern, when they notice generative semantics at all. Neil Smith's oddly titled *Twitter Machine,* for instance, mentions generative semantics sporadically, most prominently in a chapter entitled "Linguistics as a Religion," which spatters the usual mud on generative semanticists for descriptive profligacy—its practitioners were like saints who "used miracles to cure headaches as well as to raise people from the dead: an effective technique, no doubt, but one that seems a little excessive when less drastic measures would be adequate." Chomsky, as expected, is credited with saving linguistics from this pernicious tendency: "At the same time as the Generative Semanticists were expanding the domain and power of transformational rules, Chomsky was suggesting stringent constraints on any rule of grammar." And the power of rationality led to the movement's death: "Counter example . . . led to the demise of Generative Semantics" (1989:201). Raphael Salkie's *The Chomsky Update* similarly indicts generative semantics for unscientific excesses (1990:116–19). In brief, generative semantics (1) is wrong, (2) was proven wrong, (3) in the most egregious of its wrongs, became descriptively wanton, and, by the way, (4) its practitioners were too muttonheaded to see that they were wrong.

Ross is the most interesting case of these muttonheads, because there are so many appallingly negative and misinformed notions floating around linguistics about him. On the desiccated academic level, he shows up as little more than a curiosity—for instance, in van Riemsdijk and Williams's *Introduction*. Ross's thesis on island constraints is not only one of the critical defining texts of the grammatical theory van Riemsdijk and Williams champion in their book, it is hugely influential in linguistics generally; every syntactic model is forced to come to terms with the phenomena and principles Ross explored in his thesis, or it simply wouldn't be taken seriously. Yet (as John Goldsmith shows in a penetrating review) the image of Ross that emerges from their book is of a lumpen functionary who had only the vaguest of clues about what he stumbled onto. Most curiously, there is a strong suggestion in their *Introduction*—a book which, by its very nature is meant to define future generations of linguists—that Ross was only concerned with a few petty details of English, when in fact "Ross's thesis was the first (and, at the time, mind-blowing) massively crosslinguistic study of an abstract grammatical property, and his conclusions were stated at the level of theory, not that of [a] language-particular property" (Goldsmith, 1989:151).[6] On a more personal level, in many interpretivists' retrospections, and in some generative semanticists' retrospections, Ross is regarded, unkindly, as an intellectual tragedy, the fair-haired boy who did some promising work under the watchful eye of Chomsky, but who fell into bad company and went astray, the rising star who rose too fast and burned out, a cortical suicide. Sometimes, in fact, the story goes, "He took too much acid and never recovered."

Anyone familiar with him and his current work and his cortical capacity, however, knows that these stories—at least insofar as going astray or burning out or turning into a chemically induced rutabaga is concerned—are false and, where not malicious, ignorant. He certainly went through a period of intellectual depression, trying to survive in what he calls the "Black Hole," the linguistics department at MIT.[7] Not only did his research completely fail to get a toehold among the graduate students there, the only market for ideas that really counts in science, he also encountered some disquieting harassment. One of his teaching assistants, for instance, recalls that

> at least three times a semester, the department head would call me up at home and ask me if Háj was showing up for classes on time, if he taught the curriculum, bla, bla,

bla. . . . And of course, he got no graduate courses, and they'd "forget" to invite him to faculty meetings, and on and on. The jokes about him abounded.

The harassment was reluctant, surely—Ross was well liked on a personal level by everyone, though only Kenneth Hale is singled out as someone who regularly went to bat for him—but good intentions probably didn't make the situation any less unpleasant, or debilitating.

The problem was that his interests had moved so far from MIT's center of gravity that he looked like he had just spun himself off into the outer reaches of research. How far from the center of gravity? Chomsky defines linguistics in a way that leaves recalcitrant data on the extreme periphery, and Ross is a data-monger, "a lovable bear who has found a cache of honey in a hollow part of the language tree and is continually astonished at the wonder of it" (Bolinger, 1991 [1974]:29). Chomsky defines linguistics in a way that leaves the aesthetic elements of language in some cold and distant stretch of the galaxy of rational inquiry, and Ross became convinced "that trying to do linguistics which has had all its aesthetics siphoned off is ultimately pointless." Chomsky defines linguistics in a way that draws on literary criticism as a negative example, the style of analysis and thought that linguists should avoid like a poxed wooer, and Ross has moved largely into poetics. Eventually, he left MIT in unpleasant circumstances.

There is nothing recognizably generative in his work any longer, and nothing that much resembles his post-*Aspects* forays—no huge trees, no transformations, no filters and constraints (though he still has complete command of that material)—but there is a certain inevitability to his progression through deep transformational semantics into pragmantax into poetry. There is a continuum between formal linguistics and the aesthetic use of language, and Ross has traveled it.

The progression for Lakoff, too, is a recognizable and rational one, one which has also landed him in the aesthetic dimensions of language (Lakoff and Turner, 1989), but Lakoff's path from generative semantics is a good deal easier to trace than Ross's, in large part because Lakoff has provided a road map for it, with the destination clearly labeled, "Generative Semantics Updated" (Lakoff, 1987:582). His position may, in fact, be closer to his work in generative semantics than Chomsky's current work is to his post-*Aspects* period; there is no clear-cut way to make such comparisons. Certainly—what is perhaps most relevant—his work is about as far from Chomsky's as his institution, Berkeley, is from Chomsky's (Ross, by the way, is in Singapore). The hallway scuttlebutt about him, at least in some hallways, is that "he's not even doing linguistics anymore," as if linguistics was the only way to look at language or, for that matter, as if there was only one way to do linguistics.

McCawley's current work is much more recognizably in the early generative semantics tradition than Ross's or Lakoff's; Bever calls him "the truest of the true GSers." But (continuing with the geographic theme) he is midway between Cambridge and Berkeley, still in Chicago, and linguistically he is somewhere toward the midpoint between Chomsky and Lakoff as well. He has developed a wonderfully market-driven philosophy of science, where every proposal is available for a price, no matter where or who it comes from, or what auxiliary beliefs or mechanisms its proposer regards as essential to its adoption. He has quite cheerfully adopted a version of x̄-syntax, for instance, and endorses Emonds's distinction among structure-preserving, root, and local transformations, while maintaining a full catalog of transformations, an intimate relationship between the deepest syntactic level and semantics, a reverence for data, and a steadfast refusal to concede any usefulness to the notion, grammatical. This mélange of tools, commitments and perspectives, he refuses to give a name more specific than *linguistics,* and he is viewed as too idiosyncratic to have any systematic influence on the field. Everybody reads his work, everybody recognizes his brilliance, and everyone finds something challenging and rewarding they can take from his insights, but nobody really does syntax in his image. Like Sapir, he is not the sort to sponsor a school.

Speaking of school-sponsoring, we should return for a moment to someone who is very much the sort to sponsor a school, Lakoff. He has become a leading figure in the most rapidly expanding linguistic approach of the last decade, cognitive grammar. The name should be familiar. It's a holdover from one of the mid-seventies projects he was involved in, with Henry Thompson (Lakoff and Thompson, 1975a; 1975b).[8] But a huge impetus came from Ronald Langacker's decade-later *Foundations of Cognitive Grammar* (1987; it's just as well that Lakoff and Thompson provide the label—Langacker wanted to call his framework, and still does on occasion, *space grammar*). Lakoff's book of the same year, *Women, Fire, and Dangerous Things,* moved him very quickly to the forefront of cognitive grammar, but the not-doing-linguistics-anymore scuttlebutt about him is partially true, in that he is not doing anything resembling formal language modeling; even his association with Langacker's framework is somewhat peripheral; he works alongside the framework, rather than within it.[9] He is best known for his studies of image clusters (Lakoff and Johnson, 1980; Lakoff and Turner, 1989)—which have closer ties to literary criticism, philology, and rhetorical theory, than they do to linguistics—and his most noted contribution to cognitive grammar is the investigation of these clusters as image schemas.

The Cliff's Notes version of Lakoff's career is that his reputation in linguistics suffered for a while as a result of the scapegoating campaign, of his own Brownian-motion style of theorizing, and of his thrashing so publicly against the restrictiveness tide initiated by Chomsky and the Peters-Ritchie findings, but he is back on top.

The story is different for different linguists, but, in general, those who didn't drop the life-boats and row away from the good ship *Generative Semantics* when Lakoff piped global rules aboard, are still pursuing interests that grew out of the voyage. Robin Lakoff is her own inimitable brand of sociolinguist and ordinary language philosopher. Guy Carden still regards his work as abstract syntax. Georgia Green has recently published a book on pragmatics (1989). Laurence Horn has just produced *A Natural History of Negation,* a meticulous and flexible tale of one semantic concept, and a data-monger's heaven (1989). Pieter Seuren is still carrying a generative semantics torch in his discourse work (1985:120–26). Even Jerrold Sadock, who is currently up to his elbows in a model with some deeply Chomskyan guiding principles (1985b; 1991), ended one informal paean to me about generative semantics with "That's why I am a generative semanticist. *Am? Was?* I don't know." And virtually every generative semanticist, of the *is* or the *was* variety, has a deep and abiding conviction that it is completely fruitless to do linguistics without faith in the ineluctability of meaning to horn in on any and every analysis: they take that lesson as the most important legacy of generative semantics. There are others.

To oversimplify (why stop now?), there are two general classes of generative semantics legacy, one defined in terms of wholesale opposition to Chomsky and most things Chomskyan, the other in terms of adherence to most things Chomskyan (though sometimes such adherence involves opposition to Chomsky—the clearest cases being generalized phrase structure grammar, which turns Chomsky's early commitment to precise modeling against him, and lexical-functional grammar, which turns his lexicalist hypothesis against the residual transformationalism of his later work). In the first instance, generative semantics served as the thin edge of the wedge which brought context, variation, and the slippery, pragmatical pig of moment-to-moment language-making back into the field. Its central legacy on this front is pragmatics and cognitive linguistics, though it also helped the emergence of sociolinguistics, functionalism, and other varieties of *in vivo* linguistics. For workers in these fields, when they are given to historical reflection, generative semantics serves as something of an Alamo, the honorable massacre.

In the second instance, the generative semantics legacy to *in vitro* Chomskyan linguistics, the contributions are more specific and technical. They are, consequently, also quite a bit easier to chart. Generalized phrase structure grammar, for instance, has unequivocally adopted Ross's auxiliary analysis, and has an explicit, logic-influenced level of semantic representation, as does lexical-functional grammar. But the interesting wrinkle here, is that government-and-binding theory, the direct descendant (insofar as *direct descendant* makes sense in linguistics, or any other science) of Chomsky's extended standard theory, also makes very liberal use of the technical proposals of generative semantics, including a good many proposals that extended standard theorists expended much energy attacking. For this group, generative semantics represents, as above, irrationality overcome, but once they had stormed that bastion of error, the stormers saw no contradiction in pillaging and looting and taking their prizes home—lexical decomposition, for instance, or logical form—to help build their own truths.

The Legacy of Generative Semantics 1: The Greening of Linguistics

> We are dealing with people, in real, murky, often conflictual, squishy situations, where there is rarely anything like black and white to guide us. We are dealing with negotiated, improvised, always-being-born language.
>
> Háj Ross

One of the ways some people prefer to look a Chomsky's impact on linguistics is as the last gasp of Bloomfieldianism (for instance, in the best articulation of this position, Moore and Carling 1982:19–47). This perspective is far too narrow, but, like narrow perspectives everywhere, it sees one arm of the matter very well. In Chomsky's case, the arm is attached to an octopus, so there are seven more wiggling around out of sight; still, even an eighth of his impact is a pretty big part of the story of linguistics in the second half of the twentieth century. The Bloomfieldian continuities in Chomskyan linguistics, even where there have been terminological shuffles, are fairly clear.

Bloomfield, particularly as vitrified by Bloch, Trager, and others, kept meaning at bay. Chomsky is a little more optimistic about getting at meaning than Bloomfield was, and has been responsible for a good deal of semantic headway, but he retains Bloomfield's cautiousness and he retains the general Bloomfieldian dogma that meaning will never be allowed in the driver's seat. He was arguing the Bloomfieldian party line on safe linguistics as early as 1955—that meaning can't be allowed to contaminate analyses of form—and he has only occasionally wavered from that position in the four decades since. If form will help him get at meaning, he's overjoyed at the opportunity, but the reverse is unthinkable. He will never use meaning to get at form.

The Bloomfieldians had a methodological proscription against "mixing levels," especially mixing them downward—phonology wasn't supposed to infect phonetic analyses, morphology wasn't supposed to infect phonology, syntax was supposed to keep its mitts off morphology. Chomsky argued against the proscription, and happily let all these lower levels commingle in various ways. But the higher levels, the ones up closer to meaning, semantics and pragmatics, are still out in the cold. Semantics isn't supposed to contaminate syntax; pragmatics is supposed to keep its mitts off of semantics.

Bloomfield banished the mind, making *mentalism* an umbrella term for aspects of language that couldn't be approached scientifically. Chomsky broke the taboo about discussing the mind, but he has a very similar sort of chastity belt for keeping his grammar pure, performance. Just as Bloomfield's mentalism was one way of keeping meaning away from form, by consigning it to psychology and sociology, so Chomsky's performance is a way to keep meaning and other contaminants away from form, by consigning them to "memory limitations, distractions, shifts of attention and interest" as well as to "the physical and social conditions of language use" (1965 [1964]:3; 1977:3)—to psychology and sociology.

In this dying-Bloomfieldian-wheeze interpretation, then, the Chomskyan hegemony that arose in the sixties was just a new face on the old fear of meaning and mind. Chomsky belongs, this interpretation goes, not to the true vanguard in linguistics, but to the progressive elements of the old guard; the revolution that blew the top off the Bloomfieldian mountain, clearing away Hockett and Householder and the others who had helped prepare the way for the linguistic *perestroika,* is not yet over. Chomsky contributed to the initial stages of the restructuring, but he remains a victim of his Bloomfieldian prejudices, and he is destined to follow the rest of the holdovers into the ignominious historical wastebasket of the Wrong. Recall that this was pretty much the way Chomsky looked to many observers in the early seventies, sadly retrenching, like Planck or Einstein, while the generative semanticists plowed ahead on the meaning-and-mind mission he had begun. That judgment was clearly premature, but linguistics has changed. Chomsky regained a considerable amount of his waning clout in the late seventies and the eighties, but his market share is nowhere near what it was in 1965.

Other formal competitors have sprung up (including several, again, which began with groups of fledglings who noisily left the Chomskyan nest, like Bresnan's lexical-functional grammar and Brame's base-generated syntax—both taking Chomsky's lexicalist hypothesis further than he wanted to go with it). Other explanatory frameworks have developed (in particular, the many versions of functionalism, and now cognitive grammar), defining themselves largely by their confrontations with the fuzzy, contextualized, meaning-driven data that Chomsky disregards. Linguistic subdisciplines that once sought out his work for investigation and application have decided they can get along without him now (psycholinguistics, sociolinguistics, second language acquisition). Linguistics is more vibrant, pluralistic, and daring than it has ever been. It has, in Bruce Fraser's term, undergone a greening.[10] and it is easy to assign generative semantics a major role in the greening. Certainly it deserves some retrospective credit as the thin edge of the wedge that brought into linguistics a good crop of phenomena which Chomsky was happy to ignore, and a range of methods and goals he discounted or despised.

Take pragmatics. While it may seem a little peculiar to give credit for the birth of pragmatics to a movement whose most prominent spokesperson once made the notoriously arrogant pronouncement that he had reduced the whole field (and, at the time, it was a barely turned field) to garden variety semantics, Lakoff's infectious arrogance about pragmatics probably did more to break down the barriers of context and use in linguistics than any other factor. Pragmatics is now a thriving subdiscipline; which dates largely from Ross's early performative work; which includes Sadock's extensive explorations of speech act theory and Lakoff's investigations of presuppositions and of Grice's conversational research; among whose chief landmarks is Gazdar's formal text (1979), which began life with a transderivational analysis, and Levinson's informal text, which specifically attributes the pragmatic infusion of linguistics to generative semanticists (1983:36). Both books bristle with references to Ross, Sadock, the Lakoffs, Davison, Green, McCawley, Horn, and assorted other generative semanticists; that is, they take generative semantics as their chief linguistic starting point (they also, by the way, take a good deal of philosophical work as another starting point).[11]

Take, as one of the most influential formal models, relational grammar. It headed off on an early dogleg from generative semantics, when several linguists decided that grammatical relations like subject and object belonged directly among the core of theoretical primitives (in all generative research of the period, these notions were derivative; *Aspects* derives them structurally, defining *subject* as the first NP of an S, *object* as the first NP of a VP). Its roots are clear. Most of the early relational grammar players, for instance, came out of generative semantics (Ross, Morgan, and Keenan were all instrumental in its development; Postal and Perlmutter were the driving forces), and its ranks swelled with disaffected generative semanticists in the mid-seventies (Cole, Dryer, Frantz, Lawler, Sadock). More crucial are the obvious developmental connections in the model itself. The abstract syntax discarding of the VP, for instance, made structural definitions of both subject and object much more complicated. The increasing concern for cross-linguistic principles around generative semantics led to accounts of phenomena like passive sentences in terms of grammatical relations (which are quite stable across languages) rather than, as in the *Aspects* tradition, word order (which is much less stable cross-linguistically). Cross-linguistic work also had to confront languages like

Welsh and Cebuano, with verb-subject-object word order (that is, languages where the verb and object are physically separated, so that there is no natural way to define the object as the NP under a VP). Similarly, concern with a universal base led to the conclusion that order and dominance alone couldn't cut it, that grammatical relations had to be part of the primitive stock of languages.[12]

Take, too, a thick and interpenetrating bunch of developments we have barely glanced at so far—psycholinguistics, sociolinguistics, functionalism, Greenbergian typology, and a bustling contingent of yet more alternate formal models—none of which began with generative semantics, but most of which received support and recruits from its ranks, and all of which benefited hugely by the break-up of the Chomskyan hegemony that began with the disagreements over the Katz-Postal principle and deep structure.

Take, especially, the hot new grammatical framework, cognitive grammar, which many people see as following a direction that can "fulfill the promises that generative semantics was not able to keep" (Goldsmith, 1987). Cognitive grammar would be a book in itself (in fact two books in itself, and it is—see Langacker, 1987; 1991),[13] but, very briefly, it is a genuine and thriving alternative to Chomskyan linguistics of exactly the sort generative semantics wanted to be when it grew up. Like relational grammar, it also has clear sociological roots in the schism; beyond Lakoff's enthusiastic association, the chief architect, Langacker, was a generative semanticist (albeit a fairly peripheral one), and a number of other names from our story figure prominently among its supporters or fellow travelers—Bolinger, Chafe, Fillmore, and Jackendoff, for instance, all get singled out for inspirational credit by Langacker in his first *Foundations of Cognitive Grammar* volume (1987:5). But again the important connections are theoretical and methodological, rather than sociological, and these connections link cognitive grammar more closely to the fat edge of the anti-Chomskyan wedge than the thin edge, to late generative semantics, when grammaticality was jettisoned, and concern for mushy categories, social factors, and figurative language came storming in.[14]

Since generative semantics was such a catholic movement, even the ways in which cognitive grammar seems most obviously to differ from generative semantics have roots in that approach. These differences are nowhere more obvious than in the area of language universals. Langacker says "semantic structure is not universal; it is language-specific to a considerable degree" (1987:2)—a complete rejection of one of generative semantics' defining criteria. But it is a rejection that reveals how far the trend initiated in generative semantics has moved away from Chomsky's idealization of grammar as a rigid system, with some peripheral noise that can safely be ignored. In the mid-sixties, any linguist who even hinted that there might be serious diversity among languages was an object of pity and ridicule; Joos, recall, was regularly hanged in effigy on the pages of transformational articles for his languages-can-differ-without-limit-and-in-unpredictable-ways remark. But generative semanticists began taking Chomsky's universalist pronouncements to heart, which meant investigating a variety of languages, which, in turn, led to an appreciation for that variety. McCawley even endorsed Hermann Paul's stronger-than-Joos remark that "we have, strictly speaking, to differentiate as many languages as there are individuals" (McCawley, 1976b [1968]:204).[15]

By far the sharpest difference between Chomsky's cognitive perspective and Lan-gacker's is the relative interpenetration of various aspects of mental structure; in particular, the relative interpenetration of language with everything else. This dif-ference virtually defined the late stages of the generative-interpretive schism, Chomsky drawing tight boundaries around his model to keep the contaminants out, generative semanticists smashing down all the walls they could reach with their hammers, to let everything pour into their analyses. Every other difference between Chomsky's approach and Langacker's follows from these crucial differences in their respective cognitive models, and the epistemologies behind them (that's right, the familiar war-horses, rationalism and empiricism). Langacker's model gains a good deal of its prestige from adherence to a general line of computer science known as *connectionism* (a.k.a. *parallel distributed processing,* a.k.a. *neural nets*), and con-nectionism is so flatly empiricist that it is "behaviorism in computer's clothing" (Papert, 1988:9).[16]

Again, this Chomsky-as-the-dying-gasp-of-Bloomfieldianism view is far too nar-row. For one thing, we have just been discussing how cognitive grammar returns to some Bloomfieldian interests that Chomsky has ignored or condemned, linguistic diversity and empiricism. For another, there's very little evidence that Chomsky's breath is short or his pulse weak. More importantly, this view ignores the huge boost that Chomsky's work gave to psycholinguistics, which then developed fruitfully in other directions; it ignores the use to which Labov has put transformations, spur-ring the acceptance of sociolinguistics; it ignores the interchange Chomsky sparked between linguistics and philosophy and the attention he helped bring to bear on meaning, both of which led quite inevitably to the surge of pragmatic work in lin-guistics; it ignores, in fact, that Chomsky introduced focus and presupposition into the debates, which have become cornerstones of pragmatic research (he didn't do anything more with them, of course, beyond showing how they wrought havoc with the Katz-Postal hypothesis, but he brought them up); it ignores his huge influence on formal modeling generally, including relational grammar; it ignores more than we have room to address, all concerning the massive infusion of energy, ideas, and burning focus Chomsky brought to linguistics, and the doors he blew open to other disciplines. Similarly, the complement of this last-wheeze view, the generative-semantics-sponsored greening of linguistics, obscures many other contributions to the current verdure; psycholinguistics, sociolinguistics, functionalism, and every other blooming tree in the forest has far-flung roots, only some of which reach back to generative semantics. But both views get at critical elements in the course of lin-guistics over the last two and a half decades, the most important of which is: debate's the thing. The health of the field, and the currently restricted market share of Chom-sky's research has more to do with the fact that there was a schism than with the fact that it began over the existence of deep structure and the tenability of the Katz-Postal hypothesis. Rhetoric is extremely productive.

We have ignored one aspect of the current vibrancy completely, however, the one that shows the critical role of the generative-interpretive semantics debate most clearly, the grammatical approach that has been the most successful over the last decade, Chomsky's.

The Legacy of Generative Semantics 2: The Right of Salvage

> You must not mind my being rude. . . . I have a resistance against accepting
> something from the outside. I get angry and swear but always accept [it] after a
> time if it is right.
>
> Max Born

With the wreck of HMS *Generative Semantics,* huge amounts of ideas, data, mechanisms, and perspectives were cast to the seas. Some of it was lost, probably for good, probably, in fact, for the best. But much else made its way into the holds of other theories; most notoriously, into the closely guarded hold of Chomsky's commissioned frigate, the *Government and Binding.* This fate is one of the two tragedies that ex–generative semanticists recurrently cite as having befallen their model, that their work has been stolen.

The other purported tragedy, that vast quantities of their data are now being completely and systematically ignored, is off the mark. It is certainly true that some material is gone, but there is a great deal more which has had a profound impact on the way linguists look at language; that, precisely, is what the greening of linguistics is all about. What generative semanticists seem to mean by this complaint, actually, is that Chomsky and his kith are systematically ignoring this material. But, of course, it could not be otherwise. Chomsky has never had more than a peripheral interest in pragmatics, figurative language, or functional explanations; he has been dismissive about sociological aspects of language; and he has been unrelentingly hostile to empiricist research strategies in linguistics. It is virtually inconceivable that he or any of his comrades would pen a paper on syntactic and semantic horrors you can find in your medicine chest (Sadock, 1974b). So, he and his are never going to accommodate this colligation of data, no matter how thoroughly generative semanticists think it impinges on the data that does interest him. The lost-data tragedy, in short, don't feed the bulldog.

Back, then, to the pilfered-ideas tragedy: how does it fare? Considerably better. Let's start with Newmeyer's list of generative semantic contributions to formal linguistics, a list that is often sneered at by ex–generative semanticists as dismissively brief. It is, in fact, very brief, brief enough to quote in full, but it is not dismissive:

> While generative semantics now appears to few, if any, linguists to be a viable model of
> grammar, there are innumerable ways in which it has left its mark on its successors.
> Most importantly, its view that sentences must at one level have a representation in a
> formalism isomorphic to that of symbolic logic is now widely accepted by interpretivists, and in particular by Chomsky. It was generative semanticists who first undertook
> an intensive investigation of syntactic phenomena which defied formalization by means
> of transformational rules as they were then understood, and led to the plethora of mechanisms such as indexing devices, traces, and filters, which are now part of the interpretivists' theoretical store. Even the idea of lexical decomposition, for which generative
> semanticists have been much scorned, has turned up in the semantic theories of several
> interpretivists, as Wasow (1976:296) has pointed out. Furthermore, many proposals

originally mooted by generative semanticists, such as the nonexistence of extrinsic rule ordering, post-cyclic lexical insertion, and treating anaphoric pronouns as bound variables, have since turned up in the interpretivist literature, virtually always without acknowledgement. (1980a:173; 1986a:138)

It's a little difficult to see what the ex–generative semanticists see as objectionable about this passage, except that they are probably so steamed at earlier comments in the book by the time they get to it that they can't see straight. It is a realistic impact-assessment statement of how generative semantics affected formal linguistics; in innumerable ways, virtually always without acknowledgment.

Chomsky's attitude to intellectual property is cavalier at best—his own as well as others'—and it is an attitude that rubs off very quickly on his students; sometimes, even on the students of his students. Their own work, and each other's work, is all that matters. No one else gets too much attention, let alone discussion and acknowledgment. The most notorious example of this slighting is Chomsky's adoption of logical form, which occupies a critical place in his current model (that is, LF). As far as Chomsky appears to be concerned, logical form comes from Robert May, who, not coincidentally, completed a thesis under Chomsky exploring these ideas (later revising it substantially for publication—May, 1977; 1985). May cites Lakoff only once, very briefly, to deny that there is any connection between their respective suggestions (1985:158n4), and he doesn't even mention McCawley at all, despite the central role played in his work by his rule of Quantifier-lowering—with minor wrinkles, essentially the same rule that McCawley proposed much earlier (1976b [1972]:294). Next on the list of notorious borrowings is lexical decomposition, which also started to show up in interpretivist work in the mid-to-late-seventies; then comes a host of small developments, like the global properties of the trace convention and the main-verb analysis of auxiliaries.

The interesting issue, of course, is not whether Chomsky is "allowed" to incorporate generative semantics ideas into his work—McCawley, for instance, incorporates x̄-syntax in his work, and nobody complained when Gazdar and Pullum adopted Ross's main-verb analysis of auxiliaries—or even whether he should acknowledge that incorporation.[17] The interesting issue is that he denounced generative semantics so warmly for many of the tendencies and mechanisms he now embraces equally warmly, a denunciation—curiouser and curiouser—he still maintains. Consider a recent development in his framework, Mark Baker's Universal Theta Assignment Hypothesis, which ensures that semantic roles are assigned in a uniform way at D(eep)-structure (Baker, 1988:46ff). As Chomsky notes, Baker's proposal is similar to one "explored within generative semantics"; namely, "that deep structures represent semantic structure quite broadly, perhaps cross-linguistically." The earlier proposal, however, the Universal Base Hypothesis, "proved unfeasible, in fact, more or less vacuous" because of various problems with generative semantics having to do with its vast descriptive latitude. Now, with the tremendous restrictiveness built into government-and-binding theory, the same proposal "becomes meaningful, in fact extremely strong" (1988b:66–67).[18] Presumably he has similar notions about lexical decomposition and Predicate-raising—the former of which had "little empirical content," the latter of which was "quite unnecessary," at the Texas Goals conference (Chomsky, 1972b [1969]:142–43)—now that his model has mutated in ways that accommodate them.

This denounce-then-adopt policy is a thorn in the side of many ex–generative semanticists. Lakoff complains about it bitterly, and Postal has written an acid-keyboarded guide to "Advances in Linguistic Rhetoric," whence the subhead for this section advising would-be Chomskyans on how to coopt ideas successfully:

> Suppose some proponent, like McCawley, of the unquestionably wrong and stupid Basic Semantics (BS) movement has, accidentally, hit on one or two ideas you need to use, say hypothetically, the notion that surface quantifiers are connected to logic-like representations by transformational movement operations sensitive to syntactic constraints, or something like that.
>
> When adopting this idea, assuming that you wish to do so, it would be an obvious rhetorical error to cite any proponents of BS. Not only would this waste a lot of serious linguists' time if they were persuaded to actually read such misguided stuff, it might mislead less sophisticated thinkers than you into thinking something about BS was *right*.
>
> So the correct procedure is to proclaim and get others to proclaim, over a long period, many times, that BS is totally wrong, misguided, unscientific, etc. Then, quietly, simply use whatever BS ideas you want without warning and without any tiring citational or attributional material. A well-known principle of scholarly law known as Right of Salvage guarantees that you cannot be held accountable for this. (1988b:136; Postal's italics)[19]

Predictably, Chomsky has a low opinion of squabbles about priority, which he believes are a feature of linguistics only because it is not yet a fully developed science, like physics:

> There's a kind of paranoia [in "underdeveloped fields" like linguistics]. For example, [the concern for priority] is the kind of infantilism that you get in semi-existent fields. The fact of the matter is that in any real field, people are going to be thinking about the same kinds of things at about the same time, because those are the problems that are on the agenda. If you want to worry about looking and seeing if this guy said it three months before I said it, that's just childishness.

Chomsky is wrong, of course, that priority squabbles are unscientific. They are such an endemic feature of established sciences like physics that scientists often go to court about who said what when—for example, in the current patent fights surrounding gene splicing or superconductivity—and Watson's famous *The Double Helix* (1968) is almost entirely about the race for the trappings that go with being first. Even the desperately sincere Darwin, who flagellated himself constantly about his concerns over the paternity of natural selection, wrote Lyell and Hooker to press for their aid in establishing his priority (Darwin, 1958 [1892]:196–98). But ideas don't require the potential to reap huge industrial profits or Nobel prizes or places in history to inspire protectiveness, even paranoia. Ideas are the stock-in-trade of science, and very few scientists appreciate it when the credit for them goes elsewhere.

Chomsky—who, it should be clear, is not the common-thief variety of idea-absorber; he genuinely is cavalier about intellectual property, as happy to give ideas away as he is to appropriate them—has to expect some flack for his virtually total disregard of some people's contributions, particularly when he can be quite careful to acknowledge the contributions of other people, those a little closer to his theoretical heart. Which brings us to the final question of the book.

Whither Chomsky?

> To reach the port of heaven, we must sail sometimes with the wind and sometimes against it—but we must sail, and not drift, nor lie at anchor.
>
> Oliver Wendell Holmes

Two anecdotes before we get on with this section and off with this book, one personal, one not.

A year or so back I wrote a review of *Reflections on Chomsky* which included the line "Though hardly mellow (and, we can all hope, decades from the twilight of his career), [Chomsky] is no longer very active in prosecuting the dogmatists of the past or the hotheads of the present." I expected some flak about *dogmatists* and *hotheads* from the editor, although I was mildly optimistic that the context would make it clear the terms were chosen to reflect Chomsky's attitudes, not as objective labels for such linguists as Trager and Lakoff or Postal and Gazdar. But context seems to have done its job, and the terms passed without comment.

Now comes the strange part: the editor objected, strongly, to the parenthetical we-can-all-hope remark. "I don't think that everyone would agree," he told me, "that [we all hope Chomsky] will be 'decades from the twilight of his career,'" and added the somehow accusatory, "maybe you do." That phrase, he implied, would have to go. He apparently reconsidered the snipping strategy, though, and sent the review to someone else for an independent evaluation, who recommended that it not be published at all; among the reviewer's complaints, "'we can all hope' in the first paragraph is either disingenuous or deluded." Such attitudes, unfortunately far from uncommon, indicate that there are people who genuinely wish that Chomsky would die, or retire, or move exclusively into political or philosophical domains, and just leave poor little linguists alone. The level of enmity is truly stunning—we are, after all, talking about linguistics, about the study of language, not about poverty, or disease, or imperial aggression, and we are talking about a man who has turned the discipline on its head several times, who has been, further, an extraordinary fount of ideas for well over thirty years—but, clearly, there are people who would likely prefer another heading for this section, *Wither Chomsky.*[20]

The other anecdote, probably apocryphal but representative of an important perspective all the same, involves a Western historian and a Chinese historian. They meet and start chatting. "Tell me," the Western historian asks his colleague, "what do you think of the French Revolution? Was it a good thing, or was it a bad thing?" The Chinese historian says the question is premature. "It's too soon to tell."

Despite having inspired blood-boiling animosity, Chomsky will almost certainly not wither in the intellectual history of language and the mind, but it is far too soon to tell what will become of him. To retrieve our astronomy analogy for a moment, the Copernican revolution took almost a century to come off fully; Chomsky's not even at the halfway mark yet, and if, fifty years hence, the Chomskyan model of language and mind follows the trajectory that Copernicus's model of the heavens took, it will look quite different than it does at the moment. In fact, at the moment, despite his having won back a goodly portion of the field in the late seventies and the eighties, things don't look especially bright for Chomsky.

Certainly they don't look very good in psychology, for many of the reasons we have seen. Listen to one psychologist diagnose the mood surrounding Chomsky's contributions to the field:

> Despite the developments in transformational grammar over the last thirty years, and three major stages in its evolution—the initial theory, the standard theory, and the theory of government and binding—there is no complete account of the syntax (and a fortiori the semantics) of any natural language, and no generally accepted and definitive linguistic theory—no theory sufficiently explicit to be translated into an effective procedure for acquiring a grammar for a language, given a corpus of sentences from that language. Judged by the strictest criteria of scientific achievement, the Chomskyan programme has yet to succeed; compare its course with, for instance, that of Crick and Watson's theory of the structure of DNA. (Johnson-Laird, 1987:147)[21]

Johnson-Laird is quick to point out that psychologists can't, in most areas of their work, point even to the level of success that Chomsky's program has achieved, so they aren't in much of a position to turn up their noses on those grounds alone. But add to the slow and twisted progress of his program "the wranglings of one group of grammarians with another" and "the seemingly perpetual 'treason of the clerks' as successive generations of generative linguists part company with the founding father" (Johnson-Laird, 1987:148), not to mention the sweeping and unrequited optimism of the honeymoon years of cognitive psychology and transformational grammar, and the level of disillusionment is easy to explain. And things will only get worse, at least in the short haul; with the recent advent of cognitive grammar, psychologists and cognitive scientists have a much more amenable linguistic framework to explore.[22]

Not so with philosophers, of course. They never had the same optimism; their own progress, if progress it be, is glacial; they're used to wranglings and treason; and Chomsky never fails to give them a good argument; what more could they ask for? He is more of a philosopher than almost any linguist in the history of the discipline. Certainly he is a closer intellectual kin to the scattered philosophers like Zeno and Humboldt who have had an active, informed interest in linguistics, than to the linguists like Whitney and Bloomfield who have had an interest in philosophy. So, philosophers have welcomed him, in their rough and contentious way, to their hearth. They have attacked him, celebrated him, quibbled with him, and endorsed him. One philosopher, reviewing a book in which Chomsky confronts Dummet, Quine, Putnam, Searle, and assorted other formidable members of the profession, likens the spectacle to "watching the grandmaster play, blindfolded, 36 simultaneous chess matches against the local worthies" (Hacking, 1980:47). The interactions between Chomskyans, especially the titular one, and philosophers have been continuous and fruitful and are unlikely to slow down. Chomsky's place in philosophy is assured.

And his place in linguistics? That's the diciest forecast to make. Linguistics looked to be in quite dire straits in the eighties, with "as many sects as there [were] departments of linguistics" (Green, 1981:703). Some models were fairly well established—Montague grammar, relational grammar, stratificational grammar, tagmemics, a version or two of case grammar, and the (revised) extended standard theory—but there were also emergent models, like corepresentational grammar, equational grammar, daughter-dependency grammar, role-and-reference grammar, several varieties of functional grammar, and, of course, Lakoff's grammar-of-the-moment (in 1980, experiential linguistics).[23] But, while several other challengers arose, defining their approach in opposition to Chomsky$_{1965}$ and Chomsky$_{1975}$ and Chomsky$_{1985}$, the Chomsky-of-the-moment steadily retook the formal market. In 1986, Bernard Comrie lamented that "the best minds among the upcoming generation of formal syntacticians" were gravitating toward Chomsky's model (1986:774). It is a familiar moan. New contenders still arise (Sadock's autolexical syntax, for instance), and older ones continue to fade; government-and-binding theory (or whatever) just keeps growing.

His reputation is sure to see a few more hills and dales, before it settles into a steady and considered place in the history of the discipline. And, if his past is a useful guide, and if he is, as we can all hope, decades from the twilight of his career, his program will likewise see a few more hills and dales. Just look at the evolution of Chomsky's specific model. From its isolated and defensive position in the late fifties (then, again, in the late sixties), it has gained more and more influence, but uniformity is something of an illusion. Just using his own terms, we have far more stages than the three Johnson-Laird identifies. From *Syntactic Structures,* Chomsky moved to the standard theory, to the extended standard theory (which, when the "Conditions" work entered the picture, became known as the revised extended standard theory), to government-binding theory, a designation that has been giving way in recent work to the principles and parameters theory, and now there is talk about a "minimalist program." Nor is the welter of names asymptomatic. The number of technical proposals and conceptual shifts outweighs the number of designations quite substantially, and the personnel of the models has seen a rather constant turnover. There is surely more to come. Indeed, as we go to press, the work coming out of MIT (Chomsky, 1992) suggests another quite dramatic shift, and a shift again toward some of the driving ideas of generative semantics. D(eep)-structure seems finally to be following phlogiston and ether and the dodo into oblivion; S(urface)-Structure too. The old dream of a homogeneous theory with only two representations—one of sound (PF, or phonetic form) and the other of meaning (LF, or logical form)—is being revived. The goal of a mediational grammar that links sound and meaning is returning to the foreground. And . . . who knows what wonders lurk in the mind of the brilliant, impossible Chomsky?

His reputation could well slip again. Changes always seem to alienate some of his followers. Cognitive grammar appears to be a formidable competitor, with increasing journal interest (including a journal of its own), a dedicated, rapidly growing professional society, conferences all its own, a growing presence at more general conferences, and a steady flow of all-important graduate theses. And, of course, with or without cognitive grammar, the market for Chomsky-bashers is always lively. If his reputation *does* slip, however, it won't be cause for alarm, even among his proponents. Chomsky thrives on adversity, apparently having the need to be something of an underdog. The more resistance to his work there is, the more inventive he becomes. In some comments to Riny Huybregts and Henk van Riemsdijk after the generative semantics brouhaha had died down, he said "As I look back over my own relation to the field [of linguistics], at every point it has been completely isolated, or almost completely isolated" (Chomsky, 1982a [1979–80]:42), an observation that most linguists find completely baffling. It is unquestionably overstated, but Chomsky has certainly been in the minority, with his back to the wall, more than once, and he has come through magnificently; the remarks to Huybregts and van Riemsdijk came just as his most recent reinvention of generative grammar, and, some say, his richest to date, was making its way to the surface. His *Lectures on Government and Binding* were delivered in Pisa, April of 1979, and reworked through 1980 (see Chomsky, 1981a [1979]; 1982b [1979]).

Isolation and embattlement appear to be important psychological motivators for him.[24] Many of John Stuart Mill's crackling epithets can stand as slogans for his approach to science, but the one that seems best suited speaks directly to the intellectual advantages of paranoia: "Both teachers and learners go to sleep at the post, as soon as there is no enemy in the field." As we have seen, repeatedly, Chomsky quite cheerfully drives into enemydom people with whom he could otherwise work very profitably (he still speaks quite highly of Postal's work, for instance, and admires McCawley's, and has organized a great deal of his research around Ross's island work).

Brooking tautology for a moment, what Chomsky's various models have in common is Noam Chomsky, and those models have seen a rather staggering range of variations upon his core of driving themes, a range which permits the history of Chomskyan linguistics to be characterized rather nicely by a famous exchange:

HAMLET: Do you see yonder cloud that's almost in shape of a camel?

POLONIUS: By the mass, and 'tis like a camel, indeed.

HAMLET: Methinks it is like a weasel.

POLONIUS: It is backed like a weasel.

HAMLET: Or like a whale?

POLONIUS: Very like a whale.

For *Hamlet,* read *Chomsky;* for *cloud,* read *language* or *mind;* for *Polonius,* read *a rapidly changing core of bright and dedicated linguists; camel, weasel,* and *whale* are up for grabs, but there are more incarnations in Chomsky than in Shakespeare. Among the shapes that Chomsky has reported seeing in the clouds, of course, are the two at the heart of this book—the one in which deep structure was the grammatical feature that "determines the meaning of the sentence" (1966a:35), generative semantics, and the one in which deep structure was so incidental to meaning that it faded away into D-structure, interpretive semantics.

Notice, however, that the Hamlet-Polonius characterization of Chomskyan linguistics is exactly the appropriate characterization for the history of any science.

Aristotle and Ptolemy said the earth was at the center of the cosmos, and their followers agreed. Copernicus said that the sun was at the center of the cosmos, and his followers agreed. More recent interpretations of the cosmic cloud assign earth and its sun to the periphery, and this model has extremely wide assent.

The primary difference in Chomskyan linguistics is that the time frame is far more compressed, and that the same moody Hamlet is center stage throughout. These differences have affected the sociology of the discipline profoundly. Each time Chomsky goes through one of his mini–paradigm shifts, he leaves what Jackendoff terms "disillusioned Kuhnian debris" littering his wake.[25] The field is consequently quite rancorous, and the paradigmatic models more ephemeral than most. But it is also richer than many others for exactly the same reasons. It keeps the teachers and the learners alert at the post; whatever else can be said for linguistics, it is not a sleepy field. Language and its container are far too complex to give up their secrets in one fell swoop by one fell linguist, however hawk-eyed. I, for one, am very grateful for the dynamism of Chomsky's mind. I am also happy for the friction generated by variations on his themes in other models, like generalized phrase structure grammar, and for the proliferation of alternative approaches, like functionalism and cognitive grammar.

He is at the moment, in the curious flip-side to his isolationism, more sanguine than ever about his work, extrapolating from the results of his current research (in the familiar theme of getting underneath language to its more fundamental structure) to align it with the Galilean and Newtonian epistemic explosions that have virtually defined the last three hundred years of Western civilization:

> We are beginning to see into the deeper hidden nature of the mind and to understand how it works, really for the first time in history, though the topics have been studied for literally thousands of years, often intensely and productively. It is possible that in the study of the mind/brain we are approaching a situation that is comparable with the physical sciences in the seventeenth century, when the great scientific revolution took place that laid the basis for the extraordinary accomplishments of subsequent years and determined much of the course of civilization since. (1988a:91–92)

Whew. The claims and the optimism are breathtaking. Laughably so for opponents who point to the narrow syntaxicentric focus of his vision. But lasers are narrow too, and they cut deeply, and the history of linguistics in the second half of the twentieth century makes one conclusion about Chomsky inescapable, especially for the many Bloomfieldians, generative semanticists, and assorted other victims eager to get the final laugh. Don't bet against him.

Notes

Chapter 2

1. *Philology* has also had this association in the past, though never in North America.

2. For instance, in Whitney (1910 [1867]:3), Bloomfield (1933:16), and Hughes (1962), who uses exactly these terms, *pre-scientific* and *scientific*.

3. Saussure, it is important to note, did not call for the abandonment of historical linguistics. His *Memoir on the Original Vowel System of the Indo-European Languages* is a classic of diachronic linguistics, responsible for one of the most spectacular demonstrations of the power of historical reconstruction. Saussure proposed that Proto-Indo-European must have had a certain class of sounds (sonant coefficients, later called "laryngeals"), despite the absence of direct evidence for these sounds in any of its known descendants; fifty years later, Kuryłowicz found that direct evidence when he examined freshly discovered Hittite texts, and found it in exactly the phonetic positions for which Saussure had argued. "The essential point" about Saussure's work, said Bloomfield, is that it was the first to "[map] out the world in which historical Indo-European grammar (the great achievement of the last century) is merely a single province; he has given us the theoretic basis for a science of human speech" (1923:319; 1970:108).

4. For one indication of how abrupt the history in this overview is, see Julie Tetel Andresen's *Linguistics in America 1769–1924* (1990a), which offers a quite comprehensive treatment of North American linguistics for the two centuries preceding the effective starting point in this section.

5. Sapir and Bloomfield were overlapping contemporaries; competitors in a sense, confederates in another. Hockett, paraphrasing Charles Voegelin, says that "each had deep respect for the other, but with certain reservations. Sapir admired Bloomfield's ability patiently to excerpt data and to file and collate slips until the patterns of the language emerged, but spoke deprecatingly of Bloomfield's sophomoric psychology. Bloomfield was dazzled by Sapir's virtuosity and perhaps a bit jealous of it, but in matters outside language [presumably including psychology] referred to Sapir as a 'medicine man' " (Bloomfield, 1970:539–40). For Sapir, see Koerner (1984), Darnell (1990); for Bloomfield, see Hall (1987b; 1990).

6. A corollary of the differences among languages that involves the close interdependence of thought and language is the "linguistic relativity hypothesis," which dates at least to Humboldt and includes Boas, Sapir, and Benjamin Lee Whorf among its subscribers. It is the virtual opposite of the Modistae's working hypothesis, that thought had a uniform structure for all people everywhere, which language reflected, a hypothesis that arose among pre-compar-

ativist scholars who could see Latin and Greek patterns in their native tongues. Linguistic relativity (also called the *Sapir-Whorf hypothesis*), which arose with the serious, anthropologically-minded study of non-Indo-European languages, claims that since language patterns are so different, thought patterns must be as well. Its most extreme proponent phrased it this way: "I find it gratuitous to assume that a Hopi who knows only the Hopi language . . . has the same notions, often supposed to be intuitions, of time and space that we have, and that are generally assumed to be universal," and by *gratuitious,* he in fact meant *incorrect* (Whorf, 1956 [1936]:57).

Joos (1950:702–3) provides a beautiful illustration of how this language-governed structuring of the world happens in English: consider the sentence "The linguists all put their glasses on their noses." Native English speakers invariably take this sentence to refer either to a past action, or to a present action, despite the fact that *put* is one of a small class of English verbs which is ambiguous for time of activity: "They put their glasses on now"; "They put their glasses on yesterday." In a very real sense, our language makes us think temporally; Kwakwala makes its speakers think ethically.

7. At least the strong assumption is that Sapir's structuralism is homegrown. He was not a particularly covetous scholar, nor one short of the intellectual wherewithal to develop such an approach, and he never mentions Saussure in the book codifying most of his notions, *Language.* Nor would it have been unusual for Sapir not to have heard of the *Course* before writing *Language;* even in Europe, "the book's appearance was not a bombshell (in 1916, the year of its publication, Europe was being devastated by less metaphorical explosions)" (Lepschy, 1986:189). In general, Saussure's direct influence on North American linguists was relatively slight. The *Course* was not even translated into English until 1959, and most American structuralists saw their work as much more closely allied to the anthropological tradition of Boas than to anything from Europe. McDavid (1947:30), for instance, says that American structuralist thought arose "when anthropologists studying native cultures in North America realized that the structures and sound-patterns of the languages that were the vehicles of those cultures could not be distorted into the traditionally accepted patterns of Western European languages."

8. I am collapsing history to some extent, and ignoring some very significant events— like the Second World War, and Sapir's death in 1939—which affected the influence of Bloomfield and Sapir rather substantially, but the important point for our purposes is that Bloomfield became the dominant force in linguistics, and that his approach (refracted through disciples) dominated the American stage onto which Chomsky stepped. Consider just this celebration of his *Language:*

> It is not too much to say that every significant refinement of analytic method produced in this country since 1933 has come as a direct result of the impetus given to linguistic research by Bloomfield's book. If today we see more clearly than he did himself certain aspects of structure that he revealed to us, it is because we stand on his shoulders. (Bloch, 1949:92)

9. The following discussion is more ideological than strictly historical. Of the three people I identify as crucial founding members of the society, for instance, only one (Bloomfield) had an important administrative and policy-forming role, and others who had more important roles in those respects (especially George Bolling) are ignored altogether. That is, I am only interested here in what the existence of the LSA meant to linguistics, not in the history of the organization. But see Joos (1986 [1976]) and, especially, Murray (1991), for historical accounts of the LSA and its important organs, *Language* and the Linguistic Institute; Hill (1991) takes up the story in 1950, and also included Marckwardt's 1962 LSA presidential address which traces some early history of the organization; virtually every essay in Davis and

O'Cain (1980) and Koerner (1991)—collections of brief memoirs from linguists of the period—contain anecdotes about and testimonials for the LSA.

10. The analogies are not mine: Martin Joos called Bloomfield "the Newton of American linguistics [Sapir,] its Leibniz" (1957:v), and Bloomfield's postulates are self-consciously Euclidean.

11. Bloomfield's "answer" to Sapir's argument was an address delivered to the Modern Language Association, "Linguistics as a Science" (1930; 1970:227–30).

12. There is some anachronism in this severely truncated account of behaviorism—alluding to work that developed in the forties and fifties, chiefly under the influence of Watson's intellectual successor, B. F. Skinner—but nothing material is affected by this conflation, and Skinner's brand of behaviourism is important for a later chapter in our story.

13. See Hockett's interpolation on this point in his edition of Bloomfield's papers, and the review he includes by Diekhoff of Bloomfield's first text (Bloomfield, 1970:45–50).

14. The quotation is from Edgerton (1933:295; Bloomfield, 1970:258), but all the reviews of *Language* in Hockett's Bloomfield anthology play this theme (Bloomfield, 1970:257–80).

15. His "Why a Linguistic Society?" for instance, drew general attention to the "American Indian languages which are disappearing forever, more rapidly than they can be recorded" (1925:5, 1970:112), and his specific anguish over the disintegration of a culture and language is apparent in the preface to *Menomini Texts* (1928; an abridged version which retains the poignancy is in 1970:210–11).

16. Mentalists, of course, didn't shut their traps, though their numbers and the avenues open to them were both quite limited—Morris Swadesh's eloquent defense of mentalism (1948), for instance, was published in an obscure Marxist journal—so a more accurate slogan would have been "Mechanists: Shut your ears!"

17. It is mildly controversial in some quarters to say that Bloomfield-inspired linguists were uninterested in meaning, and, as we will see, the generalization is a little too strong. But meaning was definitely out of the main purview of linguistics in that period, following Bloomfield's warning that it has the potential to lead to chaos. So, for instance, Hill uses an obscure paper by Joos ("Towards a First Theorem in Semantics") to serve "as one refutation (among many) to the oft-repeated but erroneous charge that American linguistics of the 1940's and 1950's was anti-semantic and materialist, characteristics which it supposedly owed to Leonard Bloomfield," but interpolated into Hill's paper is a lamentation by Joos that the audience "reacted almost entirely negatively" (Hill, 1991:30).

18. There are quite a few alternate labels for the generation of linguists who plied their trade in the U.S. in the wake of Bloomfield's book—*descriptivists, American structuralists,* and *taxonomists,* among the more common—but, for a number of reasons, *Bloomfieldians* is best suited for our purposes. *Descriptivists* is a label best suited to linguists inspired by Boas (the tradition Harris, 1973:252, calls "descriptive informant-based grammar"), including Bloomfield and his followers; that is, it is too broad. *American structuralists* implies, among other things, that the program which displaced Bloomfield's, Chomskyan linguistics, is not structuralist; that is, its implications are misleadingly narrow. And *taxonomists,* a term introduced by Chomsky, is flatly derogatory, at least in the context in which Chomsky introduced it. So, *Bloomfieldian* it is.

Other commentators prefer to qualify *Bloomfieldian* with prefixes like *neo-* or *post-* to recognize the contributions to this program by others (such as Bernard Bloch, George Trager, Zellig Harris, and Charles Hockett), and such prefixing has the additional virtue of acknowledging that Bloomfield was somewhat less dogmatic than some of his followers became on a few of the questions we will be looking at, particularly meaning and mentalism. (See, in particular, Fries, 1961, for some discussion of Bloomfield and his influence which notes a bit of the diversity in "his" school; and Mathews, 1986, for the best discussion of how Bloomfield's work engendered the Bloomfieldian school; Anderson, 1985:277–80, is particularly succinct at capturing the uniformity and diversity in American linguistics between Bloomfield, 1933, and Chomsky, 1957a. The most representative texts of this approach are probably the ones that came latest in its reign, since they summarize and synthesize its developments and positions; in particular, Gleason's *Introduction to Descriptive Linguistics* (1956), Hockett's *A Course in Modern Linguistics* (1958), and (especially) Hill's *Introduction to Linguistic Structures* (1958).

19. A few words are in order here about the term *Chomskyan linguistics,* which probably rankles no one so much as it rankles Chomsky. Chomsky has an ego, unquestionably, and it no doubt gives him some pleasure to see his name in print associated with Big Ideas, but he regularly inveighs against the cult of personality implied by using someone's name to signal a field of inquiry or a theoretical perspective. He has, for instance, deplored the use of *Marxism,* despite his admiration for Marx's "extremely brilliant and important ideas." Terms like *Marxism* and *Marxist,* he says, belong more "to the history of organized religion" than to open inquiry. Contrasting this usage to practices in science, he opposes *Marxism* to the hypothetical *Einsteinism* and requires people to treat Big Ideas, whatever their source, simply "as intellectual contributions and not as divine inspiration." (All quotations in this paragraph are from Chomsky, 1988a:177–78.) As for *Chomskyan linguistics,* he says the phrase is "absurd." Still, the label is appropriate, and the generally preferred term, *generative grammar,* is unavailable because it could lead to confusion with *generative semantics.* So, with the following brief apologies, I will continue to employ it (sparingly) through the remainder of this book. First, although *Einsteinism* has never been a very popular term, it is not unattested, and the theory of relativity is quite regularly referred to as "Einstein's theory of relativity." *Newtonianism* (along with such locutions as *the Newtonian philosophy*) was quite common for several centuries, and *Newtonian physics* is still a frequent synonym for *classical physics.* *Darwinism* is popular in biology (indeed, Ernst Mayr uses the suffix *-ism* as a badge for the name *Darwin,* which therefore certifies him as "the greatest of all intellectual revolution[aries]" in the history of mankind"—1988:162). And many individuals have lent their names to various techniques, instruments, and phenomena in science—for example, the Fibonacci numbers, von Neumann machines, and the ever-popular Halley's comet. *Chomskyan linguistics,* too, along with variants like *Chomskyan theory* or *Chomskyan grammar,* is attested (e.g., Stout, 1973). *Chomskyite linguistics,* which also shows up, seems unequivocally fanatical and cultlike (e.g., Hall, 1990:91–92), and that is occasionally how *Chomskyan* is used as well, as something of a sneer in the writings of his detractors, but the latter term also occurs in the comments of some of his admirers (such as Harman, 1988:260). My usage, in any event, is intended to be completely neutral.

Nor has Chomsky always refrained from using proper nouns as general terms—calling, for instance, behaviorist thought "the Skinnerian pattern" (1975b:199), and entitling one of his books *Cartesian Linguistics* (rather than, say, *Rationalist Linguistics* or *Port-Royal Linguistics*). He has even, albeit under a mild, have-your-cake-and-eat-it-too protest (1991a [1989]:4), prominently participated in a conference with the label "The Chomskyan Turn" (Kasher, 1991).

20. There was evidently some tension over the handling of syntax in Bloch and Trager's book, and the job in fact was taken over by Bloomfield:

> Bloch and Trager were in effect commissioned [by the LSA generally, Bloomfield specifically] to write the *Outline of Linguistic Analysis.* They were in agreement about everything but the syntax chapter. Bloomfield was sure that the book would not serve its purpose with a radically modern syntax chapter; and after fruitless efforts to get a sufficiently conventional syntax chapter written by the two collaborators or by either one of them alone, Bloomfield wrote that chap-

ter himself. He was able to persuade Bloch to omit adding the third name to the title page, but Trager remained unreconciled, and the repeated reprintings of the *Outline,* even today, have been without his approval. (Joos, 1967:9)

21. Boas makes a similar point, though his terms are not *syntax* and *morphology* but *grammar* and *lexicography* (n.d. [1911]:26–27). The clearest statement, however, is in Hockett's (1940:56) brief comment that the morphological intricacies of Amerindian languages and the descriptive mandate of Bloomfieldian linguistics rendered syntax "almost *terra incognita.*" Hale (1976:35) suggests a related reason for the syntactic poverty of Bloomfieldianism: syntax is much more difficult for an outsider to study than for a native speaker to study, and there have been very few linguists who spoke an Amerindian language as their first language.

22. However, this observation should not be taken, as it often is, to imply that Chomsky invented syntax. There were a number of good syntactic guidebooks before the advent of transformational analysis: for English, there were Onions (1911), Wendt (1911), Poutsma (1914), Vechtman-Veth (1928, 1942), Jespersen (1937, 1949 [1909–1940]), Krüger (1914 [1897–1911]), Curme (1931); only the last of these by an American. Virtually all of these investigations, though, were pre-structuralist, in methodology if not chronology (or, perhaps *quasi-structuralist* for some, such as Jespersen, whose methods were on the cusp). Chomsky (with considerable help from Nida, Wells, and, especially, Harris) pretty much invented *structuralist* syntax.

23. The Planck story may be aprocryphal, but its moral—the widespread and misplaced confidence in the later nineteenth century that physics was nearing completion—is not. Lord Kelvin, for instance, gave a speech right at the turn of the century to the British Association for the Advancement of Science, in which he told them that "there is nothing new to be discovered in physics now. All that remains is more and more precise measurement" (Davies and Brown, 1988:4). More generally, the finished-field story is a familiar leitmotif in the history of science; a few decades after Kelvin's remarks, similar comments were common for quantum mechanics (Hawking, 1985 [1980]:132, quotes one such claim from Born in the 1920s, and Booth, 1974:69, quotes another from Russell). Its most extreme version comes from wee Francis Crick, who was impatient to become a scientist when he was a boy, but despaired there would be nothing left to do: "By the time I grew up—and how far away that seemed!—everything would have been discovered." (His mum, however, provided solace. "Don't worry, Ducky," she told him. "There will be plenty left for you to find out."—Crick, 1988:9.)

24. See Nida (1960 [1943], Wells (1947b), Bloch (1953), and, especially, Trager and Smith (1957 [1951]), which put the phonological syntax spin on Immediate Constituent analysis that was responsible for much syntactic enthusiasm among younger linguists just before Chomsky came on the scene.

25. Bloch's comment is in a letter to Robert Lees, an important force in the rise of Chomskyan linguistics, dated 31 July 1959, cited in Stephen Murray (1980:79). Bloch—"probably the truest Bloomfieldian of them all" (Stark, 1972:387) and, more importantly, as the somewhat autocratic editor of the LSA's *Languages,* "the central neo-Bloomfieldian gatekeeper" (Murray, 1980:73)—encouraged Chomsky from the outset. He appears to have had a remarkably Feyerabendian view of scientific progress, which one of his students characterizes in the Maoist metaphor of letting a hundred flowers bloom, and he openly nourished the generative flora. He placed Chomsky's theses, and his *Logical Structure,* in the Yale library. He published Lees's remarkable review of *Syntactic Structures* in *Language,* and regularly published ill-tempered transformational appraisals of Bloomfieldian work, even inserting an editorial expression of solidarity with Postal in one of his attacks (1966b). Newmeyer says that Bloch told "at least two colleagues, 'Chomsky really seems to be on the right track. If I

were younger, I'd be on his bandwagon too' " (1980a:47). Chomsky was (and remains) very grateful for Bloch's assistance; the only article he contributed to *Language* after his early assault on Skinner (1959 [1957]) was to the Bloch memoriam edition (1967a).

Hall (disputing the Newmeyer quotation above) quotes one of Bloch's students saying that Bloch "did NOT encourage TG [transformational grammar]" (1987a:108; emphasis in the original) which—given Bloch's direct encouragement of Chomsky, indirect encouragement of Chomskyans in the pages of *Language,* and documented expressions of faith—is difficult to accept at face value. Hall leaves the informant anonymous; Newmeyer's informants are anonymous in the first edition, but given in the second as Donald Foss and Sol Saporta— 1986a:38n8).

26. Respectively, these quotations are from Trager (1968:84), Hall (1987a [1965]:5), Hockett, quoted in Mehta (1971:175), and Trager (1968:84).

Chapter 3

1. Since this statement is mildly controversial—Bloomfieldians, some accounts suggest, rejected Chomsky from the git-go—some additional evidence may be in order: Bloch published Lees' (1957) long and hugely laudatory review of *Syntactic Structures;* Nida's 1959 preface to the publication of his *Synopsis of English Syntax* endorsed transformational analysis as an advance over his Immediate Constituent analysis (1960 [1943]:iv); Hockett (1958:208) apologized for not including transformational syntax in his *A Course in Modern Linguistics,* and Householder's review complained about just this omission (1958:503, 506– 8); Gleason's (1961) revisions of his *An Introduction to Descriptive Linguistics* did include transformations.

As for the Harris connection, it suffused Chomsky's reputation from the very beginning. The proceedings of the 1958 Texas Conference are clear on this point (Hill, 1962c). Chomsky's paper, for instance, begins "The approach to syntax that I want to discuss here developed directly out of the attempts of Z. S. Harris to extend methods of linguistic analysis" (124), and in several other places, particularly during the discussion sessions, he effects strong links with Harris's work (e.g., 164, 174, 178). More tellingly, many of the participants reveal a failure to distinguish between Chomsky's work and Harris's—as in Long's comment, "I agree with Chomsky and Harris here" (167), about an issue (psychological realism) on which Chomsky and Harris differed hugely.

2. Neither of these definitions is original with Chomsky, however. Bloomfield's definition of a language in the "Postulates" is "the totality of utterances that can be made in a speech-community" (see P. H. Mathews, 1986:260ff, for some discussion of Chomsky's antecedents here), and one of Hockett's criteria for a grammar is that "it must be productive: when applied to a given language, the results must make possible the creation of an indefinite number of valid new utterances" (see Hymes and Fought, 1981 [1974]:165–73 for some discussion of Chomsky's antecedents here). See Teeter, 1969; Steiner, 1971 [1969]:102–25; Gray, 1974; Hymes and Fought, 1981 [1974]:165–85; Moore and Carling, 1982:19–47; Huck and Goldsmith, forthcoming, for general discussions of Chomsky's historical connections to Bloomfieldian thought. Chomsky's contribution in these two areas is the way in which he welded the set-of-sentences definition and the generating-device definition together and organized linguistic modeling around them.

3. See Chomsky (1988b:9) for much the same account, thirty years later.

4. The shift I have just outlined came with a very specific argument, of interest primarily to linguists. Bloomfieldian theory, as Chomsky saw it, was concerned with locating a discovery procedure for grammars—a set of principles that could be turned on a corpus of texts and produce a grammatical description of that corpus, a grammar. A weaker criterion for a set of

methodological procedures, Chomsky suggests, would be the capacity to look at the corpus, look at some proposed grammar for that corpus (the discovery of the grammar being wholly irrelevant), and determine if it was the best possible grammar of that language. An even weaker requirement would be that a set of principles look at the corpus, look at two or more proposed grammars for that corpus, and return a verdict as to which grammar was better.

Now, Chomsky had redefined a grammar in terms of a scientific theory, and, in science, even the weakest of these three alternatives, formulating a general evaluation procedure for competing theories, is a very difficult goal to attain. In fact, the difficulty of formulating such a procedure is the basis of Kuhn's discussion of incommensurability and revolutionary change, and Chomsky's comment on this matter would be very much at home in *The Structure of Scientific Revolutions:*

> There are few areas of science in which one would seriously consider the possiblity of developing a general, practical, mechanical method for choosing among several theories, each compatible with the available data. (1957a:53)

Chomsky proposed downgrading expectations about linguistic methodology from the discovery procedure to the evaluation procedure—a procedure to choose between competing descriptions of the same corpus. By the point he makes this suggestion, Chomsky has already provided an extended illustration of its worth. Although he supplies no explicit mechanical procedure, the first part of *Syntactic Structures* clearly weighs the relative merits of three grammars and returns a compelling evaluation in favor of the transformational model.

There is some question as to whether the way Chomsky saw Bloomfieldian theory on this count, chasing discovery procedures, is the way Bloomfieldian theorists saw themselves. Certainly Chomsky's view is one way to read the literature he cites to illustrate this concern (1957a:52n3). It is easy to read Harris's *Methods,* for instance, as pursuing such a goal; ideally, one points the mechanical routines in the book at a sufficiently large corpus from some language and it cranks out a description of that corpus (and, by extension, of that language). But the term, *discovery procedure,* is Chomsky's, and he freely admits that it represents an *interpretation* of tacit Bloomfieldian aims ("as I interpret most of the more careful proposals for the development of linguistic theory, . . ."—1957a:52). Hymes and Fought (1981 [1974]176–86) argue that it was a *mis*interpretation. Their argument is inconclusive, since Bloomfieldian and Chomskyan terminology overlap in ways that frustrate clear comparisons, but certainly there were Bloomfieldian theorists interested explicitly in model evaluation and comparison of the sort Chomsky champions (most notably, Hockett, 1954). However, for our purposes at least, Chomsky's interpretive abilities are not an especially useful issue (and, in any case, members of competing programs rarely see each other's goals and theories in compatible ways; it wouldn't be very surprising if Chomsky and the Bloomfieldians had conflicting interpretations of Bloomfieldian goals; it also wouldn't be surprising if Chomsky's were more revealing, since outsiders often have sharper vision than insiders). What is important, and beyond dispute, is that Chomsky's argument was compelling for a great many linguists, and that, largely as a result of this argument, he brought grammar evaluation to the center of the field.

5. Here, and throughout the book, I have simplified the technical material, sometimes rather drastically, to filter off formalisms or details that aren't germane to the point at hand. I have also introduced some anachronisms here and there (like *Aux*), to simplify current or later discussions. Similarly, the discussion below of the relative merits of phrase structure grammars and transformational grammars (this time to avoid anachronism, along with unnecessary complications) completely ignores the phrase-structure innovations of Gerald Gazdar and the linguists he has influenced (see, especially, Gazdar and others, 1985). I hereby apologize for all the ground teeth this will probably cause linguists in the audience.

6. Chomsky's primary aims, here too, are Bloomfieldian; that is, distributional. His principal motive for the passive transformation is getting the sequences right, accounting for the distribution of the *be* + *-en* complex. Eliminating the "inelegant duplication" (Chomsky, 1957a:43) illustrated in the discussion of 9 and 10 above was secondary. But Chomsky clearly recognized the rhetorical virtues of capturing the systematic relations between actives and passives: that is the only feature of this solution he mentions in the preface to *Syntactic Structures*, and the first feature he introduces in an early paper, widely circulated as an introductory work, "A Transformational Approach in Syntax" (1962a [1958]).

7. Figure 3.2 is simplified to make later discussions a little easier, and therefore paints the grammar in somewhat prettier tones than it actually had; in particular, there is a little more going on in the transformational box than Figure 3.2 indicates. While Newmeyer's Figure 3.1 (1980a:65, 1986a:58) is a little more complicated than it need be, since its purpose is specifically to illustrate some complexities in the early model, it can be used as a sort of countermeasure to the attractive simplicity of my Figure 3.2.

8. Actually, the problem would be even worse, since the generalized transformations would run amok in such a grammar, and never stop splicing sentences together; not only would no simple sentences be produced, none at all would be produced, the transformations iterating endlessly so that no input ever becomes output.

9. Compare the criterion Chomsky suggests for evaluating any level of grammatical analysis in *Syntactic Structures:* "whether or not grammars formulated in terms of these levels enable us to provide a satisfactory analysis of the notion of 'understanding' " (1957a:87).

10. Although calling Harris's work a beginning in structural semantics vastly overestimates his interest in semantics, Chomsky's notions of using syntactic structure to get at meaning are clearly inherited from Harris, who argued that investigating linguistic structure will lead to "interesting distributional relations, relations which tell us something about the occurrence of elements and which correlate with some aspect of meaning" (Harris, 1954:156; see also Chomsky, 1957a:102; 1957b:290; 1964b [1962]:936). They also owe a good deal to Rudolph Carnap, author of *The Logical Syntax of Language,* and an important early influence on Chomsky.

11. This is not revisionism: Hill (1958:3) expresses the wish that the next few decades "will see results of real value in semantics."

12. As Hockett confesses, however, he was in something of a glass house during the rise of Chomskyan linguistics, since "some of us post-Bloomfieldians came close [to adopting an eclipsing stance] in the 1940s" (1987:1). (His subsequent writings, especially in the eighties, show a much fuller appreciation of the history of linguistics generally, Bloomfield's predecessors specifically.) Stark (1972:395n8) makes the parallels more explicit: "Just as early transformational papers invariably began with 'The inadequacies of phrase structure grammar' so most Bloomfieldian textbooks began with a chapter of 'Misconceptions about language', misconceptions that had been foisted upon us by traditional grammar." See, for instance, the introduction to Fries (1952) which denounces "the views and practices of the prescientific era," comparing pre-Bloomfieldian notions to the practice of using leeches in medicine (1–2). The Voegelins (1963), who coin the term *eclipsing stance* (using it for Chomsky as well as some of his predecessors), and Hymes and Fought (1981 [1974]:233ff) discuss the eclipsing stances of Boas, Sapir, Bloomfield, and others.

13. Exactly where Chomsky's mental commitments come from is not clear. He has suggested that they were always with him, saying that they don't show up in his earliest work only because he felt it was just "too audacious and premature" in the Bloomfieldian milieu of the mid-fifties to raise his concerns about mental structure in that book (1975a [1955; preface dated 1973]:35, 1979 [1976]:114). Although Chomsky has never lacked for audacity, it

is telling that his first direct claims about the psychological implications of his work were in a series of lectures at the Engineering Summer Conference in June of 1958—for an audience that wouldn't find his speculations so audacious—but there is some question as to when Chomsky actually began to develop the psychological extensions of his theory. Certainly leaving the anti-mentalist Harris at Pennsylvania and falling in with the deeply mentalist Jakobson and his student, Halle, at Harvard, must have had some influence.

One investigation that would certainly pay off in this regard is a close examination of the 1955 manuscript of *Logical Structure* in terms of psychological interests (a painstaking task, which I hereby apologize for not undertaking). Iain Boal, in an unpublished discussion of "Chomsky and the state of linguistics" which does include some close reading of the 1955 manuscript, makes the interesting charge that, just as Chomsky continued Bloomfield's anti-meaning bias, so he followed Bloomfield's anti-mentalist bias early in his career. This is a "charge" rather than an observation or a claim, since it is in the preface to the 1975 publication of *Logical Structure* that Chomsky first says he thought it too audacious to publish his mentalist concerns (a preface Boal accuses of "writing the mentalist link back into the history of generative grammar"). Boal says not only did Chomsky have none, but the 1975 edition—which purports to be, except for some technical elisions, the 1955 manuscript—selectively snips passages which betray a strong anti-mentalism:

> In the original mimeograph [Chomsky] said that "the introduction of dispositions (or mentalistic terms) is either irrelevant of trivializes the theory", and he ruled out mentalism for "its obscurity and general uselessness in linguistic theory". In the version published in 1975, these passages are expunged, and he says that the "psychological analogue" (viz. that a grammar is a model of the speaker's knowledge) "is not discussed but it lay in the background of my thinking [Chomsky, 1975a (preface dated 1973):35]." (Boal, 1983:9–10)

It is also worth noting that Chomsky claims not to have been concerned about his audience at all in writing *Logical Structure.*

Another passage, from Roger Brown, suggests that Chomsky's interest in child-language acquisition, the linchpin of his psychological notions, dates from the year after the first manuscript of *Logical Structure.* Brown says that he heard Chomsky speak on transformational grammar at a 1956 Yale conference on "Linguistic Meaning," and following the paper there was an exchange which "went something like this":

BROWN: "It sounds to me as if a transformational grammar might be what children learn when they learn their first language."
CHOMSKY: "Oh, do you think so?" (Brown, 1970:17)

It probably makes some difference to Chomsky's biography, especially for the repeated charges that he plays fast and loose in accounts of his own intellectual development, but it is irrelevant for our purposes whether Chomsky started off as anti-mentalist, mentalist, or agnostic. By 1957, at the latest, he was developing very strong psychological ramifications of his theory, and those ramifications have changed the face of linguistics, psychology, and philosophy over the intervening three and a half decades.

14. It would be very difficult to exaggerate the importance of this review. Skinner himself never responded to it, leaving Skinnerians somewhat rudderless and troubled (Weigel, 1977:118). The closest he came was, late in his life, to understate the case dramatically with "My *Verbal Behavior* has been called controversial, and in one sense of the word perhaps it is, but most of the argumentation is due to a misunderstanding." The misunderstanding, he said, is that "the book is not about language . . . *Verbal behavior* is an interpretation of the behavior of the *speaker,* given the contingencies of reinforcement maintained by the com-

munity. . . . Those who want to analyze language as the expression of ideas, the transmission of information, or the communication of meaning naturally employ different concepts . . . but I see no point in *arguing* with those who want to do things in a different way" (1987:11–12; Skinner's italics). Chomsky can certainly be accused of missing some central points in a review which is, in the mildest possible terms, ungenerous to Skinner, but Skinner also misses the point of the criticism. Chomsky is talking about language, true, and Skinner about verbal behavior, but the point of contact is how that language or that behavior is acquired.

The most important aspect of Chomsky's review, though, was that it served as a critical rallying point for the development of a sweeping new movement in psychology. No detailed response appeared at all until the early seventies, a response which laments the review's "enormous influence in psychology" and pleads for live-and-let-live treatment by the cognitivist generation that Chomsky had helped to ignite (MacCorquodale, 1970:98). That response is actually quite effective, exposing many misrepresentations and distortions in Chomsky's case, and it has served to rehabilitate Skinner to some extent in psychology. But it was far too late, and even if it had come ten years earlier, it may well have been swept aside by the fervor of the cognitive movement.

15. Such claims, actually, have also been advanced for Chomsky's work: Otero says that his linguistics could lead "to the advancement of our understanding of ourselves and our society in time to reorient the westernized course of civilization," where such a reorientation is presented as necessary to avoid disaster (1986:192).

16. The moral component of his case, effective though implicit in his review of *Verbal Behavior,* Chomsky makes very explicit and far more personal in "The Case against B. F. Skinner," a review of Skinner's *Beyond Freedom and Dignity,* for a seventies audience (that is, for an audience even more responseive to the moral critique):

> Perhaps, as the classical literature sometimes suggests, there is an intrinsic human inclination toward free creative inquiry and productive work, and humans are not merely dull mechanisms shaped by a history of reinforcement and behaving predictably with no intrinsic needs apart from the need for physiological satiation. Then humans are not fit subjects for manipulation, and we will seek to design a social order accordingly (Chomsky, 1970:23—this, by the way, is the tame part of the review, the part without totalitarian states and concentration camps and gas ovens).

The critique also shows up in Chomsky's political writings, as in "Those who rule by violence tend to be 'behaviorist' in their outlook" (1985:33). Skinner, of course, did not recognize himself in the mirror Chomsky held up, and reacted with extreme distaste, claiming to have read neither of the major reviews very far (Weigel, 1977:115, Andresen, 1990b:162n11).

17. This paper had virtually no immediate impact, but it was rapidly perceived as a threat to Bloomfieldianism. The year after its publication, the paper "was vigorously attacked [at the Second Texas Conference on Problems of Linguistic Analysis in English] and (in the opinions of the participants in the conference), conclusively demolished" (Anderson, 1985:314). Chomsky did not respond to these attacks for several years (1966b [1964]). The phonological approach which grew out of this work is known as *generative phonology,* and while Halle's influence on it was enormous, Chomsky was evidently assuming some such framework from at least the late forties (Chomsky, 1979 [1976]:111).

18. The timing of the response is mildly controversial, one observer claiming that "Lyons suckered Householder into publishing the paper . . . and, without telling him, got Noam and Morris to reply at length." Chomsky and Halle certainly must have seen a prepublication version in order to prepare such a detailed response so quickly, but where they got the copy from is irrelevant. If Lyons didn't supply them with it, professional courtesy would have called on Householder to do so. The speed of Householder's follow-up (though certainly not

the detail; the response was extremely meagre) suggests he saw a prepublication version of the Chomsky-Halle counteroffensive as well. Page numbers in this paragraph are all to Chomsky and Halle (1965).

19. One observer is particularly important to the later story of generative semantics. Householder's very public pillorying came at exactly the time he was supervising George Lakoff's thesis (another prominent Chomskyan-turned-anti-Chomskyan; Lakoff, in fact, is identified in Householder's acknowledgments, for his "helpful comments" on an earlier version of the paper which raised Chomsky and Halle's ire). Householder didn't really fight back, which all Chomskyans and many others saw as evidence that he felt himself completely out of his league (exactly the impression that Skinner's lack of response engendered). Lakoff, however, saw this passivity more as a tactical error, which effectively condemned Householder to the scrap heap. When Chomsky's guns were trained on Lakoff a few years later, at the outbreak of the generative semantics hostilities, "being an impoverished, married, 26-year-old, on a year-to-year postdoc," Lakoff "decided I wouldn't let that happen to me. I fought back."

20. Actually, it would be more accurate here to depart here from my usual conflation of diversity into the term *Bloomfieldian* and go with the one Chomsky uses throughout the essay, *post-Bloomfieldian.* Chomsky has always made a distinction between Bloomfield's practice, endorsing it in *Logical Structure* as reflecting generative principles (1975a [1955]:78n2), later telling me that it shows "he knew what linguistics was in his *bones,*" and Bloomfield's theories, which reveal "an ideological fanaticism" with behaviorist and positivist strictures. "You find that in physics, too," he added. "You find people who call themselves *operationalists,* but of course they never let it bother their actual work." The pioneers he is referring to in this quotation, in fact, include Bloomfield (along with Trubetskoy and Sapir); the bungling misplacers of their work are the American successors who took Bloomfield's methodological injunctions more strictly than he did—especially Bloch, Harris, Hockett, Trager and Smith. For a fuller discussion of Chomsky's ICL performance, which focuses on several elements slighted in my account, see Anderson and others (forthcoming).

21. Chomsky quotes Humboldt directly; the translations here are by Peter Heath (Humboldt, 1988 [1836]:48, 50, 58). I have retained the italics in the text, which both Heath's and Chomsky's edition of Humboldt agree upon.

22. Joos here is paraphrasing what he sees as the Boasian tradition, not taking the tradition to its extreme himself. Someone who does take it about as far as it can go is Whorf—"It may even be in the cards that there is no such thing as 'Language' (with a capital *L*) at all!" (1956 [1941]:239)—but even he has an exclamation mark to indicate he knows he is awfully close to passing beyond the pale linguistically.

23. The actual passage, where Joos is criticizing Trubetskoy for straying from the physical evidence, reads "Children want explanations, and there is a child in each of us; descriptivism makes a virtue of not pampering the child" (1957:96).

24. An even more frustrating manifestation of this in-group attitude was the large number of notes referring to remarks made "in lecture at MIT," or, even, to observations made in "personal conversation." See, for instance, the endnotes to any chapter of Lees (1968 [1960]). This was not, however, an exclusive trait of Chomskyans. Bloomfieldian linguists also circulated underground literature (the proofs for Harris's *Methods,* for instance, made the rounds for several years before publication, and Nida's *Synopsis* of 1943 was available only in thesis form until 1960) and cited personal contact with one linguist or another as the source for some idea. The differences are two: (1) the field was much smaller in the thirties, forties, and fifties, making it much more likely that readers had access to the unpublished material and the opinions of the cited linguists; and (2) there was less factionalization, again making it easier to get the materials and opinions.

Notes for pages 69–78

25. *Effectively,* because, while Chomsky's "influence appears on every page," and while Chomsky "offered detailed suggestions for improvement" (Lees, 1968 [1960]:i), Halle was the official supervisor.

26. As for the conference organizer, he says the purpose "was to discuss [Chomsky's] ideas and accept or reject as seemed best. And it is pretty certainly true that the result was wide-spread acceptance." (Archibald Hill, personal communication to Stephen Murray, cited in Murray, 1980:88n78). Paul Roberts, very influential in spreading the news of transforma-tional grammar to English studies, thanks Hill in his acknowledgments to *English Sentences* and again in his acknowledgments to *English Syntax,* because he "introduced me to trans-formational grammar by inviting me to the Texas Conference on Syntax" (1962:ix; also, 1964:vii).

27. For anyone who shared Chomsky's goals and methodology, reading these exchanges has exactly the effect Newmeyer describes. For many Bloomfieldians, it was something of the reverse. "All one needs to do," Hall says in response to Newmeyer, "is read the actual text [of the exchanges] to see that the exact opposite is true. The other participants tried, in a friendly fashion, to point out to [Chomsky] the severe limitations to which his schemes were subject, but he responded (typically) with a stubborn refusal to entertain any suggestions which did not correspond to his preconceived notions" (1987a [1982]:108). For outsiders, "Nobody wins a victory. Perhaps nobody can" (Francis, 1963:321).

28. The proceedings never reached publication, and the early Chomskyans for the most part believed that the conference organizers had suppressed them. Chomsky evidently tried unsuccessfully to get Hill to release his 1959 paper for Fodor and Katz's important anthology, *The Structure of Language* (1964), which only added fuel to the complaints. There was also some grousing about the four-year delay for the publication of the 1958 proceedings (though the 1956 and 1957 proceedings, at which Chomsky was not present, took even longer—all three came out the same year. See Hill, 1962a [1956]; 1962b [1957]; 1962c [1958]).

Chapter 4

1. For other typical applications or explications of Chomskyan linguistics for English studies, see Hathaway (1962, 1967), Ohmann (1964, 1966), Levin (1963, 1965, 1967), Thorne (1965), Rogovin (1965), Eschliman, Jones, and Burkett (1966), Hayes (1966), Hunt (1966), Thomas (1965), Roberts (1967), Lester (1967), Steinmann (1967), Auerbach and others (1968), Beaver (1968), or anything from the literature on sentence-combining (beginning with Hunt, 1967; see Stugrin, 1979, for a relatively full treatment). Christensen's two articles, "A Generative Rhetoric of the Sentence" (1967 [1963]) and "A Generative Rhetoric of the Paragraph" (1967 [1965]), though popular, are peripheral. He claims that the relevant term in his title is "not derived from generative grammar; I used it before I ever heard of Chom-sky," but Chomsky's association with the term certainly gave Christensen's work a cachet that helped it win some success. For some indication of the success of these efforts, see McCawley's (1976b [1967]:15–34) review of Owen Thomas (1965).

2. The rest of the remark from Lees, after *coat tails,* is "We had to. He is so smart that any idea you came up with he had already thought of, and thought over long and deeply. I wouldn't be very happy if you published that. But it's true, so I guess I'd be happy in another way. We all rode on his coat tails to prominence."

3. See Otero's introduction to his 1986 bibliography of Chomsky-supervised disserta-tions, which quotes some of these acknowledgments and charts his influence on Ph.D. work in linguistics.

4. For his most concerted discussion in this vein, see his interview with Huybregts and van Riemsdijk (Chomsky, 1982a [1979–80]:37–58), but such remarks are increasingly scat-

tered through his writings (for instance, 1988b:3, 19, 53, 72, 73; 1991a [1989]:16; 1991b [1989]:42). The quotations are from (1982a [1979–80]:38, 41).

5. There is an interesting irony here. A substantial portion of the funds for building up the MIT linguistics department came from the U.S. military—either in fairly direct ways, such as funding activity in the Research Laboratory of Electronics, or more indirectly, through various machine translation projects in Cambridge that hired MIT professors and students. Virtually every paper or book coming from MIT faculty or students in the sixties carried acknowledgments like "This work was supported in part by the U.S. Army Signal Corps, the Air Force Office of Scientific Research, and the Office of Naval Research; and in part by the National Science Foundation" (Chomsky, 1964b [1962]:914, see Newmeyer and Emonds, 1971:288–90). That is, the success of the MIT linguistics department (and, in fact, other transformationally oriented departments, like UCLA's), from which Chomsky gained much of his authority, comes from the establishment he excoriates. So much the better, many of us think. The funds might easily have been put to quite evil purposes.

6. Other books with the main title of *Noam Chomsky* include Leber (1975), Koerner and Tajima (1986), and Mogdil and Mogdil (1987a); *On Noam Chomsky* is edited by Harman (1974), *Reflections on Chomsky* by George (1989), *Chomsky's System* by D'Agostino (1986), *Challenging Chomsky* is by Botha (1989), *The Chomsky Update* is by Salkie (1990), *The Chomskyan Turn* by Kasher (1991), and *The Noam Chomsky Lectures* by Brooks and Verdecchia (1992). He also has an impressive presence in subtitles, such as in *Modern Linguistics*, with the subtitle *The Results of Chomsky's Revolution* (Smith and Wilson, 1979), *Transformational Syntax*, with the subtitle, *A Student's Guide to Chomsky's Extended Standard Theory* (Radford, 1981). Allusions are even more common, particularly through such coinages as *generative*, as in *Generative English Handbook* (Eschliman and others, 1966), *Generative Grammar* (Horrocks, 1987), and *The Lopsided Ape: Evolution of the Generative Mind* (Corballis, 1991).

7. Chomsky (1966b [1964]:25–30), George Lakoff (1969a), McCawley (1976b [1968]:183–91), Postal (1964, 1968 [1965], 1966b, 1966c, . . .).

8. Newmeyer (1980a:92; 1986a:80) makes a similar point, also calling upon biblical polysemy.

9. See Givón's similar comments (1979:9). Notice, however, that this observation is a long way from saying that Harris's model is a variant of generative semantics (as some have claimed—Muntz, 1972:270; Plötz, 1972:1–52). Harris, for instance, is completely indifferent to mental implications, and incorporates no explicit semantics.

10. This point is a bit contentious, some of the early innovators suggesting that there was no "program" to develop deep structure, that things just fell into place. But the following excerpts from the period make it clear that, at minimum, there was at least a good deal of wishful thinking in the direction of a super-kernel, even if the name, *deep structure,* was still a twinkle in Chomsky's eye:

- By this process we manage, by and large, to "factor out" the elementary content elements of the text as underlying and very simple kernel sentences. (Chomsky, 1975a [1955]:74)

- Transformational analysis, in particular, permits one partially to reduce the problem of explaining how language is understood to that of explaining how kernel sentences are understood. (Chomsky, 1957b:291)

- It would be a great step forward if it could be shown that all or most of what is "meant" by a sentence is contained in the kernel sentence from which it is derived. (Lees, 1957:394)

- It is possible that the problem of explaining how sentences mean might be reduced to the simpler problem of the meanings of kernel sentences. (Lees, 1962 [1960]:7)

• One striking fact about transformations is that a great many of them (perhaps all) produce sentences that are identical in meaning to the sentence(s) [that is, the kernel(s)] out of which the transform was built. (Katz and Fodor, 1964b [1963]:514–15)

Most tellingly, in retrospect, Chomsky says

• This [the semantic neutrality of transformations] is the basic idea that has motivated the theory of transformational grammar since its inception [Chomsky making it clear in a note that *inception* goes back to Harris]. Its first relatively clear formulation is in Katz and Fodor (1963), and an improved version is given in Katz and Postal (1964), in terms of the modification of syntactic theory proposed there and discussed briefly earlier. The formulation just suggested [the recursive base] sharpens this idea still further. (1965 [1964]:136)

Things were certainly a little more complicated than the simple, linear, kernel-to-deep-structure migration I present here, and my discussion is perhaps a little too streamlined. Deep structure was something of a *ménage à trois* involving the participants: (1) the string underlying kernel sentences, not the kernel itself, (2) the phrase structure of the kernel, and (3) the derivational history (the transformation marker) of the derived sentence. And my use of *kernel* in this section is a little too broad (and anyone who objects can mentally edit in "kernel + underlying phrase marker + transformational marker"). But deep structure became important to the theory precisely as the kernel became irrelevant, and there are very clear causal connections between their respective wax and wane, one semantic pivot displacing the other as a substantive theory of meaning entered Chomskyan linguistics.

11. I am in somewhat dicey expository territory here, since Katz and Fodor, strictly speaking, did not introduce feature notation (the uses of pluses and minuses to describe attributes), and they use the term *projection rule* rather than (the later) *semantic interpretation rule*. But, for continuity with later discussions, I am introducing some anachronism, and the example is correct in substance. Similar distortions are perpetrated on the discussion of Katz and Postal which follows, and the underlying strings have generally been simplified throughout. For instance, tense morphemes and AUX nodes are ignored.

12. Notice that when this sentence is not redundant, it is for a different word, the *bachelor* that identifies the holder of a baccalaureate degree. Similarly, when 2a is not nonsensical different words must be involved (*Logendra* referring to a woman, and *bachelor* to a degree holder)

13. In a somewhat more restricted domain, this argument echoes exactly *Syntactic Structures'* semantic case for transformations. Chomsky says that a beneficial side effect "of the formal study of grammatical structure is that a syntactic framework is brought to light which can support semantic analysis" (1957a:108). This "accidental" expansion of descriptive scope is a very prominent feature of linguistic argumentation.

14. Their treatment of questions is based, in part, on Klima's published and unpublished suggestions (Chomsky, 1965 [1964]:132). Schachter (1964 [1962]:693) also entertains a similar proposal.

15. The *you* in 5c is necessary to account for a range of facts—principally the semantic intuition that the verb in imperative sentences (here, *eat*) has an implied subject, namely the person to whom the command is addressed; and the agreement demands of reflexive pronouns (like *myself, yourself,* and *herself*), which insist that only second person reflexives can occur in simple imperatives (*Wash yourself!* is okay, **Wash herself!* isn't).

16. This answer to the problem is anticipated by Katz and Fodor (1964b [1963]:515n27). See also Chomsky (1965 [1964]:224n9), where he endorses this solution.

17. Even including the Passive transformation. Katz and Postal also proposed (1964:72), and Chomsky endorsed (1965 [1964]:103), a trigger we don't have time to look at, *by + passive*. In fact, Katz and Postal offer a double-barreled counter-argument to the claim that pas-

sive changes the meaning of quantifier examples like the Cormorant-Island sentences. We've already looked down one barrel, namely that both sentences are equally ambiguous. The other is that the two sentences don't come from the same deep structures anyway, since the passive sentence derives from an underlying structure containing a *by* + *passive* (1964:72–73).

18. Since the thrust of the Katz-Postal hypothesis is that transformations don't change meaning, since lexical insertion rules are transformations, and since lexical insertion rules affect meaning: some operational details remain. In particular, subcategorization and selectional rules are termed *strictly local transformations* (Chomsky 1965 [1964]:99–101). They are a special class. Moreover, they apply in the base component as part of the lexicon—prior to deep structure—and cannot therefore alter the semantic reading.

19. Actually, Katz and Fodor's equation was "linguistic description minus grammar equals semantics," where *grammar* meant *phonology and syntax.* I've translated the equation into more specific terms, to avoid confusion over *grammar,* which *includes* semantics in this book, and in linguistics generally from about 1965 (see Chomsky, 1965 [1964]:16ff).

20. Some of the details of these trees (Predicate-Phrase, manner, and by + passive) may seem a little odd with only the foregoing discussion to go by, but the labels are all relatively mnemonic and the function of the nodes should fall into place with a little reflection. If not, see Chomsky (1965 [1964]:128–41) for a technical, but very lucid, discussion of how the *Aspects* grammar works with similar structures.

21. In point of fact, he is discussing the relation of the Prague Circle's *Sprachegebilde* (language-structure) and *Sprechakt* (speech), heavily infuenced by Saussure and, in effect, indistinguishable from his *langue* and *parole.*

22. Lees, for instance, "corrects" his critics in the preface to the third edition of his *Grammar* on the issue of "exactly what *does* a [generative] grammar purport to describe." Many critics, it seems, had mistakenly assumed that his grammar was a model "of a speaker's gross linguistic *behavior,*" when in fact it is only "an account of a certain kind of *knowledge*" (1968 [1964]:xxix–xxxi; Lees' emphasis). But the correction looks a little disingenuous, since his book sometimes promotes exactly the view he denounces in these remarks. For instance, it proposes a separation between optional and obligatory rules, because from this separation "we might expect to gain a deeper insight into how a grammar can be used by a speaker in the production of sentences" (1968 [1960]:3), a separation he is forced, in fact, to renounce in the 1964 preface because "there is no reason to believe that of all the rules in the derivation of a certain sentence some must come before the mind of a speaker who wished to produce that sentence before he 'thinks' of others" (1968 [1964]:xxxvi).

And Postal seems a little hazy on the nature of the model in an early sixties exchange with Paul Garvin:

MR. GARVIN: . . . If you don't have a clear-cut informant response that will tell you whether this is a sentence, then you can never judge your outcome. What do you do with the unclear cases? Do you just pretend they don't exist?

MR. POSTAL: . . . The general theory of linguistic structure should be powerful enough to tell us why there should be unclear cases. One answer is that there is a limitation on memory. It may be that in the course of derivations of unclear cases, many complex rules are involved and the informant has difficulty in tracing the path of derivation. (Dallaire and others, 1962:26)

Chapter 5

1. The following discussion, and, in fact, the book at large, ignores the generative-semantics-like proposals by people outside the immediate transformational community that sprung from MIT's loins. So, for instance, Martin Kay delivered a paper in 1967 sketching a model

which mapped semantic representations, expressed in symbolic logic, onto phrase markers; which was encased in an argument from "intellectual hygiene" akin to Postal's (1972a [1969]) "Best Theory" argument; which included "rhetorical predicates" that paralleled Ross's performatives; and which showed a concern over the competence-performance distinction that surfaced later in generative semantics (Kay, 1970 [1967]). This paper appears to have had no influence on generative semantics whatsoever. It was delivered at a major conference (the Tenth International Congress of Linguists), at which Ross (and, perhaps, a few others in our *dramatis personae*) was present, but its omission from the generative semanticists' references is not especially unusual. Transformational grammarians of the period were very parochial and quite indifferent to proposals outside their immediate framework. Similarly, Petr Sgall also "explicitly propose[d] that the basis of a generative grammar should be constructed out of a set of semantic concepts" at a 1964 conference in Magdeburg, Germany (see Brekle, 1969b:84–85, who mistakenly calls Sgall the earliest to make such a proposal, indicating just how obscure Lakoff, 1976a [1963], was). He, too, made no impression on the group under discussion; very likely, they knew nothing of the proposal at all. Some independent developments in West Germany also bloomed in *generativesemantische* ways in the mid-sixties (Vater, 1971:13; of that group, Brekle, at least, was influenced by Sgall). Several linguists at Warsaw University—most notably, Andrzej Bogusławski and Anna Wierzbicka—began exploring semantic primitives in the mid-sixties, and Wierzbicka visited MIT in the 1966–67 academic year, when she urged the emerging generative semanticists to go more deeply, more quickly, than they were prepared at the time to go. A few of her papers also made some impression on the underground circuit (Wierzbicka, 1976 [1967]; 1972:166–90, 203–20), and her work was generally well received by generative semanticists, but she was never an active member of the school; indeed, she found generative semantics rather tame and semantically halfhearted. Wallace Chafe proposed obliterating the syntax-semantics distinction in generative grammar and dispensing with deep structure, in a 1967 *Language* paper and a review of Katz (1966) the same year (Chafe, 1967a, 1967b; see also Chafe, 1970a, 1970b; Langacker, 1972). He was an outsider at the time, though he became a sympathizer of the general program of generative semantics, while rejecting many of its specific proposals.

Zellig Harris had a quite indirect influence on generative semantics, primarily through his influence on Ross. Specifically, Harris proposed a performative deletion transformation which prefigures Ross's work in that area (see Harris, 1968:79–80, 212), and made some suggestions when Ross was working under him which led Ross to the "squishy" notions most identified with his late role in generative semantics. (See, also, Plötz's 1972 preface, for an unconvincing argument that Harris's model *is* generative semantics.) Uriel Weinreich also had a small but significant influence on the development of generative semantics, by way of his suggestion that the semantic elements of lexical entries are, in some important respect, equivalent to deep structures (see McCawley's review of his 1966 [1964] "Explorations," 1976b [1968]:192–99, especially, 198–99), and by his public entertainment of the "advantages of including semantic features in the base" (1966 [1964]:466). Very unfortunately, Weinreich died prematurely and had no direct involvement in the generative-interpretive semantics debates, on one side or the other.

I also make very little attempt, except in passing, to follow the developments of generative-semantics-like models influenced by the conceptions of Lakoff, Ross, McCawley, and Postal. Liefrink (1973), for instance, developed his interesting *Semantico-Syntax* in the wake of those generative semanticists, but had virtually no interchange with them. Similarly Sgall's later work was influenced by their proposals, particularly through his commerce with German linguists working in generative semantics (see, especially, Sgall, Haličová, and Benešová 1973), and Eugene Nida claims "a dependence upon the generative-semantics approach" for his componential analysis (1975:7–8), especially under the influence of Chafe's work.

2. Most of these quotes and paraphrases are from discussions or letters. Newmeyer's comments, however, are also in print (1980a:93f; 1986a:82f), as are McCawley's (1976a:159), and Ross gives the lion's share of credit to Postal in the acknowledgments to one paper (1974b:122).

3. The only remotely similar paper is an unpublished study from Bever and Rosenbaum, which worked out a blueprint for incorporating semantics into the base (heavily under Klima's influence, as was Ross, and to a lesser extent, Lakoff)—a paper that Chomsky cites in *Aspects'* closest brush with generative semantics (1965 [1964]:159). Bever has even sometimes remarked, with considerable chagrin, that he bears a measure of original sin for generative semantics. But the study came several years after Lakoff's paper, only briefly entertained a structural reorganization such that deep structure and semantic representation were identical, and rejected it summarily. It also (as with Lakoff's mimeographed paper) had very little impact on the Chomskyan community. Chomsky, for some reason, makes direct claims for the patrimony of Bever and Rosenbaum in *Language and Responsibility,* citing "some work by Thomas Bever and Peter Rosenbaum, in which a virtual obliteration of the distinction between syntactic and semantic rules was proposed, an idea that led finally to generative semantics" (1979 [1976]:151). This claim, however, is way too strong, substantially distorting the genesis of the theory. Lakoff, Ross, Bever, and Rosenbaum all discussed early versions of generative semantics and *Aspects*-type semantics in 1963, but Lakoff had already written "Toward Generative Semantics," and all the participants continued to work with interpretive assumptions until Lakoff and Ross broke away formally from the *Aspects* model in 1967 by rejecting deep structure. The relevant Bever and Rosenbaum paper has proven extremely difficult to find; my characterization of it comes from Bever's recollections.

4. The chronology is a little tight here. In the *ICL Proceedings* version, Chomsky's "Logical Basis" paper says that "a generative grammar contains a *syntactic component* and a *phonological component*" (1964b [1962]:915). In the two "Current Issues" versions of that paper, he says "the generative grammar of a language should ideally contain a central *syntactic component* and two *interpretive components,* a *phonological component* and a *semantic component*" (1964c [1963]:51–52, 1964d [1963]:9). The phonological component always embraced a goodly amount of morphology, though real morphology didn't enter the theory until the latter part of the sixties.

5. As is the case with most movement labels, the movers weren't especially pleased with it once it caught fire, and everyone proposed alternatives, ranging from Postal's *Homogeneous I* to Green's *pragmantax* to Seuren's *semantic syntax;* Lakoff alone has had upwards of a half dozen labels for his work (which he still regards as true to the tenets of what was originally generative semantics). Even its opponents objected (Sinha, 1977a:35n7). But here is McCawley, at about the midpoint of the dispute, trying to make the label workable:

> I will describe a set of rules as "generative" if they specify what is possible or impossible at one specific stage of a derivation and "interpretive" if they specify how two stages of a derivation may or must differ. It is in this sense of "generative" that it makes sense to refer to this version of transformational grammar as involving "generative semantics." This distinction between "generative" and "interpretive" presupposes that a grammar is a set of contraints on allowable derivations; it would make no sense in conjunction with, for example, a Turing machine. It is related but not identical to the distinction that Chomsky draws in *Aspects* (pp. 136, 143) between "creative" and "interpretive": Chomsky admits as "deep structures" only those structures generated by his base component that can form part of a complete derivation, and he calls "interpretive" those parts of the grammar that do not affect the class of deep structures. (in Lakoff, 1970a [1965]:vii)

6. *Irregularity in Syntax* is the title under which it was published several years later (with an important preface by McCawley); it was accepted at Indiana under the title, "On the

Nature of Syntactic Irregularity," and that is how it shows up in much of the generative semantics literature.

7. Indeed, even the word *abstract* had always been approbatory in the Chomskyan vocabulary, closely associated with transformational grammar. Consider this passage from *Logical Structure of Linguistic Theory:*

> A grammar is justified by showing that it follows from a given abstract theory of linguistic structure. This abstract theory must provide a practical and mechanical evaluation procedure for grammars. The abstract theory must have the property that for each language, the highest-valued grammar for that language meets the external criteria of adequacy set up for the given language. . . . We are far from having an abstract theory which is not hopelessly *ad hoc,* and that leads to an adequate grammar of even one language. Our goal is to construct an abstract theory that is not *ad hoc.* (1975a [1955]:65; underscoring added)

The word recurs frequently in his early work, and Chomsky regularly employs degree of abstraction as an evaluation metric for comparing theories. In particular, increased abstraction was one of the dominant themes of his address to the Ninth International Congress of Linguists, an important metric separating transformational work from the "far simpler, more 'concrete' " taxonomic model of the Bloomfieldians (1964c [1962]:916–17).

8. Both 6 and 10 have been elided somewhat, to streamline the discussion.

9. If you're raising your hand to point out that 11a and 11b don't mean exactly the same thing, you're right, which two examples raised by Robin Lakoff show even more clearly:

(i)   Dianna doesn't need to go to the bathroom.
(ii) Dianna needn't go to the bathroom.

But you'll have to wait a few chapters to get some justice. The *Aspects*-period notion of meaning, remember, was quite restricted, limited essentially to truth values. It was only the serious exploration of meaning that grew out of the semantics wars that linguists realized that even prototypically equivalent sentences like (iii) and (iv) don't mean exactly the same thing:

(iii) Dianna walked the duck.
(iv)   The duck was walked by Dianna.

10. The phrasing, "the same sequence of underlying P-markers" might appear to imply that the relevant sentences must have exactly the same derivational history, which was not Katz and Postal's intention, nor how the community construed their heuristic. Since the theory still had generalized transformations, this phrase refers to the set of each kernel's deepest underlying representation. With the advent of the term *deep structure,* phrasing the heuristic became much easier: (relevant) paraphrases should be analyzed such that they share the same deep structure. Though his point is somewhat different, Sampson's (1975:160) remarks identify this heuristic as a step onto a slippery slope toward generative semantics:

> If we allow semantic facts, such as synonymy, to count as evidence for syntactic transformations and hence for the deep structures of sentences, we can hardly be surprised if the deep structures we posit turn out to reflect the semantic properties of the sentences!

The only necessary qualification here is that Katz and Postal were not advocating the use of semantic evidence to prove a given transformation, so much as advocating the use of semantic facts as a signpost for syntactic transformations (possibly even as a symptom of them)— semantic facts were neither necessary nor sufficient to posit a transformation—but the distinction between evidence and indication is an easy one to lose sight of.

11. There are some anachronisms and simplifications in this discussion—in particular, I am taking the decompositional analysis further than Lakoff does in his thesis, and using

abstract verbs rather than feature bundles, but this stretching simply goes in the direction these analyses later followed in generative semantics.

12. More specifically, 15b is a transliteration into English, with the abstract morphemes ignored, of an "underlying linguistic structure" in the generative semantics mode, but this aspect of generative semantics (long, unwieldy, underlying structures) is part and parcel of the abstract syntax program. For the somewhat arbitrary purposes of this book, the cutoff between abstract syntax and generative semantics is March 1967, the date of a mimeograph paper by Lakoff and Ross (1976 [1967]) which is something of a declaration of independence from the *Aspects* model. But there are a good many continuities.

13. PM-2 is taken from Abraham and Binnick (1972:41), who cite an underground Ross paper from 1968. Other versions of this tree are in Newmeyer (1980a:95; 1986a:84) and Shenker (1972); Ross (1974b) has some fairly wild trees along with considerably more supporting argumentation than in Newmeyer or Shenker; Gruber (1976 [1965]) also has some many-branched trees.

14. Actually, there was some disagreement about what the three ultimate categories were; in particular, if the primitive nominal category was N (noun), as in PM-2 above, or NP (noun phrase). S (sentence) and V (verb) were okay in everyone's books. I have followed McCawley (1976b:199n*h*), who explored these issues most thoroughly, and gone with NP (see also Dowty, 1979:19). Brief histories of the proposals that brought linguistics to this inventory and its relation to symbolic logic are given in McCawley (1976b [1968]:136–39), Newmeyer (1980a:148–50; 1986a:100–101), and R. Lakoff (1989:946–53). Additionally, not every abstract syntactician was happy with only three base categories; Ross, for instance, retained the VP in much of his work.

There is a curious parallel to Sapir in this work which no one at the time seems to have noticed. In *Language,* Sapir says "It is well to remember that speech consists of a series of propositions," and that propositions have two essential ingredients, nouns and verbs. As much as he rejects a "logical scheme of the parts of speech," Sapir goes on to observe that "no language wholly fails to distinguish noun and verb" (1949a[1921]:119). That is, he finds a common core of language to be propositions, verbs, and nouns.

15. If some of the categories appear unbracketed, it's because they are, for simplicity's sake; a more accurate labeled and bracketed string for 18 would represent all the lexical categories, as in

$$(((\text{the})_{Det}(\text{man})_N)_{NP}((\text{abuse})_V((\text{the})_{det}(\text{duck})_N)_{NP})_{VP})_S$$

16. I hasten to add that it was not just the semantic virtues of logic that led to its enthusiastic adoption. Syntax was still king in transformational grammar, and logic lent a hand in some thorny syntactic problems. In particular, there was a very significant class of sentences discovered by several people in 1967 (Susumu Kuno, William Woods, Emmon Bach, and Stanley Peters, in various degrees of collaboration and independence), illustrated by (i), which had a critical impact on replacement theories of pronouns:

(i) The reporter who chases it will get the scoop she deserves.

This sentence looks pretty ordinary, but what makes it and others like it (called Bach-Peters sentences) so significant is the trouble they cause for the Lees-Klima approach to pronouns that was an exemplar of the early transformational theory. This approach derives (ii) from (iii), (iv) from (v), and so on:

(ii) The reporter is as lovely as *she* is smart.
(iii) the reporter is as lovely as *the reporter* is smart
(iv) The lovely, smart reporter is excited by the story *she* is chasing.
(v) the lovely, smart reporter is excited by the story *the lovely, smart reporter* is chasing

This approach is doomed with (i), since *it* must derive from "the scoop she deserves" and *she* must derive from "the reporter who chases it," leading to an infinite regress. Consider what happens when you work backwards from the surface toward the deep structure. The first step back, replacing each pronoun with its "antecedent" yields (vi) (the brackets indicate the replacement domains):

(vi)   the reporter who chases [the scoop she deserves] will get the scoop [the reporter who chases it] deserves

The second step back yields (vii):

(vii)   the reporter who chases [the scoop [the reporter who chases it] deserves] will get the scoop [the reporter who chases [the scoop she deserves]] deserves

One more time gives you:

(viii)   the reporter who chases [the scoop [the reporter who chases [the scoop she deserves]] deserves] will get the scoop [the reporter who chases [the scoop [the reporter who chases it] deserves]] deserves

And so on. You can never "get back" to a point in the derivation where there is neither a *she* nor an *it*. Associating the deepest structure with logical form, however, McCawley (1976b [1970]:145) was able to offer an elegant solution, illustrated by the following derivation, with (ix) as the deep structure/logical form.

(ix)   (x will get y) (the reporter chases y) (the scoop x deserves)
(x)   [the reporter who chases y] will get [y]
(xi)   [the reporter who chases it] will get [y]
(xii)   [the reporter who chases it] will get [the scoop x deserves]
(xiii)   [the reporter who chases it] will get [the scoop she deserves]

See Karttunen (1971), McCawley (1976b:152nl; 1981:182-85, et passim; 1989:327f, et passim), and references therein, for subsequent developments. The interpretivists, partly under the pressure of the Bach-Peters paradox, abandoned the Lees-Klima replacement solution to pronouns altogether, generating deep structures with pronouns already present, so that their deep structure for (i) would be (xiv):

(xiv)   the reporter who chases it will get the scoop she deserves

Superficially, the interpretive solution may be more attractive (see Jackendoff, 1972:109f, where he calls the generative semantics solution "drastic"). But it still needed rules similar to McCawley's transformation, only "in reverse," to ensure the appropriate semantic representation. Moreover, as Lakoff discovered, investigating these sentences (1976b [1968]:329-33), not all pronouns can participate in Bach-Peters sentences—indicating that a uniform treatment of all pronouns (for instance, all are derived from deep structure noun phrases, the Lees-Klima solution, or all are present in deep structure, the post-*Aspects* interpretive solution) is not necessary.

17. Indeed, since before his career began. Harris, remember, had ensured that Chomsky read widely and studied deeply in modern logic. Harris and Chomsky both stayed away from a direct incorporation of logical syntax and semantics into linguistics, but they borrowed from the mechanisms and terminology for their models rather eagerly. For attitudes about logical syntax, see, especially, Chomsky's (1955b) response to Bar-Hillel in *Language,* occasioned by the latter's (1954) criticism of Harris. For faith in logical mechanisms, see Harris (1951 [1947]:18). Chomsky's position on logic became a little more expansive than Harris's, though. Early on, he speculated that his program might "constitute a bridge between vernacular language in all its variety and complexity and the restricted language of the logician," in a way that he linked closely with one of the key selling points of his grammar, "the problem of understanding how language is understood" (1957b:291). As his grammar became more

explicit semantically, the terms of logic often began to attach to deep syntactic analysis. Even adopting tree formalisms for symbolic logic may have had its roots directly in Chomsky's work: in *Aspects* he says that "it is interesting to note," in connection with the Katz-Postal principle, "that the grammars of the 'artificial languages' of logic and theory of programming are, apparently without exception, simple phrase structure grammars" (1965 [1964]:163).

18. "Concerning the Base Component of a Transformational Grammar" (McCawley, 1976b [1967]:35–58); see Anderson (1976 [1966]:114ff), Lancelot and others (1976 [1968]:258), Bach (1968 [1967]:114), for commentary.

19. Greenberg showed up only sporadically in discussions of the universal base (as in Bach, 1967; Lakoff, 1968b [1966]), but R. Lakoff (1989:950) suggests he had a good deal to do with the generative semanticists' notions about constituent order.

20. See Newmeyer (1990:170; 1991 [1989]:208–14, et passim) for a fuller discussion.

21. The assumptions, the technical machinery, and the complicated interaction of Lees's many rules, are far too detailed to even summarize here, but, very roughly, the two transformations serve to get from an underlying string like (i) to a sentence like (iv), in the following way: rule 21 deletes the first *for* (i ⟹ ii); rule 20 deletes the *her* (ii ⟹ iii); rule 21 kicks back in to delete the second *for* (iii ⟹ iv):

(i)   Alexia pleaded for for her to have another Guinness ⟹
(ii)   Alexia pleaded for her to have another Guinness ⟹
(iii) Alexia pleaded for to have another Guinness ⟹
(iv) Alexia pleaded to have another Guinness

22. Rather incredibly, early transformational grammar derived a sentence like (i) from any number of kernels like (ii)–(v).

(i)   Pop reads.
(ii)   Pop reads books.
(iii) Pop reads newspapers.
(iv) Pop reads skywriting.
(v)   Pop reads soup-can labels.

And so on, with no end in sight. This approach was amended first in favor of an amorphous *something* in the object position, and then in favor of a dummy symbol, both of which were "recoverable," where *books, magazines, skywriting, soup-can labels,* and so on, are not. See Newmeyer (1980a:69–71; 1986a:62–63).

23. Ross called it the *A-over-A-principle* because it prohibited the movement of a constituent dominated by another constituent of the same category, indicated in Chomsky's proposal by the variable *A*. As in the following illustration (from Ross, 1986 [1967]:9), it prohibited moving a category A out of a constituent dominated by another A—transformations under the principle can only move the topmost A (the specific configuration of the subtree isn't especially important, except for the relation of the two As—the Z-branch needn't be there, nor the W-branch; conversely, there could be several other branches under the topmost A, or the lowest A, or Z, and so on).

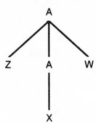

So, for instance, the rule Topicalization moved NPs to the beginning of the sentence to increase their salience. But Topicalization out of a conjunction makes for ungrammatical sentences: (i) is okay, from an underlying structure like (ii), but neither (iii) nor (iv), both of which are equally easy for a transformational grammar to produce, is a legitimate English sentence.

(i)   Dahl and rice, Barrie likes.
(ii) Barrie likes dahl and rice
(iii) *Rice, Barrie likes dahl and.
(iv) *Dahl, Barrie likes and rice.
      (where *dahl* and *rice* are assumed to go together, not to be two separate items that Barrie likes individually, which would come from a different deep structure; "Good jokes, Kenny likes, and playing horseshoes" would come from "Kenny likes good jokes and Kenny likes playing horseshoes.")

In the highly specific approach of Lee's *Grammar*, Topicalization would have to include a condition prohibiting it from applying to members of conjunctions. But with Chomsky's A-over-A principle, which applies to all movement rules, including Topicalization, the right grammaticality predictions follow because both *dahl* and *rice* are NPs dominated by another NP:

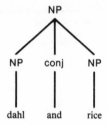

Hence, the topmost NP can be moved (yielding i), but neither of the lower NPs can be moved (preventing *iii and *iv).

24. Actually, the story is a little more complicated, since [+Nominalization] only implies that AGGRESS *can* undergo Nominalization, when, in fact, it *must* undergo Nominalization (or else the grammar generates illegitimate sentences like *Hitler and Stalin aggressed against Poland*), a condition Lakoff calls "an absolute exception" (1970a [1965]:60), and McCawley (1988.1:137) describes as "superobligatory."

25. Among the vast number of notions ignored in this epitome is that Austin also classed the functions of sentences in terms of their locution and their perlocution; the former, very roughly, corresponding to the coding of the content, and the latter to the emotional and epistemic consequences of an utterance. Sadock (1988:184) characterizes the difference thusly:

> The illocutionary act is central to the speech event in something like the way that killing an official is central to an assassination. Performing a locutionary act is more like pulling the trigger, while performing a perlocutionary act is like causing the government to fall.

As this passage indicates, however, speech act linguistics (effectively born out of the work by Ross discussed in this section) is concerned almost exclusively with illocutionary force. Locution is the province of mainstream linguistics (phonology, morphology, syntax, and truth-conditional semantics). Perlocution is the province of several disciplines in the humanities, particularly rhetoric.

26. This point requires a little elaboration. Chomsky was very impressed with ordinary

language philosophy, and it has had at least two obvious consequences for the way he and his do linguistics. First, the switch that his approach had on linguistics with respect to data collection probably comes in part from that school. The Bloomfieldians were in large measure methodologically confined to their corpora. They looked for distributional and co-occurrence patterns in the data they collected from informants. Ordinary language philosophers had no qualms about adduced data, collected off the top of their own heads. That is, as the wont of philosphers, they just mused about their language, mined their intuitions about the appropriateness of saying "I do" in given circumstances. Minus the circumstances, this is also the principal Chomskyan technique: thinking about sentences like *John walked the dog* and *The dog was walked by John*, or *John is easy to please* and *John is eager to please*, and musing about their differences and similarities. (One of the most prominent effects of the musing approach is that ordinary language data and Chomskyan data has been heavily biased toward English.)

Second, Chomsky viewed his theory as embedded in a larger, as-yet-to-be-fully-specified, theory of language which included a Wittgensteinian use-theory of meaning. In *Syntactic Structures,* for instance, he talks about "some more general theory of language that will include a theory of linguistic form and a theory of the use of language as subparts" (1957a:102), and in the preface to *Logical Structure* he comments that "it is assumed . . . that the theory is to be embedded in a broader semiotic theory which will make use of the structure of [a language], as here defined, to determine the meaning and reference of expressions and the conditions on their appropriate use" (1975a [1973]:5). Chomsky, of course, has never pursued this theory of use or broader semiotic very far (his collaborations with Miller marking some notable, if narrow, exceptions); his driving concerns have always been with the theory of linguistic form or the structure of language. Starting with Ross, generative semanticists became very interested in use.

27. The *I* in the quotation is the first person singular pronoun, of course, not the imperative trigger morpheme. Interestingly, Chomsky appears to endorse this solution in *Cartesian Linguistics* (1966a:46, 103n86), though it is difficult to discern in that book what he approves of in the rationalist tradition, and what he is only commenting on.

28. Robin Lakoff (1968:170) and McCawley (1976b [1967]:84) both credit Ross with the proposal, and it was one of the highlights of the elaborate Floyd talks he gave on the early lecture circuit. In addition to Lakoff and McCawley, Boyd and Thorne (1969) and Sadock (1969) also beat Ross to print with underlying performative treatments of illocutionary force. Sadock's work was clearly inspired by Ross, but Boyd and Thorne's proposal appears to have been an independent development. Their notes acknowledge the comments of Ross on earlier versions of their proposal, but don't indicate any influence beyond useful criticism. In any case, Ross does not mention their work in his paper and it had virtually no impact on the performative debates, which is unfortunate because it is in some ways more subtle, cutting finer semantic distinctions. The paper, published in England, appears to have been a casualty of American chauvinism.

29. Ross says only that the underlying abstract verb is "like *say*" (1970b [1968]:238), and that it is associated with certain syntactic features, like [+performative] and [+declarative]. Since nothing of consequence hangs on representing the abstract verb as SAY or DECLARE, as most representations of his argument do, I have opted for the more congruous TELL. Ross (1975:249n19) adopts this convention.

30. Notice then, that, in addition to facing the normal problems of justification, Ross's specific proposal for declarative sentences is also not required by the incorporation of illocutionary force. That is, one could easily have a grammar that codes illocutionary force in terms of covert performative verbs and still not need such a verb for declaratives. Indeed, much of Austin's work, where he enforces a distinction between performatives (which do not

have truth values) and constatives (which do have truth values) suggests that the most appropriate mechanism for coding declaratives (constatives) is the *absence* of performative verb. Moreover, to the extent that transformational grammar had developed a typopology of sentences, it had followed exactly this format (declaratives were those sentences without an underlying Q or I). That is, Ross chose the most difficult aspect of the performative analysis to defend.

31. According to Ross (1975b:71), Predicate-raising began life as Lakoff's transformation, Plugging-in; presumably this is from unpublished work which generalized Inchoative and Causative transformations.

32. Similarly, of course, other lexical insertion rules, like the following, were implicitly part of the package:

33. In some ways, generative semantics dates more accurately from this paper than from Lakoff and Ross's, since it is the first compelling positive proposal in the framework. Although it is a decade after the fact, the following paean from Vroman (1976:38) captures the reception that greeted McCawley's proposal:

> McCawley's pre-lexical analysis fits in with prior work in syntax, is more constrained in that it reduces the number of assumptions (substa[n]tive universals like selectional features, semantic interpretation rules, and the like); i.e., criterion of simplicity, and automatically expands the set of data that can be accounted for: possible meanings of words, adverbial scope ambiguities, referential/non-referential ambiguities, etc., i.e., criterion of generality. His model unifies a theory of syntax with one of semantics and formally says the rules which govern the distribution of forms in sentences are those which govern the distributions of meanings in words.

34. See also another Lees paper from the same period, where he offers evidence that "the deepest syntactic structure of expressions is itself a more or less direct picture of their semantic descriptions!" (1970b [1967]:185).

Chapter 6

1. Bolinger's comments here are spliced from two sources; some come from letters to me, and some from his "First Person, Not Singular" memoir (1991 [1974]:29).

Newmeyer (1980a:93, 1986a:82) says that the Lakoff-Ross seminars were "devoted to challenging analyses then favored by Chomsky," which both Lakoff and Ross deny. There is substance for both interpretations. Some participants recall antagonism toward Chomsky, others recall reverence; the two attitudes are not exclusive and were probably equally well represented. Indeed, even relatively late in the wars, generative semantics papers could cite *Syntactic Structures* or *Aspects* like scripture and dismiss Chomsky's contemporary work rather curtly.

2. Chomsky's leave was for the fall term of 1966. He returned to teach two courses in the spring term of 1967, "Structure of English I" and "Intellectuals and Social Change." These lectures (the "Remarks" lectures) took place in fall 1967.

3. Jackendoff's better-known version is "transformations do not perform derivational morphology" (1972:12–13), which means pretty much the same thing.

4. He voiced the suspicions somewhat vaguely, as we will see, but one Katz-Postal consequence of the lexicalist hypothesis should be immediately apparent, namely that the deep structure (subtree) of phrase (i) would no longer be the same as the deep structure underlying sentence (ii).

(i) Chomsky's refusal of generative semantics.
(ii) Chomsky refused generative semantics.

Since (ii) is a rather crucial component of the meaning of (i), deriving them both from different deep structures swam directly against the spirit of the Katz-Postal principle (and the corresponding heuristic that governed abstract syntax).

5. Chomsky and Halle, for instance, call using analogy "of dubious validity" (1968:356n12; it is difficult to put a "real" date, as opposed to a publishing date on this book. It was under development for close to a decade, first being cited as "to appear" in 1959, and generative works appealed to it frequently in the sixties; Lakoff's 1965 thesis adapts one of its proposals). In *Cartesian Linguistics,* Chomsky used Bloomfield and Hockett's appeal to analogy as a symptom of their misguidedness (1966a:12, 81n21, 82n22). Nor has his position on analogy, except for the blip of "Remarks," changed in any noticeable way: "Analogy seems to be a useless concept, invoked simply as an expression of ignorance as to what the operative processes and principles really are" (1988a:24). Ikeuchi (1972) disputes the generalization expressed in the following example by *2b; see also McCawley (1982b [1973]:116n24, 1988.2:408ff).

6. The reference to *Time* has to do with Lakoff's recollections: he, Chomsky, and Ross had only one meeting upon Chomsky's return, which was interrupted by a call from the magazine which took up most of the scheduled time; subsequent meetings were cancelled. Chomsky getting the call is certainly easy enough to believe. Demands for his time were extremely pressing (see Shenker, 1971, for instance, who discusses these demands on page 105, and four pages later notes "Characteristically, Chomsky had to interrupt our conversation to attend a meeting of dissident professors at M.I.T. His colleague Morris Hale . . . replaced him on the dilapidated chair").

7. Lakoff wrote a (November, 1967) letter to Chomsky on the "Remarks" lectures which he then mimeographed for underground circulation. It is fascinating for a number of reasons. Principally, in this context, the tone is very illuminating—haranguing and superior, but cheerful. Lakoff is not the type to be deferential, at least not when there are ideas at stake, but the difference between his and Chomsky's statures was immense. Lakoff was a junior lecturer. Chomsky, in addition to having founded the framework which underlay the debate, and having become the unquestioned leader in North American linguistics, had gained (and was continuing to gain) considerable fame outside of linguistics—as a proponent of rationalism, as the sharpest critic of behaviorism, and as a tireless, trenchant opponent of the Vietnam War. He was rapidly ascending to the reputation he now widely holds, as one of the foremost and influential thinkers of the twentieth century. Lakoff, in fact, says of his antagonist's position in the late sixties, "Chomsky was a Holy Presence."

But holiness apparently did not present Lakoff with much of an obstacle. His principal criticism in the letter is that Chomsky is dodging all the real issues, and he uses a venerable tool of needling, the rhetorical question, to hector him about facing up to them (though, of course, since it is a quasi-public letter, the function of this insistence is more in the nature of an insult than an invitation):

Is there anything to stop this sort of [absurd] argument? If not, why don't you embrace it? (4)

A personal question: Do you believe this? Do you use [this principle] as a heuristic principle in doing grammar? If not, what exceptions can you think of, or envision? (8)

If you have some linguistically relevant sense of the word 'natural' in mind, please say what you mean. While you're at it, you might explain why your "by Δ" analysis of the passive is 'natural' by this criterion. (12)

In the discussion of causatives, why don't you answer my S-deletion argument (at the end of the causative paper)[?] It's crucial that you do. (13)

The harping tone is important to notice, since it probably has some bearing on the stance that Lakoff adopted in his oral clashes with Chomsky and helps to explain much of the antagonism (just as it is important to notice that Chomsky singles out Lakoff in "Remarks," since it probably reflects his general attitude toward Lakoff's work and helps to explain much of the antagonism). It is also important to note that "Remarks," whatever elaboration its ideas got in class, is a very sketchy paper, unclear and hesitant about substantial, wide-ranging alterations to the *Aspects* model; Lakoff *did* have a lot of questions to ask.

8. Chomsky's use of *determine* in such contexts earned him some rebukes from generative semanticists about his finesse with terminology. See Lakoff (1971b:236–37); McCawley (1982b [1973]:66).

9. See also Chomsky (1975b:239n8), where Jackendoff's role is somewhat muted, though the chronology of his influence is still wrong:

My own version of the standard theory was qualified in that I suggested that some aspects of meaning are determined by surface structure. By the time that [*Aspects*] appeared, I had become convinced that this was true to a significant extent, in part, on the basis of work by Jackendoff.

10. The title under which Chomsky's paper was read is only slightly less neutral, "Credo, 1969."

11. The pagination for the quotations from "Some Empirical Issues" (1972b [1969]) is: "uninteresting" (137), "vacuous" (133), "totally obscure" (148n22), "no substance" (146), "permitting any rule imaginable" (141), "at best dubious rules" (152), "a terminological proposal of an extremely unclear sort" (137), "not only unmotivated but in fact unacceptable" (150), "is probably correct, in essence," (151) "more natural" (187), "somewhat more careful" (188), "well-supported" (165), "to be preferred" (196), "again to be preferred" (197), "more restrictive, hence preferable" (197).

12. As, indeed, Postal found them at the time. Whenever citing key proposals or documents in generative semantics, he always gave McCawley's work priority. For instance, in his "Best Theory," he describes generative semantics as developing from proposals "by Bach, Gruber, and most extensively, McCawley" (1972a [1969]:134); in his *remind* paper, he gives a catalog of publications containing "the significant proposals" of generative semantics: one is by Bach, one by Gruber, two by Lakoff, and nine are by McCawley (1971b [1969]:248).

13. In a letter to Stephen O. Murray, 6 September 1977, quoted by Murray in his forthcoming revision of his 1983 book.

14. The paper has been republished with a brief but illuminating postscript (in Schiller and others, 1988:25–45).

15. Lakoff explains his comments this way:

A book had just been published about Freud and his circle, by Paul Roazen, *Brother Animal* [1986 (1969)]. And somehow Shenker had been talking to people and had heard about us. He or somebody else had gotten the idea that Chomsky was sort of like Freud and we were like Jung and Adler. Háj was like Tausk. You know? He called up to ask about this, and I said, "Well, look, that's pretty stupid. It's very silly and over-simplistic. Let me tell you what's going on."

I talked to him for 45 minutes, and he took several quotes out of context. For one quote he took two halves of sentences from different parts of the interview and spliced them together, so as to be utter nonsense. And then he took this thing about Chomsky fights dirty, which was a description of what had happened in those classes [when Chomsky was attacking generative semantics] and in that article about McCawley [regarding his *respectively* argument]. So, it wasn't just "Chomsky fights dirty in general." It was "given this particular instance, in that case, what we found was Chomsky fights dirty."

I should also say that it was the first time I was ever interviewed by a reporter, and the last time I will ever do anything like that, because I discovered that they take things out of context and piece quotes together that you don't say.

Lakoff's motivation for trying to clarify his comment is understandable, since the quotation has stigmatized him to some degree. But the remark is not at all uncommon. Indeed, its sole claim to uniqueness is that it showed up in *The New York Times*. Published accusations of rhetorical dishonesty against Chomsky (largely in politics, but also in linguistics) are numerous, and off-the-record accusations of dishonesty are even more numerous.

Chapter 7

1. The quotation is from Nuel Pharr Davis (1968:314). The introduction to the quotation, as follows, does not give the name of the person alleging the remarks, but they fit the temper of the times and Alvarez's feelings well (he is on record, along with Teller, Latimer, Pitzer, and Griggs as swearing that he thought Oppenheimer to be a security risk):

> One of the leaders of the atomic establishment says that he was appalled by an intimation he caught in 1954 of the way anger and frustration had affected Alvarez' mind: "I remember a shocking conversation I had with Alvarez. It was before the Hearings [into Oppenheimer's patriotism]. I want to make it clear that I am not giving his words but trying to reconstruct his reasoning. What he seemed to be telling me was 'Oppenheimer and I . . .' ".

2. For an earlier version of this argument form, when Chomsky was test-marketing it against a Bloomfieldian, see Chomsky (1964d [1963]:54n18), where Bolinger is either (1) wrong, or possibly (2) saying the same thing as Chomsky. For a more recent version, see his comments to Gazdar that generalized phrase structure grammar is, loosely, a notational equivalent of transformational grammar, except for its "needless complexity" in places, which makes it inferior (Longuet-Higgins and others (1981:[64f])).

3. At a theory-comparison conference in Milwaukee (Moravcsik and Wirth, 1980), for instance, where a mimeograph paper made the rounds, concerning specious evaluation arguments such as "the claim that 'Your theory is a notational variant of mine, and it's wrong' " (Lawler, 1980:59n14, quoting or paraphrasing K. Whistler and others).

4. The lack of clarity had mostly to do with Chomsky's presentation, which was something of a mess. In particular, he failed to be very specific about what *standard theory* meant. Reconstructing Chomsky's argument (always a very risky enterprise), *standard theory* meant not just the immediate *Aspects* model, but also several of the abstract syntax extensions of that model—especially, Ross's island constraints, Perlmutter's deep and surface structure output conditions, Postal's crossover constraints, and perhaps a few other innovations. With this construal, the standard theory does indeed look fairly close to generative semantics in its empirical predictions. But the notational variants charge still doesn't hold very convincingly. Chomsky did not mean *all* of the abstract syntax extensions—in particular, he seems to have excluded Lakoff's rule features, lexical exceptions, and lexical decomposition work—and it was in the area of lexical decomposition that most people thought the differences between the two models could be best adjudicated. He also apparently included his own lexicalist and x-bar proposals in the standard-theory package, which the generative semanticists repudiated, and which, presumably, Chomsky wouldn't have proposed had he thought they made

no empirical difference to the model (the proposals have to do with what he calls "trading relations" among various grammatical components, and he says that "the proper balance between various components of the grammar is entirely an empirical issue. We have no a priori insight into the 'trading relation' between various parts"—1972b [1967]:13).

Chomsky was still publicly insisting in the mid-seventies (for instance, 1975b:238n2), and continues to insist privately, that the two models were essentially notational variants. But the argument was a nonstarter then (again, excepting Katz), and unconvincing now. Some Chomskyans appear to have bought it. Talmy Givón, for instance, was a generative-semantics sympathizer until the 1969 Texas conference, when

> it finally dawned on me (I'm kinda slow) that the "great debate" involved no empirical issues, let alone theory; only formalisms and egos. So I said "plague on both your houses." . . . [I no longer have] any use for either position.

And, considerably later in the debate, Ronald Langacker (also aligned for a time with generative semantics) complained that "however different the theories may look, in most respects they are essentially equivalent" (1976a:99). But no one, not even Katz, found in the argument any reason to adopt the model Chomsky was using the argument to promote. Indeed, the clearest impact of the notational variants argument, as we will see somewhat later, was among nonlinguists (and, to some extent, among linguists far from the action), with whom it reinforced the idea that all the bad blood was over something very trivial, and the argument therefore damaged Chomsky at least as much as it damaged his opponents—both houses seemed plague ridden.

5. Three other arguments against deep structure predate this one: Lakoff and Ross's unpublished and extremely allusive "Is Deep Structure Necessary?"; Bach (1968 [1967]:119–21), which is only briefly sketched and attracted very little attention (see Janet Dean Fodor, 1980:124ff, for discussion); and McCawley (1976b [1967]:103–17), which is more properly an argument *for* the unity of syntax and semantics than an argument *against* deep structure. None of them have the simple formal structure of McCawley's *respectively* argument. Additionally, Fillmore (1968 [1967]:88) comments that "it is likely that the syntactic deep structure of the type that has been made familiar from the work of Chomsky and his students is going to go the way of the phoneme," but offers no argument.

6. He says that McCawley's argument depends on the derivation of (iv) from (i) in the following sequence:

(i)   ($\forall$x) (x ϵ {Larry, Tom} & x loves x's wife)
(ii)   Larry loves Larry's wife and Tom loves Tom's wife.
(iii) Larry and Tom love Larry's wife and Tom's wife, respectively.
(iv)   Larry and Tom love their respective wives.

Now, Chomsky says, the move from (i) to (ii) is certainly semantic, but (ii) $\Rightarrow$ (iii) $\Rightarrow$ (iv) are syntactic moves: the phenomenon is not unitary, and the semantic action is partitioned away from the syntactic action in a way completely compatible with the *Aspects* model. He also says, incorrectly, that the *Respectively* transformation derives 2b from 3b, and neglects to mention that a representation like 4b is part of the picture, so McCawley's derivation here looks strictly syntactic (and, therefore, irrelevant). Beyond the distortion, he also accuses McCawley of equivocation (on the term, "*Respectively* transformation") frequently enough in the short discussion to cast considerable doubt on either McCawley's integrity (if intentional) or his acumen (if not).

Lakoff, not known for turning the other cheek, took up the public gauntlet for McCawley, and accused Chomsky thusly:

Chomsky describes [this position] as the position McCawley *accepts* rather than the one he rejects. He then proceeds to point out, as did McCawley, that such a position is untenable because of its inadequate handling of plurals. On the basis of this, he claims to have discredited McCawley's position in particular and generative semantics in general. (1971b:275)

Adding, for good measure, an appraisal of Chomsky's case that mirrors identically Chomsky's appraisal of McCawley's case:

Chomsky's claim . . . that McCawley has not proposed anything new in this paper is based on an equivocation in his use of the term 'deep structure' and collapses when the equivocation is removed.

Chomsky replied:

Lakoff. . . claims that the position I tried to reconstruct from McCawley's scattered argument is actually a position that McCawley rejects, rather than one he proposes. Since Lakoff gives no argument at all for this claim (specifically, no reference to McCawley's text) and does not indicate in what my reconstruction, which was based on cited comments from McCawley's text, is inaccurate, I cannot comment on his claim—though it may be correct, for as I noted there explicitly, it is quite difficult to reconstruct McCawley's argument. (1972b [1969]:147n22)

Chomsky did not in fact say explicitly in the original discussion that McCawley's argument is difficult to reconstruct, or even that his presentation is a reconstruction, though there are two textual clues to these ends ("Presumably, then, McCawley intends. . ." and "McCawley seems to have in mind . . ."—1972b [1968]:78); the presentation, actually, suggests that everything comes pretty directly from McCawley's paper. Adding to the confusion is Dougherty (1975:268n4), who somehow finds Lakoff to agree with Chomsky's claim that McCawley's *respectively* argument is an equivocation. Dougherty, by the way, does get closer to the real problems of McCawley's argument—as, in fact, McCawley does, both in correspondence with Chomsky and in a review article of the book in which Chomsky published his reconstruction (1982b [1973]:39–41).

7. See McCawley (1976b [1967]:121–32; 1972:535–38; 1982b [1973]:39–41; 1988.2:536–41); see also Green (1974 [1972]:7, 27–28), who indicates that even if generative semanticists were not willing to acknowledge the full force of McCawley's argument, they felt it at least indicated something was wrong with deep structure. In responding to Chomsky's attack, Lakoff gives a fairly clear presentation of McCawley's argument (1971b:273–77), but ends up only with a lukewarm endorsement. Lakoff also offers some Hallean-type arguments with other constructions (1972b:547–59).

8. Chomsky's remark was in a letter to McCawley (20 December 1967) over deep structure; see also his similar comments in *Logical Structure* (1975a [1955]:93), though they don't concern deep structure.

9. That McCawley actually recognized where the burden of proof lay is clear from his frank comments about a presentation of the *respectively* argument to the LSA:

I had an ulterior motive in writing and presenting 'The respective downfalls [of deep structure and autonomous syntax]' at the Dec. 1967 LSA meeting, namely corruption of the young, specifically stimulating them to do research based on a model which involved generative semantics. . . . My description of this as 'corruption of the young' should not be taken as indicating any disapproval on my part of the model just sketched, which I in fact consider to be correct. It rather indicates that I intended to generate more confidence in that sketch of a model than is warranted by current knowledge, which I consider a desirable goal in that it will not be possible to get a sizeable body of conclusions bearing on whether there is a linguistically significant level between semantics and surface structure [that is, whether there is a deep structure] until a sizeable body of research is done in which the researcher tried to do without such a level. (1976b [1967]:129–30)

Despite the generative semanticists' protests to the contrary, by the way, virtually everyone else apparently also realized that the burden of proof fell on the challengers, not the defenders, of deep structure. Generative semanticists mounted a number of negative arguments against it, but there were only a few positive arguments on its behalf (Chomsky, 1972b [1968]:85–86, [1969]:151–62).

10. The STRIKES-LIKE analysis is due to a famous paper by Postal ("On the Surface Verb 'Remind' "—1971b [1969]—which Ross still calls "the best articulation of what generative semantics is about . . . a beautiful argument." Postal uses Predicate-raising to argue that the most revealing analysis of the word *remind* is not as a three-place simplex verb (x reminds y of z), but a complex of two two-place atomic predicates ([x LIKE z] STRIKES y). See also Lakoff's (1971b:246–52) derivation of *dissuade* from x CAUSE y INTEND NOT, Postal's (1988a [1969]:85–86) derivation of *pork* from MEAT FROM PIGS, and Binnick's (1971) derivation of *bring* from x CAUSE y TO COME.

The most famous support arguments are Morgan's scope-of-ambiguity argument and the possible-word argument (available in various versions from Morgan, McCawley, and Postal).

Morgan (1969b:62–65) argued that (i)-a is several ways ambiguous, corresponding to (i)-b through (i)-d.

(i) a   Ralph almost killed Alice.
     b   Ralph almost caused Alice to become not alive.
         [I.e., he almost did something that would have killed her.]
     c   Ralph caused Alice to almost become not alive.
         [I.e., he did something which almost caused her to die such as narrowly missing her while driving his bus.]
     d   Ralph caused Alice to become almost not alive.
         [I.e., he did something which caused her to almost die such as hitting her with his bus and injuring her near fatally.]

The possible-word argument is a little more subtle, resting on how Predicate-raising combines with constraints on movement rules of the sort Ross had pioneered in his dissertation. To illustrate, Ross had argued for a movement prohibition he called the *coordinate structure constraint,* which says that movement transformations can't extract material from conjoined phrases. Consider (ii)-a and (ii)-b, which transformational grammar related with a rule variously called *Topicalization* and *Y-movement:*

(ii) a   I like horseshoes.
      b   Horseshoes, I like.

But there are no corresponding pairs of sentences like those in (iii):

(iii) a   I like Horseshoes and hand grenades.
       b   *Handgrenades, I like horseshoes and.

The phenomenon is extremely general, and very well established, as are several other movement constraints. Combining the research on movement constraints with the proposal that lexical items are the result of movement transformations (in particular, Predicate-raising), allowed the generative semanticists to make substantive claims about the content of the lexicon. In particular, they could claim that certain types of words were impossible in any language (see, e.g., Morgan, 1969b:52–53, McCawley, 1976b [1971]:327–29). The possible-word argument proceeds very simply: the hypothesis that words arise from movement rules predicts there can be no words which violate movement constraints; for instance, that there can be no word, *splarm,* such that (iv)-a means (iv)-b:

(iv)   a   I splarm handgrenades.
        b   I like horseshoes and handgrenades.

That is, there can be no word which means something like "to like horseshoes and."

After the dispute was pretty much over, a former generative semanticist who had begun working in a new framework called *Montague Grammar* raised formidable difficulties for the possible-word argument (Dowty, 1979:237–38).

11. For his counter-arguments to the scope-of-ambiguity argument, see 1972b [1969]:150, for his counter-arguments to the possible-word argument, 1972b [1969]:143.

12. This was something of a red herring, since McCawley also found evidence that, within his assumptions, some of the more established transformations sometimes must precede lexical insertion, but these arguments had less of an impact than the one involving Predicate-raising, since all of the other transformations were under steady revision in this period and several of them were subsequently discarded by the interpretivists.

13. Of course, transformational theory had already changed markedly as a result of the issues surrounding meaning preservation (which didn't hold of transformations in *Syntactic Structures* but did in *Aspects*). The difference was that now transformational linguists had forked into two mutually hostile channels.

14. The Katz-Postal principle, that is, was only a working hypothesis in the *Aspects* milieu; one which Chomsky expressed mild reservations about, and which S.-Y. Kuroda and Paul Chapin, both of whom were Chomsky's doctoral students, had explicitly rejected (the former because of quantifier scope issues, the latter because of nominalizations—Kuroda, (1969[1965]; Chapin, 1967); and which Fillmore was prepared to relax in his move toward case grammar (1966, 1968 [1967], 1969 [1966]). It was generally regarded as a welcome addition to transformational grammar, but it was a long way from dogma. It only became dogma with the advent of generative semantics.

15. McCawley also had an elegant way around this violation in terms of logical form (1976b [1967]:108–9).

16. *Shallow structure* was Postal's term for the level after the application of all cyclic transformations and before any post-cyclic transformations.

17. Darwin, though not given to Chomsky's level of agonism, had the similar view that "truth can only be known by rising victorious from every attack" (letter to Sedgwick, 26 November 1859), and, though not given to Chomsky's level of acerbity, also had rather harsh things to say about some of his critics, using in his attacks misrepresentation (as in his characterizations of Matthew and Blyth—see Clark, 1986 [1984]:142f) and ridicule (as in his comments to Lyell, 5 December 1859, about FitzRoy and the dismissal of Buckland, which his family expurgated from his autobiography).

18. The fleshing-out comes especially in two important books, *Semantic Interpretation in Generative Grammar* (1972), and *X-bar Syntax* (1977), and an important paper, "Morphological and Semantic Regularities in the Lexicon" (1975).

19. The one possible contender in Chomsky's oeuvre as an articulation of the extended standard theory is his *Studies on Semantics in Generative Grammar* (1972b), but it just collect his anti–generative semantics trilogy, "Remarks on Nomalization," "Deep Structure, Surface Structure, and Semantic Interpretation," and "Some Empirical Issues in the Theory of Transformational Grammar."

20. His first major work, for instance, adopts $\bar{x}$-notation for some phrase structure rules, but seems embarrassed to extend the conventions to other rules (e.g., 1972:60ff). In partic-

ular, he uses x̄-syntax only for noun phrases, as had Chomsky—a move that must have reinforced the impression that the notation was an *ad hoc* convenience Chomsky had drafted to
get himself out of the explanatory pickle he was in for having dropped a transformational
account of derived nominals. It would take us too far afield here to discuss either these proposals or Jackendoff's treatment of them, which diverged significantly from Chomsky's original suggestions, but both lexical redundancy rules and x̄-notation mark important recon-
ceptualizations of grammatical theory, helping to define formal linguistics since the mid-
seventies, and Jackendoff is a major reason for their success.

21. In some construals, Jackendoff had one semantic representation with four parts,
which amounts to the same thing. See Jackendoff (1972:3) for the clearest account of his
model's architecture.

22. The ugliness also corresponded to the ambition, however: Jackendoff (1972), in addition to applying much finer grained semantic criteria than Katz and Fodor (1964b [1963]),
Katz and Postal (1964), or Chomsky (1965 [1964]), also accounts for many more types of
semantic phenomena, like focus, presupposition, and modal structure.

23. Jackendoff, too, has fared quite well in the retrospections of generative semanticists.
When feelings ran high, he was considered little more than a loud-mouthed parrot, repeating
Chomsky's bile-infused contra-generativism, but he is now almost always singled out as an
honorable exception among the scabrous horde of interpretive semanticists. McCawley
(1982b:8), for instance, warmly acknowledges Jackendoff's help and insights. Indeed, despite
the several very fierce confrontations between Jackendoff and the generative semanticists (or
perhaps because of them), they tend to regard him as something of a comrade, and there are
certain similarities of form. As Kac puts it, "the most conspicuous failing of [*Semantic Interpretation*] is that it has too much in common with what it purports to oppose" (1975:23), and
the resemblance was not just formal: drugs, politics, and general goofiness (the stylistic proclivities of the generative camp) play a role in his sample sentences; he is the breeziest stylist
of the intepretive semanticists by far (and matched in lucidity only by McCawley); he is frank
about gaps in his analyses; and he is almost as relentless as the generative semanticists about
citing his data sources. Nor is he above the practice of in-jokes, often co-opting the favorite
proper nouns of generativist papers (Max, Irving, Seymour), and even managing to find a
legitimate way to cite Quang's best grammaticality juxtaposition:

(i)   Drown the fucking cat.
(ii) *Drown that cat which is fucking. [wrong meaning]

More importantly, he shared most of their driving concerns about meaning. This call, for
instance, could have come from any generative semanticist: "[We need] a much more serious
study of semantics—and not semantics reduced to syntactic terms—than has been fashion-
able" (Jackendoff, 1971:142).

24. I have snipped out Lakoff's citations to the relevant texts (the quotation is from Lakoff,
1970b:627).

25. I have reversed the order of the two sections of this quotation (separated by the elision
dots). Chomsky's general points here follow on a discussion of a global proposal by Ross,
but he obviously regards the vagueness problem to hold for Lakoff as well (e.g., 1972b
[1969]:141); Postal gets marginally better treatment (p. 140).

26. The term "wild cards" in connection with Chomsky's shell-and-pea data game comes
from Lakoff's (1967) response to "Remarks on Nominalization." Postal's comments here
have been edited slightly. A longer passage containing these remarks is quoted in my dissertation (Harris, 1990:381–82).

27. Chomsky didn't pull the virtues of complexity out of his hat at this point in the debate.
It had been an important undercurrent in his argumentation for some time. Transforma-

tional grammar, for instance, was an advance over taxonomic grammar because it is "far more complex and highly structured" (1964b [1962]:917), and Skinnerian behaviorism was bankrupt because it failed to handle the "extremely complex mechanism for generating a set of sentences" in the child's head (1959:57). Indeed, complexity was connected with a buzz-word of early theory construction that Postal had taken more seriously than anyone else, *abstractness*. But complexity took a decided backseat to simplicity. Now, confronted with a conceptually much simpler theory, Chomsky stitched complexity to restrictiveness, which now became the primary virtue of any linguistic theory.

Oddly enough, though, when Chomsky says in the Goals conference paper that the merit of complexity is "a point that has been noted repeatedly" (1972b [1969]:126), he gives only one citation, a section of *Aspects* which discusses the notion rather obliquely, calling only for "abstract statements and generalizations . . . enriching the theory and imposing more struc-ture on the schema for grammatical description" (1965 [1964]:46)—raising yet another frus-tration many people find with Chomsky's arguments. His references are usually poorly cho-sen, even to his own work. He is not, so far as I can tell, guilty of fabrication in such references (as some have charged); when he claims to have said something, he has said it, but the ref-erences he offers as support for these claims do not always advance his case as clearly as they might, and often there are much better passages he could reference.

28. This discussion is somewhat misleading. The Peters-Ritchie results concern what is known as *weak generative capacity,* the ability of a rule system to specify strings of symbols. Chomsky argues in *Aspects* (1965 [1964]:60–62) and elsewhere that this is not the goal of linguistic theory, which should concern itself with *strong generative capacity*—the assign-ment of structural descriptions to strings of symbols (for instance, assigning phrase structure trees to English sentences). Additionally, he argues in *Aspects* that "the real problem [of grammatical theory] is almost always to restrict the range of possible hypotheses by adding additional structure to the notion 'generative grammar' " (1965 [1964]:35), where he is also talking of (among other criteria) strong generative capacity. That is, restrictiveness for Chom-sky does not dovetail with Peters and Ritchie's work in any direct way. Nonetheless, the results and Chomsky's call for restrictiveness (which was far more urgent in "Some Empirical Issues" than in *Aspects,* or anywhere else) were almost always mentioned in the same breath. Newmeyer (1980a:176), for instance, says

> The Peters-Ritchie findings served as silent witness to almost all of the significant work in syn-tax in the 1970s [where *significant* means *significant in the interpretivist framework*]. There was hardly a paper written that did not appeal to the increased restrictiveness of the theory that followed as a consequence of the adoption of the proposals in its pages. Constraint after con-straint was put forward to limit the power of the grammar.

Newmeyer tempers these remarks somewhat in the second edition (cf. 1986a:189), written when he was a little further from the action, but still ascribes a great deal of influence to the proofs.

29. There is more potential for confusion here. In particular, it is not necessarily the case that a theory which includes constraints is a more constrained theory by virtue of that inclu-sion. If a theory allows any conceivable form of constraint (as the interpretivists charged of the generativists), it is not thereby a more restrictive theory. In fact, it might be much more powerful. The easiest way to see this is in terms of Perlmutter's work. He had argued that no conceivable sequence of transformational rules could handle pronominal clitics in Spanish, and therefore a surface-structure constraint had to be brought in. By adding his constraint, he expanded the descriptive power of the model (see his epilogue, 1971:123–34, where he worries about just this issue). Nonetheless, the necessity for constraints is almost always dis-cussed in the literature of the period in terms of restricting or constraining the descriptive

powers of the grammar. And, most importantly, the early filters-and-constraints work of the generative semanticists was very much of a piece with the later work of the interpretivists, which was portrayed relentlessly as opposed to generative semantics.

30. See also McCawley's remarks: "When Lakoff proposed that you need global rules, that did not carry with it a proposal that every imaginable global rule is a possible rule. You can raise the question of what global rules are needed and set about restricting the class" (in Parret, 1974 [1972]:268).

31. Lakoff's transderivational constraint paper, much like his global rules paper, is largely a collection of data that suggests the need for such constraints, and it acknowledges work by Grinder, Postal, and Perlmutter as pointing the way to transderivational constraints (1973a [1970]:442). That is, the concept is only partially his. Brame (1976:69), however, goes too far when he says that transderivational constraints are due only to Perlmutter and Hankamer, citing Hankamer's 1973 discussion of his transderivational "No-Ambiguity Condition," which seems to have been an independent development. Hankamer does not cite Lakoff, and his research into the condition dates to 1971 (nor does Lakoff cite Hankamer).

32. See Bach (1977:135ff), Postal and Pullum (1978), and Lightfoot (1980:155ff) for some discussion on the globality of the trace convention.

33. Trace theory can quite safely be called Chomsky's; it bears his indelible stamp. But it has immediate roots in work by Postal (1970), by Baker and Brame (1972), and by Selkirk (1972). The *want-to/wanna* data was first noticed by Lakoff, in the paper that launched global rules, and used as one of the justifications for such rules (1970b). See also Baker and Brame (1972), Lightfoot (1980), Chomsky and Lasnik (1977; 1978), Postal and Pullum (1978), Pullum and Postal (1979), McCawley (1982b [1973]:126n87) for some ups and downs and sidesteps in the *wanna* debate.

34. McCawley (1982b [1973]:29), however, argues that Chomsky had no compunction about appealing to transderivational constraints as needed—in particular, that his "rules of analogy" ploy in "Remarks" was an appeal to a transderivational rule—but that he avoided the term like poison. He makes a similar point with respect to the use of global rules in Chomsky and Halle (1968); see McCawley (1974c:73).

35. Actually, the term in Chomsky is usually *grammaticalness* (e.g., 1965 [1964]:3, for *acceptability,* see, e.g., 1965 [1964]:11).

36. See especially Sadock, 1969, 1970, 1971, 1972, 1974a, 1985a; Davison, 1970, 1972, 1973; Gordon and Lakoff, 1988 [1971]; Robin Lakoff, 1972a, 1972b, 1973a, 1973b, 1973c, 1977; George Lakoff, 1972a, 1972b, 1972c, 1977b.

37. The word itself, as a foil to *semantics,* is from Charles Morris, who used it to lump together "all the psychological, biological, and sociological phenomena which occur in the functioning of signs" (1938:108), and, as the scope of the definition indicates, philosophers generally used it to refer to something they *weren't* going to deal with either, like Bloomfield's *mentalism,* Chomsky's *performance,* and, in slightly different degrees, to what both Bloomfield and Chomsky meant by *meaning.* So, pragmatics remained ill-developed for a long time, *qua* pragmatics.

The use theory of meaning which developed out of Wittgenstein into the speech-act philosophy of Austen and Searle, and the conversational logic of Grice, was seen in the seventies to embrace largely the sorts of issues that Morris had in mind. The word accordingly came to designate something that linguists (some of them at least) *were* going to study. Generative semanticists, however, had something of a constitutional fear of boundaries, and refused to set up a principled dividing line between semantics and pragmatics and, of course, syntax. One of the words (this one due to Georgia Green) for what they were studying in the seventies was *pragmantax.* See Kempson (1975), who treats pragmatics as a performance phenomenon, Gazdar (1979), who treats it as competence phenomenon, and Green (1989), who just treats it, for some worthy accounts of how pragmatics fits into linguistics.

38. To remind, Postal's principle was meant to account for facts like:

(i)   It is tough for Jeff to shave himself.
(ii) *Jeff is tough for himself to shave.

39. See Jackendoff (1972:120ff) for a quite detailed argument supporting this syntactic-to-semantic reanalysis.

40. For those who appreciate historical ironies, the following quotation is offered, since it comes from Michael Brame, one of Chomsky's staunchest supporters in this period, and one of Lakoff's nastiest enemies, who later found Chomsky as exasperating as Lakoff had: "as time runs out on trace theory, one sees ever more far-fetched devices proposed to accommodate counterexamples that genuinely follow from more realistic approaches" (1979:13). There was also a sense in which generative semanticists were open to a similar charge, at least according to Talmy Givón. He was one of the few people to indict the generative semanticists for data shifting, though with an implication of capriciousness rather than of dishonesty: he snorts about a theory "where adjectives were proclaimed to be verbs one day (Ross and Lakoff, 1967) and nouns the next (Ross, 1969[a])" (1979:14).

41. Surveying the arguments, Chomsky says they establish their case "quite persuasively, in my opinion" and says that "structuralism and Lakoff's semantics . . . are in fact rather similar" (Chomsky, 1979 [1976]:154). See also Brame (1976:26n1).

42. See McCawley (1975), Kuiper (1975), Dougherty (1975) for further discussion of the issues.

43. Chomsky's earliest comments on feedback are fairly strong: "there is neither empirical evidence nor any known argument to support any SPECIFIC claim about the relative importance of feedback from the environment" (1959 [1957]:44). But, as the emphasis suggests, he didn't rule out a general role, and Chomsky (1962b [1960]:530) explicitly grants it some such function: "Other data [than degenerate input] that the child has available to him may play an essential part in language learning. Thus he may have available a set of nonsentences (that is, corrections by the speech community). He may need information about repetition of utterance tokens." The arrows Chomsky adds to the diagram of his acquisition device, though, are quite different from the one McCawley adds. Chomsky suggests a refinement simply of the notion "primary linguistic data," not of the model itself in any serious way:

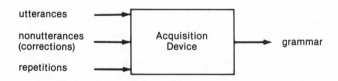

See Chomsky (1988a:60) for a recent indication that even this sort of refinement doesn't interest him very much: it is the character of the acquisition device, not of the input, that is his principal concern.

44. Ross and McCawley are mentioned only in passing by Katz and Bever; Postal is cited only once, as co-author of a two-page squib with Ross; and the only other generative semanticist to make an appearance (in connection with a sociolinguistic article) is Robin Lakoff. Katz and Bever do, however, make one interesting concession that they are narrowing all of

generative semantics to the work of one individual, with their title for one section, "A Case Study: Generative Semantics, Lakoff Style" (1976 [1974]:30), but they hold him to be representative, citing sympathetic work by Ross and McCawley (p46), and generally discussing Lakoff's work as if it was universally endorsed by generative semanticists.

45. Of course, such a position fed the interpretivist argument for generative semantics as an empiricist backslide. Witness Dougherty (1975:154) surpassing Ross's absurdity here: "Generative semantics has been developed internal to Harris's transformational taxonomic system."

46. See also Parret (1974 [1972]:250). McCawley here uses *generative grammar* in the strict *Syntactic Structures* sense of describing sets of sentences, hence the quotation marks. He now holds a more expansive notion of generative grammar, which describes sets of "something other than sentences"; namely, "complexes of sentence, meaning, context, and style/register" (1988.1:6). Lakoff also said "generative semanticists are not doing generative grammar" at about the same time (Parret, 1974 [1972]:152), though his rejection was far more sweeping; he does not, for instance, feel called to use quotation marks, and uses *generative semantics* and *generative grammar* as virtual antonyms throughout the interview.

47. For similar discussions, see Lyons (1970c:137–38); Lehmann (1972:221–22); Elgin (1973:134–35); Wardaugh (1977:171–77); Hayes, Orenstein, and Gage (1977:102–5), Simpson (1979:237–43). In several of these discussions, case grammar is side by side with generative semantics.

48. There are, as in chapter 5 above, some interpretations under which Harris's grammar is a fundamentally semantic one, especially in light of statements like: "Almost everything that there is to say about the meaning of a sentence can therefore be obtained directly from the meaning and the positions of the components ψ, K. Hence, given this theory . . . there is little need for an additional semantic theory" (Harris, 1968:211; ψ designates *base transformations,* K designates *kernels*). But Muntz, a collaborator with Harris at the University of Pennsylvania on the Transformation and Discourse Analysis Project there, and Plötz (1972:1–52), in a related argument, are clearly Harris supporters attempting to hitch his work to the generative semantics bandwagon, which was in full gallop at the time.

49. For examples of generative semanticists at psychology conferences, see Ross, 1974b; McCawley, 1974b. For examples at philosophy conferences, see Lakoff, 1972b, 1973d, 1975; McCawley, 1972, 1973b.

Chapter 8

1. In large part, Robin Lakoff ascribes the generativist style and its impact to the temper of the times, "when experimentation with lifestyle and personality was encouraged" (1989:977). See also Darnell's Introduction in the 1992 reprint of Zwicky and others (1971).

2. It is important, however, to make several important caveats. First, there were many linguists who bought into generative semantics—who, for instance, published dyed-in-the-wool generative semantics papers like "On the Alleged Boundary between Syntax and Semantics" (Newmeyer, 1970), and "On the Syntax and Semantics of the Atomic Predicate CAUSE" (Dowty, 1972)—but whose work shows only faint traces of this style. Second, there are linguists who employed this style only in very restricted subsets of their writing. Postal, for instance, tended to write in a very formal, almost Chomskyan style, but frequently juxtaposed offbeat sample sentences to this prose, and his publications carried a much more serious tone than his *samizdat* papers. Third, some interpretive semanticists, like Jackendoff and Akmajian, also used elements of this style—considerably more elements, for instance, than Newmeyer or Dowty. And, most importantly, a great deal of work which showcases this style is to be found in the various proceedings of the Chicago Linguistic Society; that is, one

of its principal aims is to be effective when delivered orally. Indeed, this chapter might easily have been entitled "The CLS Ethos."

But the identification of generative semantics with the CLS was an extremely close one. The society was the organizational center of generative semantics, publishing at least half the movement's papers in its proceedings, serving as a swap meet for the latest ideas, and throwing the best parties. The CLS exuberance has abated somewhat in more recent, more staid times, but not much. The dating scheme of the preface to *The Best of CLS* (Schiller and others, 1988), for instance, is given in terms of "the Nixon years" and "back when Johnson was president." A Quang Phuc Dong paper is included "to capture the flavor of CLS" because it deals "with idioms which were ignored by more straightlaced linguists." And the preface ends with "The Supreme Court recently re-affirmed our right to speak freely. At CLS, you can always do so!"

3. For readers to whom this allusion is obscure: there was an elaborate and bizarre rumor, characteristic of the paranoia and mysticism of the period, that Paul McCartney was dead, that there had been a massive cover-up of this by evil-minded money-mongerers, but that the genuine remaining Beatles were trying to get the truth to the faithful in the form of cryptic lyrics, symbolic album covers, and, strangest of all, by encoding messages that could only be discerned by playing certain sections of their records backwards.

4. Ved Mehta, for instance, was concerned enough about Chomsky's expository dullness to ask him how it was that someone with such a penetrating understanding of the technical features of language could write so poorly (1971:192), and thought the trait so definitive that he entitled the article containing the interview, "John Is Easy to Please." When he later collected the article and some of his other *New Yorker* pieces into a book, he retained his pleasure with the title (the book is *John Is Easy to Please*).

5. To keep this paragraph from getting too bogged down by parenthetical references, I've saved them all for here: *Flip,* Ross recalls, was coined by Postal in his early sixties classes at MIT, showing up in print in Lakoff's thesis; *Slifting* (Ross, 1973a); *Sluicing* (Ross, 1970a [1967]:252); *Stuffing* (Ross, 1972a:162); *Irving* (Morgan, ms., cited in Horn, 1970:325); *Ludwig* (Neubauer, 1970:403); *Richard* (Rogers, 1971); *Apparel Pronoun Deletion* (Grinder, 1970:300); *Euphemistic Genital Deletion* (cited in Borkin, 1984 [1974]:105–6).

Stockwell (1977:131n2) criticizes the generativists for choosing such rule names, "which are neither mnemonic nor transparent in their meaning" but, curiously, of the three names he indicts—*Pied-Piping, Tough-movement,* and *Sluicing*—only the last is especially troublesome. *Pied-Piping* is very appropriate and mnemonic, since it describes the situation where one word can follow after another in movement analyses, and *Tough-movement* names a very common word subject to the processes it labels. See Robin Lakoff's partial response to Stockwell (1989:975n11), and Postal's backhanded advice to would-be Chomskyans: "To appear serious, of course, one should avoid . . . names like 'Rumplestiltskin' or 'Debby Does Dallas'. Select something like 'The Contraction Determination Condition' or 'Recoverability'" (1988b:131).

6. The role of forum, and consequently of audience, is an underappreciated factor in the generativist ethos. Newmeyer's largely negative discussion of the generative semantics style, for instance, lists three titles, three rule names, three sample sentences, and three prose extracts as evidence for his remarks on "the whimsical style of presentation that pervaded so much written in [the generative semantics] framework" (1980a:171; 1986a:136); all but four examples come from CLS publications. This ratio is perhaps not unrepresentative; putting aside the difficulty of clearly identifying all the generative semantics papers, the majority likely appeared in CLS volumes, though probably not as many as two-thirds. Newmeyer accurately attributes the style reflected in his examples to "youthful enthusiasm (the average age of generative semanticists in 1970 was well below 30) and [to] the rambunctious person-

alities of several prominent generative semanticists" (1986a:137), but he fails to comment on the contribution of venue.

The CLS gatherings were primarily oral events, and effective oral techniques do not always work equally well on paper. Ross and Lakoff had set the standard with the energetic Floyd presentations that had recruited so many generativists, raising wit and flash to the top of the rhetorical currency market. But it is noteworthy that the Floyd shows—though epitomes and data-sheets from them were in circulation—never made it to press. More importantly, the CLS assemblies were largely gatherings of the faithful. In Wayne Booth's terms, its presentations were principally directed at a community of the blessed, and the style reinforced the group ethos: "You could always tell a GS paper: by its title, its breezy style, its funny examples. You knew who belonged, who your people were. It was cozy comfort in a heartless world" (R. Lakoff, 1989:977).

7. Even the generally forgiving editors of CLS, among whom Lakoff is a favorite, felt compelled to add an editorial comment on this one, quoting Snoopy's comment: "Actually, after you ace someone, you really shouldn't say 'Nyahh, Nyahh, Nyahh!'" (Lakoff, 1973b:290). This was not the only example where Lakoff's aggressiveness was not especially well concealed. His sample sentences also took occasional potshots at the other side, such as juxtaposing *lexicalists* and *whores* in parallel sentences (1971c:333), and offering up this gem of agonism, "Chomsky is the DeGaulle of linguistics" (1973d:235).

8. He has subsequently published books in which the blending of his formal linguistics and his politics is considerably more overt (e.g., 1979 [1976]; 1986; 1988a), and ironically, been taken to task for it by some former generativists.

9. There is a curious parallel to Lakoff's remarks here in some later comments by Chomsky, at a 1984 colloquium—though, of course, Chomsky's conclusions are very different: "Maybe someday experiments will be useful, but right now if you sit and think for a few minutes, you're just flooded with relevant data that you can't explain" (Chomsky, 1984:44).

10. This preoccupation with problematic data marks another point of disaffection that many generative semanticists have with Newmeyer's treatment of the movement. He terms the preoccupation "data fetishism," and castigates generative semanticists for their "nihilistic outlook" (1980a:168; 1986a:133). Lakoff calls the phenomenon "data love" and "honesty," and, completing the connubial metaphor, celebrates generative semanticists for their "fidelity to the facts." See McCawley (1980a:917–19) for an argument to this effect—where, using the inverse metaphor, he compares Newmeyer's attitude to "the traditional Christian attitude towards sex: the pleasure of gathering data is proper only within the confines of holy theory construction and when not carried to excess; recreational data-gathering is an abomination"—and Newmeyer (1980b:932–34) for his response. Robin Lakoff (1989) continues the discussion.

11. All quotations in this paragraph are from the preface to Borkin and others (1968). The first quotation is a paraphrase of Burt's position, presumably by Borkin. The others are quotations directly from Burt.

12. The IULC was a mimeograph distribution house of great importance, which circulated many groundbreaking theses and working papers, on both sides of the debate, and continues to serve a similar purpose.

13. Into this cheerful dissensual stew of rhetoric, Lakoff throws a clear countercultural appeal, some humor, and the unmistakable theme that the generative semantics program is the road to salvation, in part because it embraces the sort of unflinching skepticism the old school can't abide, and in part because it simply makes more sense than shallower brands of linguistics ("We know much more about the meaning of a sentence than we know about its surface structure"). More interestingly, all of it, from the snide preface to the urgent foreword, is informed by the attitude that true science involves embracing the artificiality of models and

the muddiness of data. The preface demands the right "to be confronted with the realities of science." The foreword calls for linguistic students to be responsible, like "any student of the physical sciences," and recognize fudges for what they are.

14. Nobody acknowledges even a kernel of truth to Newmeyer's can't-you-see-we-were-joking? characterization, and Ross in fact addressed it as early as his thesis, commenting that although one of his technical proposals (the Pied-Piping convention) got its "terminology from the realm of fairy tales," this use should not be interpreted as "a disclaimer on my part of psychological reality" of the convention (1986 [1967]:126n3). Newmeyer has had the chance to recant, or at least bury, this observation. The generative semantics–linguistic war sections of his book were rather severely pruned for the second edition (1986a), after he was well aware of the vitriolic reaction to his, and his informant's, accusation of comic hedging, but the passage survived the cuts (1986a:137). (One of the few areas in the discussion of generative semantics that was actually expanded, however, was the treatment of style—indicating that he was sensitive to complaints about his handling of the topic). See also these parallel comments—suggesting something of an "urban myth" status to the just-joking position— embedded in Raimo Anttila's rant against all things generative: "a proponent of generative phonology says that McCawley intended his generative phonology as a joke to see how soon his benighted colleagues would see through it. A similar opinion has been expressed by another linguist concerning Lightner's work" (Anttila, 1975:174). The sentence following this remark reveals one of the motives underlying Anttila's angst: "As it is, the jobs certainly went to the jokesters" (and, in a letter quoted by Hagège, he gives a quantitative basis to his plaint, saying "a non-TG person has to publish about three times more and still not get the 'normal' raises"—1981 [1976]:167).

Chapter 9

1. Newmeyer (1979), however, draws exactly the opposite conclusion about the book. He apparently admires it a great deal, but finds the framework it articulates seriously wanting.

2. See Bever (1988:123–24) and Wanner (1988:149) for wistful recollections of success. See McMahan (1963), Mehler (1963), Miller and McKean (1964), Savin and Perchonock (1965), and Slobin (1966) for some of the successes of the derivational theory of complexity. See Fodor and Garrett (1966), Bever (1970 [1968]), Glucksberg and Danks (1969), Baker, Prideaux, and Derwing (1973), and Fodor, Bever, and Garrett (1974) for some of the failures. See Fodor and Garrett (1966) and Prideaux (1985:100–23) for general discussions of the derivational theory of complexity.

3. For instance, MacKay and Bever (1967) investigated structural ambiguity, which the *Aspects* model accounts for by assigning one surface structure two deep structures, each representing a different meaning, and found that subjects were relatively slow recognizing structural ambiguity with respect to lexical ambiguity (as in *Superman was high,* where he could be either at a great altitude or in an altered state of consciousness, depending on the meaning of *high*). The deep structure hypothesis suggests these results, since listeners would have to reconstitute two distinct deep structures for the first type of ambiguity, and only have to make a lexical decision for the second. However, sentence length, surface clausal complexity, and location of ambiguity were all possible contributors to these results as well, and indeed, when they were controlled, Prideaux and Baker (1976) failed to replicate MacKay and Bever's findings. See Wanner (1974, 1988:147–49) on testing the deep structure hypothesis.

4. In a retrospective essay on the impact of linguistics on psychology, Bever comments

The wars in linguistics highlighted another problem: Linguistic theory changes like quicksilver. Psychologists think they have their hands on it, but it slips through. . . . It takes a month to develop a new syntactic analysis of a phenomenon; it takes a year to develop an experimental investigation of it. All too often, the psychologist is in the position of completing an arduous series of empirical studies, only to discover that the linguistic theory underlying them is no longer operative. (Bever, 1988:130)

See Chomsky (1988b:18–19) for similar comments, and for comments on the derivational theory of complexity. There is, however, evidence the disgruntlement dates from at least the time of abstract syntax. Fully sympathetic psychologists like Roger Brown and Camille Hanlon, for instance, lamented the lack of consensus, in these terms: "Unluckily for us there is considerable disagreement even among transformational publishing around 1965" (Brown and Hanlon, 1970 [1968]:170).

5. The novel is a good deal of fun for linguists, providing a deer park of allusions to chase down, some of which seem to be (but one can't be entirely sure) to generative semantics. For instance, a former colleague of Sandground's, one Liedlich, motivated primarily by anti-Sandground sentiments, is behind Oh's training: Oh and Liedlich commune in "special semantikaramas"; Liedlich has written *Brother Creatures,* a title recalling Roazen's *Brother Animal,* a depiction of Freud's treatment of heresies in which many saw strong Chomskyan parallels.

There were, incidentally, a good many ape-experimenters gunning for Chomsky's species-specific hypothesis in this period, the most famous being the Gardners' attempts with Washoe (Gardner and Gardner, 1969; 1975), Premacks' with Sarah (Premack and Premack, 1972), Terrace's with Nim (Terrace, 1979), and Patterson's with Koko (Patterson, 1978; Patterson and Linden, 1981). None of the apes were quite in Oh's league (that is, none of them came close to compromising Chomsky's resilient hypothesis) and they came under immediate criticism from linguists (Bronowski and Bellugi, 1970; McNeill, 1970; Sebeok, 1982), but some remarkable progress was gained all the same; optimistic estimates put the apes' linguistic abilities in the range of a three- or four-year-old child (e.g., Brown, 1970:224). See Chomsky's comments to Huybregts and van Riemsdijk (Chomsky, 1982a [1979–80]) or to Rieber (1983:58–90) for succinct accounts of his views on this research, Putnam (1983) for some counter-argument and Chomsky (1983) for additional remarks. For a popular account of this material, including recent work with the amazing Kanzi, see Ingram (1992:218–37).

6. See Newmeyer (1986b:120–26), and the references therein.

7. There are many discussions of Chomsky's Cartesianism and the issues surrounding it. For arguments parallel to Aarsleff's (by far the majority), see Zimmer (1968), Breckle (1969a), Salmon (1969), Hall (1969), or R. Lakoff (1969b); for more favorable appraisals (which still usually include some finger-wagging at Chomsky's cavalier scholarship), see Prideaux (1967), Kampf (1967), Harman (1968), Bracken (1970; 1972; 1984). The most thorough and dispassionate discussion is probably Kretzmann (1975).

8. I have elided Derwing's examples of these three argument types in Chomsky, all of which he documents in the book: for the fully specious, Derwing cites Chomsky's attack on the discovery procedure; for the mainly irrelevant, his and Halle's campaign against the phoneme; for the out-and-out false, Derwing quotes Chomsky and Halle's (1968:49; Derwing's italics) "It is a *widely confirmed empirical* fact that underlying representations are fairly resistant to historical change."

9. For some of the history of this model—which includes Ross, Morgan, and Keenan in its early stages, Lakoff somewhat later and more peripherally—see Johnson and Postal (1980:15–19). Ross's only real publication dealing with relational grammar is in his "Three Batons for Cognitive Psychology," where he argued that rule application was controlled by a hierarchy of grammatical relations (1974b:106–10); Keenan's only direct participation in the relational grammar program was his investigation of universals concerning passive sentences (1975a; see also Keenan and Comrie, 1977 [1972]). Morgan never published in the framework at all.

10. In particular, see Ross, 1972c, 1973b, 1973c, 1974a, and 1975. Ross's work developed from "offhand remarks of Zellig Harris" (Ross, 1973c:231) that had been puzzling him in various ways since his classes with Harris ten years earlier, and was the result of frustration that generative semanticists had built up as they tried to get the on-or-off notions of grammaticality in transformational grammar to work with increasingly complex data sets. A major force on this endeavor (and on the work of Lakoff we will see a few paragraphs hence) was the research of Dwight Bolinger, a long-standing critic of discreteness in linguistics.

Chomsky had been long aware of the difficulties of forcing the flux of language into strict binary categories. *Logical Structures,* for instance, contains a goodly amount of fretting about the various degrees of grammaticality and acceptability before settling on an argument that the most profitable focus of linguistics should be on "sentences of the highest degree (first order) of grammaticalness" (1975a [1955]:154), and Chomsky revisited the topic on several occasions, most notably in a paper entitled "Degrees of Grammaticalness" (1964a [1961]). In transformational semantics, Katz (1964) had discussed at length phenomena he called *semi-sentences,* ungrammatical sentences, like *I have overconfidence in you,* which are nonetheless perfectly intelligible.

But the generative semanticists took the notion to heart, and began to use graded grammaticality as linguistic evidence. McCawley argues in his foreword to Lakoff (1970a [1965]), that (1) the Postal-Lakoff notion of relative grammaticality is distinct from the Chomsky-Katz notion, and (2) that Lakoff's use of this notion in his thesis was a contribution to grammatical argumentation of "great importance." Lakoff, for instance, used differences in grammaticality to motivate his abstract verb proposals (1970a [1965]:62), arguing that (i)-a is less grammatical than (i)-b (and, hence, double-starred), in precisely the same way that (ii)-a is less grammatical than (ii)-b:

(i)   a **The lawnmower critiqued the book.
      b *The lawnmower's critique of the book.
(ii) a **The tuba aggressed against India.
      b *The tuba's aggression against India.

Lakoff's double asterisk was just the tip of the iceberg. Generative semanticists developed a whole array of signs, called *stigmata* by McCawley, to indicate that sentences weren't up to grammatical snuff (where *grammatical* is much more widely construed than in Chomsky's usage), including the question mark (?), for sentences that might be grammatical for some people, or some circumstances, but not for others; the exclamation mark, singly or in pairs, as the "double shriek" (!!), for screamingly bad sentences; the percentage sign (%), to signal dialectal variation; even the star of David (✡), courtesy of McCawley, to signal English sentences which were grammatical if spoken with Yiddish intonation.

Ross, though, was the most systematic in his use of grammatical gradation markers. In his thesis, he adopted a scheme of combining question marks and asterisks to signal several levels of grammaticality, and he was the transformational linguist most resistant to binary notions of grammaticality, basing quite forceful theoretical arguments on very subtle shades of grammatical goodness. In the early seventies, however, with his squishiness work Ross began to explore these dimensions of grammaticality virtually for their own sake, offering very little in the way of an account for them.

11. In fact, Ross made several finer distinctions, including the continuum Present > Perfective > Passive to indicate the squish among participles, and introducing the term *adjectival noun* between *Preposition* and *Noun.* He also placed a parenthetical question mark after *Preposition,* to suggest some uncertainty about its place on the continuum. None of these details are relevant to the present discussion.

12. Though it was just developing when Lakoff began importing its concepts, Rosch's

work is now quite extensively published; see, for example, the papers and references in Rosch and Lloyd (1978 [1976]).

13. Zadeh (1965) is his initial paper on fuzzy set theory and logic. See also Zadeh (1987), and the references therein, for subsequent developments in his work.

14. "I don't want to give the impression that I take the proposals in section IV to be correct in all or even most details. Hedges have barely begun to be studied and I have discussed only a handful. I have no doubt that the apparatus needed to handle the rest of them will have to be far more sophisticated. In fact, it is easy to show that far more sophisticated apparatus will be needed merely to handle the hedges discussed so far" (Lakoff, 1973d:246).

15. The first appearance of this term, an invocation of the idea that logic should be an empirical pursuit, appears to be in Lakoff (1971b:277), but tendencies in this direction date back at least to McCawley (1976b [1967]:106). Natural logic, then, was not solely Lakoff's project, though he was on this front, as on most fronts, the move vociferous. At one conference, for instance, McCawley made some modest suggestions to stimulate logicians "to study the logical properties of items [they] generally ignore (e.g., is it valid to argue 'goddamn all imperialist butchers; Nixon is an imperialist butcher; therefore, goddamn Nixon'?)" (1976b [1972]:319). At the same conference, Lakoff delivered a long, forceful sermon (1972c) to the effect that logicians had painted themselves into a corner by allowing their formalisms to shrink the subject matter of logic until it contained only the narrowest subset of facts about human reason and its vehicle, natural language; generative semantics, he suggested, would be their salvation.

16. Newmeyer's first reaction was noncommittal, avoiding VSO deep structures mostly for expository reasons (1970:179n2), but later explicitly rejected (1971). See also Postal (1972e; a reply to Newmeyer), Berman (1974; a reply to McCawley), and Anderson and Chung (1977; which includes a reply to Berman) for some discussion. McCawley no longer sees the need for any word order at levels much below the surface (e.g., 1988.1:42), and regards his original VSO arguments as "very weak" (1982b:7).

17. As should become clear shortly, this quotation is slightly misleading. That is, "On Generative Semantics" is not a flat-out parody of Chomsky, but it does contain a good many parodic elements, particularly in the first few pages.

18. Another, more recent, example of Chomsky catching it on the chin for failing to get a joke is his reaction to Pullum's satirical "Formal Linguistics Meets the Boojum" (1989; 1991:48–55), which berates Chomsky for his abandonment of formal linguistics and ends with a doomsday scenario for "the few formal linguists who survive, slightly crazed as a result of isolation and inbreeding," with them "taking to the hills in places like Montana and northern Idaho. . . . Perhaps sometimes a lonely old madman with stringy gray hair and wild eyes will be found seizing people by the arm at an LSA meeting and haranguing them about precise definitions of formal underpinnings, until he is taken away by hotel security" (1989:43; 1991:55). In short, "it gets crazy at the end, as the reader of TOPIC . . . COMMENT [Pullum's column at the time in *Natural Language and Linguistic Theory*] should expect. What the reader should *not* expect is that anyone should take the piece as stone-cold serious and write a serious response to it," Pullum said a few years later. "But unfortunately, this has actually happened with the appearance of Noam Chomsky's 'On formalization and formal linguistics' [1990]" (Pullum, 1991:47).

19. Lakoff sets up a "basic theory" in parallel to Chomsky's "standard theory," and defines it so broadly as to encompass virtually any conceivable theory of language, thereby making every theory a notational variant of the basic theory. Much of the style is recognizably mock-Chomskyan, with rather tortuous syntax, numerous uppercase variables, subscripts, and superscripts, and the tendency to traffic for a long while in abstractions before offering specific proposals or analyses.

20. Compare the quoted formula with the one Lakoff offers in his interview with Parret: "The abstract objects generated [in generative semantics] are not sentences but quadruples of the form (S, LS, C, CM) where S is a sentence, LS is a logical structure associated with S by a derivation, C is a finite set of logical structures (characterizing the conceptual context of the utterance), and CM is a sequence of logical structures, representing the conveyed meanings of the sentences in the infinite class of possible situations in which the logical structures of C are true." Lakoff also suggests, in parallel with the ellipsis of the 1971 formulation, that "even this is inadequate," since "one must take into account much more than conceptual contexts" (Parret, 1974 [1972]:163). Such "place holders" were not unique to Lakoff, however. On a smaller scale, for instance, Jackendoff (1972:40–41) includes the variable *Y* in his dictionary entries for *buy* and *sell,* defining it as representing "simply any semantic residue that has not yet been expressed" in the rest of the entries.

21. Compare Chomsky's "I will refer to any elaboration of this theory of grammar as a 'standard theory', merely for convenience of discussion and with no intention of implying that it has some unique conceptual or empirical status" (1972b [1968]:66). There are a number of commentaries on Chomsky's deft use of labels. One that is particularly germane to this discussion is Postal's (1988b:136) discussion of "the unquestionably wrong and stupid Basic Semantics (BS) movement." Postal's use of *BS* is obvious enough to ensure the parody is not missed; Lakoff's *basic theory* is just too subtle.

22. Maclay says Lakoff

mistakenly assumes that Chomsky wishes to differentiate these positions [the semantic positions of the standard theory and *Aspects*]. He presents a quotation from Chomsky's paper which seems to indicate that the form of semantic representations in the standard theory is different from such representations in *Aspects*. In fact, this quotation is Chomsky's description of an alternative to the standard theory which he rejects. (1971:177n)

The quotation from Chomsky (1972b [1968]:71–72) is a brief discussion of reformulating semantic representation as a phrase marker which can embrace lexical decomposition, and otherwise function as generative semanticists claim semantic representations function; Chomsky does not explicitly reject this formulation of semantic representation, but implies that it is not a "genuine alternative" to the standard theory. The discussion in Lakoff (1971c:268–70) is an attempt to show that Chomsky has defined the standard theory in "Deep Structure" expansively enough to be a notational variant of (some aspects of) generative semantics, but that the reformulation of semantic representation that Chomsky effects is not possible for an *Aspects* theory—hence, that the standard theory and *Aspects* are not one and the same.

23. Cited in Hagège (1981 [1976]:33n29); translated by Robert Hall; my interpolation.

24. This was a widespread but not universal trend. Dowty (1979:18), for instance, takes several representative early papers as the central defining documents. He certainly doesn't ignore Lakoff, but he pays very little attention to his mid- and late-seventies work, and focuses on the clear early claims.

25. Postal's influence extends to the titles as well. His Anarchy Notes use titles like "Horrors of Identity," "Temporal Monstrosities," and "Coordinate Mind Snappers" (1976 [1967–70]:203–4).

26. Montague grammar started in the late sixties with the work of logician-philosopher, Richard Montague, and began attracting the attention of linguists after his death, particularly as reworked by (the Chomskyan) Barbara Partee (see especially Partee, 1976). Among the disaffected generative semanticists who signed up to investigate Montague's model were Karttunen, Dowty, and Bach. Relational grammar began with Postal and Perlmutter's work in the early seventies and gained a great deal of momentum after a very well attended Lin-

guistic Institute they team-taught in Amherst, 1974 (see especially Perlmutter, 1980; Perlmutter and Rosen, 1984; Blake, 1990). Among its ex–generative semantics customers were
Dryer, Cole, and Frantz. The generative semanticists who joined Chomsky's program seem
to be limited to Newmeyer, Baker, and Lightfoot.

27. *Cognitive grammar* comes from some papers he developed with Henry Thompson in
the mid-seventies (Lakoff and Thompson, 1975a; 1975b); *experiential linguistics,* from Lakoff (1977a). Newmeyer also notes Lakoff's fondness for new labels—"By coming out almost
yearly with a newly named theory . . . Lakoff has not presented himself to the linguistic world
as a consistent theoretician" (1980a:172)—though two of the labels he cites (*global transderivational well-formedness grammar* and *dual-hierarchy grammar*) are from underground
sources. The matter of nominal fondness is an interesting one, which has much to do with
presentation, since the other major linguist with the same proclivity, Chomsky, *does* come
off as a consistent theoretician. Lakoff presents new developments as breakthroughs which
dramatically change the way we now have to look at language; Chomsky presents them as
inevitable stages in his march on truth.

28. If this discussion resembles the account of dummy symbols in the context of *Aspects*
in chapter 4, it is for good reason. The dummy symbol, Δ, was the first of a good many phonologically null pronouns in Chomskyan linguistics; among the more prominent devices to
follow in its noiseless footsteps are the trace *(t)* and PRO.

29. Chomsky repudiated the term *deep structure,* but in an extremely halfhearted way. In
his 1975 Whidden Lectures at McMaster University, he pledged to avoid the term, because,
among other reasons, people were unfortunately thinking there was something necessarily
profound about it (1975b:81–83); more importantly for his research at the time, which was
beefing up the other end of the *Aspects* derivational schema with traces and dedicated semantic interpretation rules, people were thinking that there was something necessarily trivial or
superficial about surface structure. He had a very brief fling with *initial phrase marker,* but
he was soon back with his old friend, kept provocatively behind a very thin veil—*D-Structure*
(1980b [1978]:145)—the use of which is really quite perverse, as almost everyone realizes.
Even Chomsky's students, for instance, adopt such terms as "D(eep)-Structure" (May,
1985:3), and Chomsky himself almost always draws attention to the term *deep structure;* the
index entry in *Lectures on Government and Binding,* for instance, is "Deep-structure (D-
structure)" (1981a [1979]), *Concepts and Consequences* introduces it as "*D-structures* ('deep
structures')" (1982b:5), and the first mention in *Knowledge of Language* includes the footnote "Called 'deep structures' in earlier work" (1986:205n8).

30. The double-edged adjective may have come from the translator, Robert Hall, whose
contempt for the entire Chomskyan program is well known and vastly surpasses that of Hagège (see, for instance, Hall, 1987a); my point, however, is unaffected by the origin of the
insult. Notice, too, that *serious* is the word O'Donnell uses in similar circumstances
(1974:75).

31. This elevation of frankness was another counterculture trait. See, for instance, Rubin's
(1971:124–26) discussion of the hypocrisy of the Chicago police testifying in court, who
insisted they could not repeat some of the defendants' words in front of women jury members, but who boasted to one another in the back room about how they busted the heads of
"the fucking little fagots"; yippies, on the other hand, don't hide anything.

32. Attributing arguments and data to other linguists has a long history in transformational grammar, and was certainly present in interpretivist papers of the period. My point
here is not that it was exclusive to generativists, but that they took the practice to new levels,
and that, in part, the extremism followed from their attitudes toward Chomsky's attributive
practices.

33. Chomsky's contrast with generative semantics here is even more direct than these quotations indicate, since he specifically singles out generative semanticist issues (like the "use of . . . cognitive structures"—138), as falling in the mystery camp.

Chapter 10

1. Notice that the general answer here partially nullifies the question, revoking its chief presupposition; that is, the dispute was not especially acrimonious as scientific clashes go. Hurling obscenities at one another in very public forums may seem atrocious behavior, but within its context—in particular, within the sixties—it is no different than a geological debate among Victorian gentlemen turning "topsy turvy without scruple" (Edward Turner, in Rudwick, 1985:99–100), or Cuvier's post-revolutionary eulogy of Lamarck, so vitriolic it wasn't even published until after *Cuvier's* death, or Huxley publicly ridiculing a bishop, or _____ (fill in the blank). Science is often a very agonistic process, and the vituperation follows the style of the times.

2. This penchant for not delivering on cited publications was so prominent that it became something of a leitmotif in generative semantic self-mockery. One of Ross's acknowledgments, for instance, goes "to George Lakoff, who and I ([improbably] forthcoming) are said to be writing a book about abstract syntax" (1974b:123). Postal (1972a [1969]:168) cites "a no doubt never-to-be-written paper, Lakoff, Postal, and Ross (forthcoming)." But such commentary was not always reflexive to the author: McCawley changed many of his citations when he collected his early essays into *Grammar and Meaning* (1976b) from the format "Lakoff (to appear a)" to "Lakoff (abortion a)," and the references for a pseudonymous Zwicky paper include the multibarbed entries:

Coughlake, Gorge. (To appear a) Natural logic and unnatural linguistics.
Coughlake, Gorge. (To appear b) On the factual inadequacy of all theoretical positions in linguistics, this one included.
Coughlake, Gorge. (To appear N_0) The straight truth about quantifiers, and other heavy facts about high predicates. (Zwicky, 1970:149)

3. This is not to say, however, that no promises were addressed and none fulfilled. McCawley's work since the early seventies is largely an attempt to satisfy promises made in the late sixties. For instance, his abstract analyses of the seventies largely addressed the simplicity promises of early generative semantics; *Everything That Linguists Have Always Wanted to Know about Logic* (1981) is the most detailed exercise in natural logic, by far; and *The Syntactic Phenomena of English* (1988) is the most comprehensive application of generative semantics principles to linguistic description. Levi's *The Syntax and Semantics of Complex Nominals* (1978) satisfies much of the promise of a transformational analysis of nominalization that generativists kept alleging was possible in response to Chomsky's "Remarks." Lakoff's *Women, Fire, and Dangerous Things* (1987) addresses many of his late generative semantics promises. But, where these projects were not too little, they were certainly too late.

4. It was fairly common in the seventies to call Chomsky the "Einstein of linguistics" (Leber, 1975), and Dougherty had just published "Einstein and Chomsky on Scientific Methodology" (1976b), so Brame's epigraph was clearly meant to cut both ways.

5. The declaration of Lakoff as Public Enemy Number One has its roots in Chomsky's "Remarks," where he is set up as the chief proponent of the transformationalist position, and laying the sins of generative semantics at his feet continues as a prominent theme through all of Chomsky's anti–generative semantics papers; by "Some Empirical Issues" Lakoff is a com-

plete whipping boy, and "Lakoff's revised generative semantics" (1972b [1969]:140)—that is, Homogeneous II—is the nadir of linguistics, the worst possible theory. Even after he had pretty much stopped addressing generative semantics arguments, Chomsky could still take a couple of pages out to attack Lakoff (1977 [1975]:52–54), and his responses to Lakoff's direct criticism are rife with personal insults. I don't mean to suggest (or, for that matter, preclude) that Chomsky stirred the troops into a Lakoff-bashing frenzy, but, at minimum, there was a gathering-momentum-on-the-downslope effect to his animosity, of the sort Karl Vogt remarked on in the midst of the social revolutions of mid-nineteenth-century Germany: "The brutality which is present in higher circles filters down, and this brutality which above lives only in thoughts, below takes the form of action" (Robertson, 1952:413).

6. Ross's thesis is not in the Greenbergian leagues in its cross-linguistic virtues, but in early-Chomskyan terms it is remarkable—while concentrating on English, it considers data from a variety of languages, including Danish, Finnish, French, German, Japanese, Latin, Russian, and Serbo-Croatian—and Ross is very clear that his work concerns universal principles, not English-specific properties.

7. An interesting contrast here is Perlmutter, who, like Ross was in the MIT linguistics department, and who also had a great deal of trouble influencing research there. He left fairly early, and has had a powerful impact on the field; Ross stayed, and his influence on the field dwindled to almost nothing (though, of course, his earlier work continued to be quite instrumental in the development of Chomsky's model).

8. See Lawler's comments on the Lakoff-Thompson version of cognitive grammar, which he called "dead" and "stillborn" and "only a steppingstone in a history of theoretical research which proceeds through 'Linguistic Gestalts' (Lakoff, 1977[a]) [to unpublished work of Lakoff's on "experiential linguistics" with Mark Johnson]" (Lawler, 1980:54, 54, 51). Lawler also says, invoking the metaphor of a "dead metalanguage" in a particularly poor but reasonable prophecy that "it does not look like the Miracle of Modern Hebrew will be repeated for CG [cognitive grammar]" (1980:54).

9. Although Kirsner (1991:171) talks about "the Lakoff wing of Cognitive Grammar," there really isn't one. Langacker has a definite, articulated framework, and Lakoff is no more eager to embed his wide-ranging concerns in a definite framework than he was in the seventies. He is a sympathizer and fellow traveler, and an important idea man, but not a co-articulator, of cognitive grammar. As the Kirsner quotation (and many other quotations throughout this book) indicates, linguists are fond of using uppercase letters for models (transformational grammar is TG in the vernacular, relational grammar is RG, generative semantics is GS), and they talk about "doing RG" or "doing GS," and one cognitive grammarian puts Lakoff's work into perspective this way: Lakoff is, unquestionably, a cognitive grammarian, and a prominent, exciting one at that, "but he doesn't do big-cee, big-gee, Cognitive Grammar." See Langacker's (1988a) review of *Women, Fire, and Dangerous Things*, for the biggest-cee, biggest-gee, Cognitive Grammarian's view of Lakoff's role.

10. Bruce Fraser, an early MIT linguistics graduate who managed to avoid the fray better than most (though occasionally letting his generative semantics sympathies show), organized a lecture series at Boston University in 1978, just as it was becoming unavoidably clear that generative semantics was collapsing around the ears of its few remaining adherents, and his own contribution to the series (patterned on a seventies bestseller—Reich, 1970) was "The Greening of Linguistics." Fraser argued that the Chomskyan tradition had narrowed language to overly abstract concerns, and that the tradition was undergoing a fractionation into a more diverse group of more empirical concerns, such as sociolinguistic and functional research.

11. Another good indication is Cole's early collection of essays in volume 9 of the *Syntax and Semantics* series (1978), seven of whose twelve articles are by generative semanticists

(four are by philosophers, and the remaining one is by Givón, who had an early affiliation with generative semantics, but, by 1969 had turned his back on the dispute).

12. Postal's word on the origin of relational grammar, by the way, is that there is no interesting way in which it grew out of his participation in generative semantics (he also says, however, that he is not a particularly good authority on his own intellectual development, that ideas just present themselves and he follows them for a while, without, apparently, any sharp sense of where they come from), and Perlmutter isn't talking. My own suspicion, aside from the obvious parallels just sketched, is that Postal looked over the brink of generative semantics for a while, saw that meaning was such a morass of unsolved perplexities that it was going to cause more problems than it was going to solve, at least in the short run, and decided to go back to hardcore syntax. In any case, his version of relational grammar (arc pair grammar) is so hardcore syntactically that even Chomsky approves of it, and there is virtually no semantic discussion at all in his and Johnson's seven-hundred-page formalization of relational grammar (Johnson and Postal, 1980), aside from some treatment of a few standard logical syntax problems of the early seventies, like quantifier scope—vastly less discussion than in the work of the remaining generative semanticists, for instance, or in the work of Jackendoff, or in contemporary models like Montague grammar.

13. See also, Langacker (1986; 1990; the latter containing the former) for more compact presentations, Rudzka-Ostyn (1988), for a collection of exploratory papers in the framework (including another succinct overview by Langacker—1988b), and virtually anything in the new journal, *Cognitive Linguistics.*

14. Langacker's words on the roots of cognitive grammar begins with "Cognitive grammar is not in any significant sense an outgrowth of generative semantics, but it does share with that conception a concern for dealing explicitly with meaning, and for providing a unified account of grammar and lexicon" (1987:4), which is certainly accurate, but understates the influence considerably—perhaps because Langacker is using "generative semantics" to refer to the early, neater version of the theory. Certainly he warmly acknowledges the influence of many aspects of Lakoff's later and fuzzier research.

15. McCawley quotes the original. The translation is by H. A. Strong (Paul, 1891 [1886]:21). This emphasis on language diversity, of course, harkens back to the Amerindian imperative by way of the Boas-Sapir-Whorf thread of linguistic relativism, and Langacker has worked extensively on Amerindian languages, initially finding some evidence for various generative semantics proposals (1973; 1976a; 1976b), and later taking a more straighforwardly descriptive approach, as in the three volumes of Uto-Aztecan studies he edited (1977; 1979; 1982).

16. In this connection, see Andresen's (1990b) reappraisal, "Skinner and Chomsky 30 Years Later," originally presented to the 1989 LSA with the psychoanalytical subtitle, "The Return of the Repressed."

17. The issue of citation etiquette, at any rate, is a very complex one, far too complex to delve into here, but it is quite easy to come up with parallel cases from the generative semantics literature, or from virtually any scholarly pursuit. One interesting recent example is Ross and the relation his squishiness work bears to cognitive grammar. In a great many ways, cognitive grammar is founded on the sorts of insights Ross was struggling with in the mid-seventies, and if that model continues to grow in influence, Ross will look absolutely visionary to anyone with any eye on the historical record (which also includes Chomsky, Harris, Sapir, and Bolinger—all of whom have been concerned with gradations), but he shows up rarely in the cognitive grammar literature, even in Lakoff's work. See Pullum (1988; 1991:147–58) for some (unrelated but, as always with Pullum, fascinating) remarks on citation in linguistics.

On a more general note, salvaging is a widely established practice in science. When two programs clash the victorious one frequently, and often covertly, incorporates solutions, data, and methods from the defeated one. Even in relatively uncontentious circumstances, proposals are up for grabs in science. See Pullum (1991 [1983]:14), for instance, on the rapid coopting of material from relational grammar in the seventies.

18. In other interesting wrinkles, Baker's hypothesis depends on compromising one of the earliest generative semantics beating-sticks, the lexicalist hypothesis, and adopts a version of Predicate-raising for lexical incorporation.

19. See also Bach (1977:140–41), Ruwet (1991:xxi–xxii), Huck and Goldsmith (forthcoming), and Gazdar's wonderfully snide "As has often been remarked, the 'Aspects' view of language leads inexorably to Generative Semantics. What is surprising about REST [the Revised Extended Standard Theory] is that is has taken Chomsky so long to get there" (1982:472). In addition to the complaints of ex–generative semanticists about the increasing resemblance of Chomsky's model to generative semantics, and the observations of neutrals, many interpretivists, especially the ones who sat through his early condemnations, were very perplexed; see, in particular, Brame (1979).

20. If anyone cares, the review (of George, 1989) was requested by Konrad Koerner for his *Historiographia Linguistica*. He had second thoughts when he saw my review and sent it to a colleague for his advice, who recommended rejection. Koerner then wrote me that my review was unsuitable, for a variety of reasons that had all been confirmed for him by another reader, and kindly suggested some other journals to which I could submit it. But he inadvertently enclosed a copy of his letter to the reader which spilled some of the beans (in particular, who the reader was, Stephen Murray, someone whose opinion of Chomsky, as Koerner knew, is extremely low; complicating matters, my review also included a characterization of an earlier work of Murray's which could have been, and was, taken as hostile). Of course, we all exchanged a few mildly incensed letters, and that was that. The review editor for *Word* (Sheila Embleton) subsequently accepted it for publication without comment. It came out, but someone somewhere along the editing chain snipped out the offending "we can all hope" (Harris, 1991:328), so that the comment now looks predictive rather than hopeful. All told, I suppose, justice was served. "We can all hope" clearly didn't belong, since a good many people would so enthusiastically opt out of the *we*.

21. I have elided Johnson-Laird's exemplars for each of the three stages of Chomskyan evolution he identifies—*Syntactic Structures* (1957a) for the initial theory, *Aspects* (1965 [1964]) for the standard theory, *Lectures on Government and Binding* (1981a [1979]) and *Some Concepts and Consequences of the Theory of Government and Binding* (1982b) for government-and-binding.

22. One debate about cognitive grammar that has been fairly heated is its right to take the defining adjective of its name, since there is a barely covert implication in the name that other linguistic frameworks (especially Chomsky's framework) are not cognitive. The submerged claim in this debate is the relevance of the research in either framework for psychologists and other cognitive researchers. The debate is interesting, but no one should lose sight of the fact that Chomsky's model is unquestionably cognitive; as we have seen, it was largely his work that introduced cognitive concerns to linguistics. The principal difference from cognitive grammar, however, is that his general model of the mind is modular,

in the sense that it consists of separate, at least partially separate systems, each with its own intrinsic structure, each designed specifically to handle a particular kind of problem, with the whole system interacting in such a way as to create a very intricate complex of highly special structures. Now, on that . . . view, which I believe to be correct, the mind would be more or less analogous to the body. How do we think of the human body? Well, it is basically a complex of organs. One doesn't expect to find principles of functioning which are going to involve the heart as a special case, the spleen as a special case, and so on; there is a level at which they all fall together, namely, the level of biology, but if you really want to study the structure of the body, you will ask how the specific organs function, what their structures are, what their principles are, how their development is genetically determined, how they interact with one another, and so on. The physical body, the human body, is an intricate and delicate system of interacting subsystems, each of which has very special characteristics and special modes of development. Well, the modular approach to the mind takes essentially the same view; it assumes that we are going to find in the brain—which is perhaps the most complicated system we know of in the universe, and maybe the most complicated system that exists in it—what we are going to find is qualitatively like what we find in any other biological system known to us, namely, a high degree of modularity and specific structure, and that there will be an array of cognitive faculties, call them mental organs if you like—one might think of them as analogous to the physical organs—and that each of these cognitive faculties, each of these mental organs, will have its own very specific properties, its own specific structural properties, its specific physical representation, specific mode of development. It'll mature along a course that is predetermined. The result of the flowering of all these systems will be mental representations of a high degree of richness and intricacy, but quite different from one another and interacting in ways which are also biologically determined by the basic genetic structure of the system. (Chomsky and Saporta, 1978:308)

Langacker's position is very different:

Language is an integral part of human cognition. An account of linguistic structure should therefore articulate with what is known about cognitive processing in general, regardless of whether one posits a special language "module" (Fodor, 1983), or an innate *faculté de langage*. If such a faculty exists, it is nevertheless embedded in the general psychological matrix, for it represents the evolution and fixation of structures having a less specialized origin. Even if the blueprints for language are wired genetically into the human organism, their elaboration into a fully specified linguistic system during language acquisition, and their implementation in everyday language use, are clearly dependent on experiential factors and inextricably bound up with psychological phenomena that are not specifically linguistic in character. Thus we have no valid reason to anticipate a sharp dichotomy between linguistic ability and other aspects of cognitive processing. Instead of grasping at any apparent rationale for asserting the uniqueness and insularity of language, we should try more seriously to integrate the findings of linguistics and cognitive psychology. . . . To put it contentiously, language has appeared special and unassimilable to broader psychological phenomena mainly because linguists have insisted on analyzing it in an inappropriate and highly unnatural fashion; once the many layers of artifact are removed, language starts to look much more natural and learnable in terms of what we know about other facets of human cognitive ability. (1987:12–13)

You can decide which argument carries more weight (both are pretty potent, but both depend on a network of tacit auxiliary premises), but there is a clear way in which Langacker's general model makes his brand of linguistics "more cognitive," since it builds on a wider array of mental mechanisms and processes. Which is more *correct* is an empirical question, one which won't be answered for a while, possibly a very long while. But now ask yourself what action these arguments would lead you to if you were a psychologist. An insular model of language like Chomsky's means, among other things, that any results or ideas you have from exploring other mental structures are essentially useless for looking at language, and any insights you come away with from looking at language are essentially useless for the rest of your work. The potential for generalizable findings is much greater with Langacker's heavily integrated model.

23. This list is taken pretty much from the proceedings of a 1979 conference on Current Approaches to Syntax at the University of Wisconsin-Milwaukee, sometimes called *the syntax sweepstakes* (Moravscik and Wirth, 1980; Kac, 1980). Lakoff's paper (on cognitive grammar) was not published (though see Lawler, 1980); Brame's paper was published in his *Essays toward Realistic Syntax* (1979:19–63).

24. For an interesting parallel, see Holton's (1988 [1973]:240) comments on Mach, and for some intriguingly similar comments about Chomsky's relation to society as a whole, perhaps suggesting some overlapping motivations for his political concerns, see Otero's introduction to the Chomsky anthology, *Language and Politics,* paraphrasing some autobiographical remarks Chomsky has made in various places about his youth:

> He always felt completely out of tune with almost everything around him . . . he was always either alone or part of a tiny minority [in his political beliefs] . . .
>
> He was always on the side of the losers . . .
>
> [After the bombing of Hiroshima] he just walked off by himself into the woods and stayed alone for a couple of hours. He felt completely isolated. (Otero, 1988:22)

25. Newmeyer (1980a:206–7) catalogs this debris somewhat, and Wasow (1985) makes some similar remarks. I went a little overboard in an earlier paper (Harris, 1989:106), characterizing the series of related disputes as "Chomsky$_1$ vs. Chomsky$_2$ vs. Chomsky$_3$ . . . Chomsky$_n$," but only a little. Among the most obvious symptoms of debrisdom is the choice of buzzword. Postal, Ross, the Lakoffs, Gruber, McCawley, and many others came out of the early-to-mid-sixties, when Chomsky was pushing for deeper and more abstract analyses, and generally trumpeting the virtues of abstractness. The next wave, from the period when Chomsky was surfacing, came out on the side of concreteness and "realism." So, Brame's Milwaukee paper was "Realistic Grammar," and its containing anthology (1979) was *Essays toward Realistic Syntax;* Bresnan's best-known breakaway paper was "A Realistic Transformational Grammar" (1978).

Works Cited

Aarsleff, Hans. 1970. The history of linguistics and Professor Chomsky. *Language* 46:570–85.

Aarsleff, Hans. 1971. 'Cartesian linguistics': Fact or fantasy? *Language Sciences* 17:1–12.

Abraham, Werner, and Robert I. Binnick, editors. 1972. *Generative Semantik.* Frankfurt: Atheneum Verlag.

Anderson, Stephen R. 1976 [1966]. Concerning the notion "base component of a transformational grammar." In McCawley (1976a:113–28).

Anderson, Stephen R. 1985. *Phonology in the twentieth century: Theories of rules and theories of representation.* Chicago: University of Chicago Press.

Anderson, Stephen R., and Sandra Chung. 1977. On grammatical relations and clause structure in verb-initial languages. In Cole and Sadock (1977:1–26).

Anderson, Stephen R., Sandra Chung, James McCloskey, and Frederick J. Newmeyer. Forthcoming. Chomsky's 1962 program for linguistics: A retrospective. *Proceedings of the XVth International Congress of Linguists.*

Anderson, Stephen R., and Paul Kiparsky, editors. 1973. *A festschrift for Morris Halle.* New York: Holt, Rinehart and Winston.

Andresen, Julie Tetel. 1990a. *Linguistics in America 1769–1924: A critical history.* London: Routledge.

Andresen, Julie Tetel. 1990b. Skinner and Chomsky 30 years later. In Dinneen and Koerner (1990:145–66).

Anttila, Raimo. 1975. Revelation as linguistic revolution. In Makkai (1975:171–76).

Aristotle. 1991. *On Rhetoric: A theory of civic discourse.* Translated by George A. Kennedy. New York: Oxford University Press.

Arnauld, Antoine, and Claude Lancelot. 1975 [1660]. *General and rational grammar: The Port-Royal grammar.* Edited and translated by Jacques Rieux and Bernard E. Rollin. The Hague: Mouton.

Arnauld, Antoine, and Pierre Nicole. 1963 [1662]. *The art of thinking: The Port-Royal logic.* Edited and translated by James Dickoff and Patricia James. New York: Bobbs-Merrill.

Auerbach, Joseph, Philip H. Cook, Robert Kaplan, and Virginia J. Tufte. 1968. *Transformational grammar: A guide for teachers.* Rockville, MD: English Language Services.

Austin, J. L. 1962 [1955]. *How to do things with words.* Oxford: Oxford University Press.

Baars, Bernard J. 1986. *The cognitive revolution in psychology.* New York: The Guilford Press.

Bach, Emmon. 1967. *Have* and *be* in English. *Language* 43:462–85.

Bach, Emmon. 1968 [1967]. Nouns and noun phrases. In Bach and Harms (1968:91–124)

Bach, Emmon. 1974. Explanatory inadequacy. In Cohen (1974:153–71).

Bach, Emmon. 1977. Comments on the paper by Chomsky. In Culicover, Wasow, and Akmajian (1977:133–55).

Bach, Emmon, and Robert T. Harms, editors. 1968. *Universals in linguistic theory.* New York: Holt, Rinehart and Winston.

Bain, Alexander. 1879. *Logic.* 2 vol. London: Longmans, Green, and Company.

Baker, C. L., and Michael Brame. 1972. Global rules: A rejoinder. *Language* 48:51–77.

Baker, Mark. 1988. *Incorporation: A theory of grammatical function changing.* Chicago: University of Chicago Press.

Baker, William, Gary D. Prideaux, and Bruce L. Derwing. 1973. Grammatical properties of sentences as a basis for concept formation. *Journal of Psycholinguistic Research* 2:201–20.

Bar-Hillel, Yehoshua. 1954. Logical syntax and semantics. *Language* 30:230–37.

Bar-Hillel, Yehoshua. 1967. Review of Katz and Fodor (1964a). *Language* 43:526–50.

Beach, Woodford A., and others, editors. 1977. *Papers from the thirteenth regional Chicago Linguistic Society meeting.* Chicago: Chicago Linguistic Society.

Beaver, Joseph C. 1968. A grammar of prosody. *College English* 29:310–21.

Belletti, Adriana, Luciana Brandi, and Luigi Rizzi, editors. 1981. *The theory of markedness in generative grammar: Proceedings of the 1979 GLOW conference.* Pisa: Scuola Normale Superiore di Pisa.

Berlin, Isaiah. 1980. *Personal impressions.* London: Hogarth Press.

Berman, Arlene, 1974. On the VSO hypothesis. *Linguistic Inquiry* 5:1–37.

Bever, Thomas G. 1970 [1968]. The cognitive basis of linguistic structures. In J. Hayes (1970:279–362).

Bever, Thomas G. 1988. The psychological reality of grammar: A student's eye view of cognitive science. In Hirst (1988:112–42).

Bever, Thomas G., and others, editors. 1976. *An integrated theory of linguistic ability.* New York: Thomas Crowell.

Bierwisch, Manfred, and Karl Erich Heidolph, editors. 1970. *Progress in linguistics: A collection of papers.* The Hague: Mouton.

Binnick, Robert I. 1971. 'Bring' and 'come.' *Linguistic Inquiry* 2:260–65.

Binnick, Robert I., and others, editors. 1969. *Papers from the fifth regional Chicago Linguistic Society meeting.* Chicago: Chicago Linguistic Society.

Blake, Barry J. 1990. *Relational grammar.* London: Routledge and Kegan Paul.

Bloch, Bernard, 1949. Leonard Bloomfield. *Language* 25:87–94.

Bloch, Bernard. 1953. Linguistic structure and linguistic analysis. In Hill (1953:40–44).

Bloch, Bernard, and George L. Trager. 1942. *Outline of linguistic analysis.* Baltimore: Linguistic Society of America.

Bloomfield, Leonard. 1914. *An introduction to the study of language.* New York: Henry Holt and Company.

Bloomfield, Leonard. 1923. Review of Saussure's *Course on general linguistics. Modern Language Journal* 8:317–19.

Bloomfield, Leonard. 1925. Why a linguistic society? *Language* 1:1–5.

Bloomfield, Leonard. 1926. A set of postulates for the science of language. *Language* 1:153–64.

Bloomfield, Leonard. 1928. *Menomini texts. Publication of the American Ethnological Society* 12. New York: G. E. Stechert.

Bloomfield, Leonard. 1930. Linguistics as a science. *Studies in Philology* 27:533–57.

Bloomfield, Leonard. 1931. Albert Paul Weiss [Obituary]. *Language* 7:219–21.

Bloomfield, Leonard. 1933. *Language*. New York: Holt, Rinehart and Winston.

Bloomfield, Leonard. 1970. *A Leonard Bloomfield anthology*. Edited by Charles F. Hockett. Bloomington: Indiana University Press.

Boal, Iain A. 1983. Chomsky and the state of linguistics. Unpublished manuscript. (My copy is dated January, 1983; others have cited a 1984 version.)

Boas, Franz. No date [1911]. *Introduction to the Handbook of American Indian languages*. Washington: Georgetown University Institute of Languages and Linguistics.

Boas, Franz, editor. 1938. *Handbook of American Indian languages*. Vol. 3. New York: J. J. Augustin.

Bolinger, Dwight L. 1950. Rime, assonance, and morpheme analysis. *Word* 6:117–36.

Bolinger, Dwight L. 1957. *Interrogative structures of American English*. Huntsville: University of Alabama Press.

Bolinger, Dwight L. 1991. [1974; 1988]. First person, not singular. In Koerner (1991:19–46).

Boole, George. n.d. [1854]. *An investigation into the laws of thought*. New York: Dover Publications.

Booth, Wayne C. 1974. *Modern dogma and the rhetoric of assent*. Chicago: University of Chicago Press.

Borkin, Ann. 1984 [1974]. *Problems in form and function*. Norwood, NJ: Ablex.

Borkin, Ann, and others. 1968. *Where the rules fail: A student's guide. An unauthorized appendix to M. K. Burt's* From deep to surface structure. Bloomington: Indiana University Linguistic Club.

Boslough, John. 1985. *Stephen Hawking's universe: An introduction to the most remarkable scientist of our time*. New York: Quill.

Botha, Rudulph P. 1989. *Challenging Chomsky: The generative garden game*. Oxford: Basil Blackwell.

Boyd, Julian, and J. P. Thorne. 1969. The semantics of modal verbs. *Journal of Linguistics* 7:57–74.

Bracken, Harry M. 1970. Chomsky's variations of a theme by Descartes. *Journal of the History of Philosophy* 18:188–92.

Bracken, Harry M. 1972. Chomsky's Cartesianism. *Language Sciences* 22:11–17.

Bracken, Harry M. 1984. *Mind and language: Essays on Descartes and Chomsky*. Dordrecht, Netherlands: Foris.

Brame, Michael K. 1976. *Conjectures and refutations in syntax and semantics*. New York: North-Holland.

Brame, Michael K. 1979. *Essays toward realistic syntax*. Seattle: Noit Amrofer.

Brekle, Herbert E. 1969a. Review of Chomsky (1966a). *Linguistics* 49:74–91.

Brekle, Herbert E. 1969b. Generative semantics vs. deep syntax. In Kiefer (1969:80–90).

Brekle, Herbert Ernst. 1970 [1968]. *Generative Satzsemantik und transformationelle Syntax im System der englischen Nominalkomposition*. (= *Der Internationalen Bibliothek für Allgemeine Linguistik* 4.) München: Wilhelm Fink.

Brekle, Herbert Ernst. 1978. *Generative Satzsemantik im System der englischen Nominalkomposition*. (Second edition of Brekle, 1970 [1968].) München: Wilhelm Fink.

Brentari, Diane, Gary N. Larson, and Lynn A. Macleod. 1992. *The joy of grammar: A festschrift in honor of James D. McCawley*. Philadelphia: John Benjamins.

Bresnan, Joan W. 1978. A realistic transformational grammar. In Halle, Bresnan, and Miller (1978:1–59).

Brooks, Daniel, and Guillermo Verdecchia. 1992. *The Noam Chomsky lectures: A play*. Toronto: Coach House Press.

Bronowski, J., and U. Bellugi. 1970. Language, name and concept. *Science* 168:669–73.

Brown, Roger. 1970. *Psycholinguistics: Selected papers.* New York: Free Press.

Brown, Roger, and Camille Hanlon. 1970 [1968]. Derivational complexity and order of acquisition in child speech. In Brown (1970:155–207).

Bruner, Jerome. 1983. *In search of mind.* New York: Harper & Row.

Bruner, Jerome. 1988. Founding the Center for Cognitive Studies. In Hirst (1988:90–99).

Bursill-Hall, G. L. 1971. *Speculative grammars of the Middle Ages: The doctrine of the* partes orationis *of the modistae.* The Hague: Mouton.

Burt, Marina K. 1971. *From deep to surface structure: An introduction to transformational syntax.* New York: Harper and Row.

Butterfield, Herbert. 1957. *The origins of modern science.* Rev. ed. New York: Free Press.

Bynon, Theodora, and F. R. Palmer, editors. 1986. *Studies in the history of western linguistics: In honour of R. H. Robins.* Cambridge, Cambridge University Press.

Carden, Guy. 1968. English quantifiers. Aiken Computational Laboratory Technical Report NSF-20-IX:1–45. Cambridge, MA: Harvard University.

Carden, Guy. 1970. The deep structure of *both.* In Chicago Linguistic Society (1970:178–89).

Carnap, Rudolph. 1937 [1934]. *The logical syntax of language.* London: Routledge and Kegan Paul.

Carnap, Rudolph. 1942. *Introduction to semantics.* Cambridge, MA: Harvard University Press.

Carr, John W. III, editor. 1958. *Computer programming and artificial intelligence.* Ann Arbor: Michigan University College of Engineering.

Carroll, John B. 1953. *The study of language.* Cambridge, MA: Harvard University Press.

Cassirer, Ernst. 1945. Structuralism in modern linguistics. *Word* 1:97–120.

Catwell, N. R. 1966. *The new English grammar.* Cambridge, MA: MIT Press.

Chafe, Wallace L. 1967a. Language as symbolization. *Language* 43:57–91.

Chafe, Wallace L. 1967b. Review of Katz (1966). *International Journal of American Linguistics* 33:248–54.

Chafe, Wallace L. 1970a. *Meaning and the structure of language.* Chicago: University of Chicago Press.

Chafe, Wallace L. 1970b. *A semantically based sketch of Onondaga.* Memoir 25. Bloomington: Indiana University Publications in Anthropology and Linguistics.

Chafe, Wallace L., editor. 1976. *American Indian languages and American linguistics.* Lisse, Netherlands: Peter de Ridder Press.

Chapin, Paul. 1967. *The syntax of word-derivation in English* [*Technical publication 16*]. Bedford, MA: MITRE Corporation.

Chicago Linguistic Society. 1970. *Papers from the sixth regional Chicago Linguistic Society meeting.* Chicago: Chicago Linguistic Society.

Chicago Linguistic Society. 1971. *Papers from the seventh regional Chicago Linguistic Society meeting.* Chicago: Chicago Linguistic Society.

Chicago Linguistic Society: See also Beach and others (1977), Binnick and others (1969), Corum and others (1973), Darden and others (1968), Grossman and others (1975a, 1975b), LaGaly and others (1974), Mufwene and others (1976), Peranteau and others (1972), Schiller and others (1988).

Chomsky, Noam, 1955a. Semantic considerations in grammar. *Georgetown Monograph Series in Linguistics,* 8:140–58.

Chomsky, Noam. 1955b. Logical syntax and semantics: Their linguistic relevance. *Language* 31:36–45.

Chomsky, Noam. 1957a. *Syntactic structures.* The Hague: Mouton.

Chomsky, Noam. 1957b. Logical structures in language. *American Documentation* 8:284–91.

Chomsky, Noam. 1958. Linguistics, logic, psychology, and computers. In Carr (1958:429–54).

Chomsky, Noam. 1959 [1957]. Review of Skinner (1957). *Language* 35:26–58.

Chomsky, Noam. 1962a [1958]. A transformational approach to syntax. In Hill (1962c:124–86).

Chomsky, Noam. 1962b [1960]. Explanatory models in linguistics. In Nagel and others (1962:528–50).

Chomsky, Noam. 1962c. The logical basis of linguistic theory. *Preprints of the ninth International Congress of Linguists.*

Chomsky, Noam. 1964a [1961]. Degrees of grammaticalness. In Fodor and Katz (1964:384–89).

Chomsky, Noam. 1964b [1962]. The logical basis of linguistic theory. In Lunt (1964 [1962]:914–77).

Chomsky, Noam. 1964c [1963]. Current issues in linguistic theory. In Fodor and Katz (1964:50–118).

Chomsky, Noam. 1964d [1963]. *Current issues in linguistic theory.* The Hague: Mouton.

Chomsky, Noam. 1965 [1964]. *Aspects of the theory of syntax.* Cambridge, MA: MIT Press.

Chomsky, Noam. 1966a. *Cartesian linguistics.* Cambridge, MA: MIT Press.

Chomsky, Noam. 1966b [1964]. *Topics in the theory of generative grammar.* The Hague: Mouton.

Chomsky, Noam. 1966c [1964]. Topics in the theory of generative grammar. Sebeok (1966:1–60).

Chomsky, Noam. 1967a. Some general properties of phonological rules. *Language* 43:102–28.

Chomsky, Noam. 1967b. The formal nature of language. In Lenneberg (1967:397–442).

Chomsky, Noam. 1970. The case against B. F. Skinner. *New York Review of Books* (30 December):18–24.

Chomsky, Noam. 1972a. *Language and mind.* Enlarged edition. New York: Harcourt Brace Jovanovich.

Chomsky, Noam. 1972b [1967–1969]. *Studies on semantics in generative grammar.* The Hague: Mouton.

Chomsky, Noam. 1973a [1971]. Conditions on transformations. In Anderson and Kiparsky (1973:232–86). Also in Chomsky (1977).

Chomsky, Noam. 1973b. Chomsky replies [letter to the editor]. *New York Review of Books* (19 July):33.

Chomsky, Noam. 1975a [1955; preface dated 1973]. *The logical structure of linguistic theory.* New York: Plenum Press.

Chomsky, Noam. 1975b. *Reflections on language.* New York: Pantheon.

Chomsky, Noam. 1977. *Essays on form and interpretation.* New York: North-Holland.

Chomsky, Noam. 1979 [1976]. *Language and responsibility.* [Conversations with Mitsou Ronat.] Translated by John Viertel. New York: Pantheon.

Chomsky, Noam. 1980a. Author's response: The new organology. *Behavioral and Brain Sciences* 3:42–58.

Chomsky, Noam. 1980b [1978]. *Rules and representations.* New York: Columbia University Press.

Chomsky, Noam. 1981a [1979]. *Lectures on government and binding.* Dordrecht, Netherlands: Foris.

Chomsky, Noam. 1981b [1979]. Markedness and core grammar. In Belletti, Brandi, and Rizzi (1981:123–46).

Chomsky, Noam, 1982a [1979–80]. *The generative enterprise: A discussion with Riny Huybregts and Henk van Riemsdijk.* Dordrecht, Netherlands: Foris.

Chomsky, Noam. 1982b. *Some concepts and consequences of the theory of government and binding.* Cambridge, MA: MIT Press.

Chomsky, Noam. 1983. On cognitive structures and their development. In Piatelli-Palmarini (1983).

Chomsky, Noam. 1984. *Modular approaches to the study of mind.* San Diego: San Diego University Press.

Chomsky, Noam. 1985. The manufacture of consent in democracy. *Philosophy and Social Action* 11.1:21–39.

Chomsky, Noam. 1986. *Knowledge of language.* New York: Praeger.

Chomsky, Noam. 1987. *The Chomsky reader.* Edited by James Peck. New York: Pantheon.

Chomsky, Noam. 1988a. *Language and problems of knowledge: The Managua lectures.* Cambridge, MA: MIT Press.

Chomsky, Noam. 1988b. *Generative grammar: Its basis, development, and prospects.* A special issue of *Studies in English Linguistics and Literature.* Kyoto: Kyoto University of Foreign Studies.

Chomsky, Noam. 1988c. *Language and politics.* Edited by Carlos Otero. Montreal: Black Rose.

Chomsky, Noam. 1990. Topic . . . comment: On formalization and formal linguistics. *Natural Language and Linguistic Theory* 8:143–7.

Chomsky, Noam. 1991a [1989]. Linguistics and adjacent fields: A personal view. In Kasher (1991:3–25).

Chomsky, Noam. 1991b [1989]. Linguistics and cognitive science: Problems and mysteries. In Kasher (1991:26–55).

Chomsky, Noam. 1992. *A minimalist program for linguistic theory [MIT occasional papers in linguistics* 1]. Cambridge: MIT Department of Linguistics.

Chomsky, Noam, and Morris Halle. 1965. Some controversial questions in phonological theory. *Journal of Linguistics* 1:97–138.

Chomsky, Noam, and Morris Halle. 1968. *The sound pattern of English.* New York: Harper and Row.

Chomsky, Noam, Morris Halle, and Fred Lukoff. 1956. On accent and juncture in English. In Halle and others (1956:65–80).

Chomsky, Noam, and Howard Lasnik. 1977. Filters and control. *Linguistic Inquiry* 8:425–504.

Chomsky, Noam, and Howard Lasnik. 1978. A remark on contraction. *Linguistic Inquiry* 9:268–74.

Chomsky, Noam, and Sol Saporta. 1978. An interview with Noam Chomsky. *Linguistic Analysis* 4:301–19.

Choseed, Bernard, and Allene Guss. 1962. *Report on the eleventh annual round table meeting on linguistics and language studies.* Washington: Georgetown University Press.

Christensen, Francis. 1967. *Notes toward a new rhetoric.* New York: Harper and Row.

Christensen, Francis. 1976. *A new rhetoric.* New York: Harper and Row.

Clark, Ronald W. 1986 [1984]. *The survival of Charles Darwin: A biography of a man and an idea.* New York: Discus Books.

Cline, Barbara Lovett. 1987. *Men who made a new physics.* Chicago: University of Chicago Press.

Cohen, David, editor. 1974. *Explaining linguistic phenomena.* New York: Wiley and Sons.

Cole, Peter, editor. 1978. *Pragmatics*. Syntax and Semantics, vol. 9. New York: Academic Press.

Cole, Peter, and Jerrold Sadock, editors. 1977. *Speech acts*. Syntax and Semantics, vol. 8. New York: Academic Press.

Cole, Peter, and Jerry L. Morgan, editors. 1975. *Speech acts*. Syntax and semantics, vol. 3. New York: Academic Press.

Comrie, Bernard. 1986. Relational grammar. Whence, where, whether. *Linguistics* 24:773–90.

Corballis, Michael C. 1991. *The lopsided ape: evolution of the generative mind*. New York: Oxford University Press.

Corum, Claudia, and others, editors. 1973. *Papers from the ninth regional Chicago Linguistic Society meeting*. Chicago: Chicago Linguistic Society.

Coseriu, Eugenio. 1970. *Sprache: Strukturen und Funktione*. Tübingen, Germany: Tübingen Beiträge zur Linguistik.

Craft, Ebbing [Arnold M. Zwicky]. 1970. Up against the wall, fascist pig critics! In Zwicky and others (1970:147–50).

Crick, Francis. 1988. *What mad pursuit: A personal view of scientific discovery*. New York: Basic Books.

Culicover, Peter, Thomas Wasow, and Adrian Akmajian, editors. *Formal syntax*. New York: Acadamic Press.

Curme, George O. 1931. *Syntax*. Volume 3 of *A grammar of the English Language*. Boston: D. C. Heath and Company.

D'Agostino, Fred. 1986. *Chomsky's system of ideas*. Oxford: Clarendon.

Dallaire, Raimonde, and others. 1962. The transformational theory [A panel discussion]. In Woodworth and DiPeitro (1962:3–50).

Darden, Bill J., and others. 1968. *Papers from the fourth regional Chicago Linguistic Society meeting*.

Darnell, Regna. 1990. *Edward Sapir: Linguist, anthropologist, humanist*. Los Angeles: University of California Press.

Darwin, Charles. n.d. [1872]. *The origin of species* and *The descent of man*. New York: Modern Library.

Darwin, Charles. 1958 [1892]. *The autobiography of Charles Darwin and selected letters*. Edited by Francis Darwin. New York: Dover.

Davidson, Donald, and Gilbert Harman. 1972. *Semantics of natural language*. Dordrecht, Netherlands: Reidel.

Davidson, Donald, and Jaakko Hintikka, editors. 1969. *Words and objections: Essays on the work of W.V.O. Quine*. Dordrecht, Netherlands: D. Reidel.

Davies, P.C.W., and J. Brown, editors. 1988. *Superstrings: A theory of everything?* Cambridge: Cambridge University Press.

Davis, Boyd H., and Ramond O'Cain, editors. 1980. *First person singular*. Studies in the History of Linguistics, vol. 21. Philadelphia: John Benjamins.

Davis, Nuel Pharr. 1968. *Lawrence and Oppenheimer*. New York: Da Capo Press.

Davison, Alice. 1970. Causal adverbs and performative verbs. In Chicago Linguistic Society (1970:190–201).

Davison, Alice. 1972. Performative verbs, adverbs, and felicity conditions: An inquiry into the nature of performative verbs. Unpublished dissertation for the University of Chicago.

Davison, Alice. 1973. Words for things people do with words. In Corum and others (1973:114–22).

Dean, Janet. See Fodor, Janet Dean.

Derwing, Bruce L. 1973. *Transformational grammar as a theory of language acquisition.* Cambridge: Cambridge University Press.

Dinneen, Francis P., S. J., editor. 1966. *Report of the 17th annual round table meeting on linguistics and language studies* [*GURT 1966*] Washington: Georgetown University Press.

Dinneen, Francis, P., S. J., and E. F. Konrad Koerner, editors. 1990. *North American contributions to the history of linguistics.* Amsterdam Studies in the Theory and History of Linguistic Science, series 3, volume 58. Philadelphia: John Benjamins.

Dixon, R. M. W. 1963. *Linguistic science and logic.* The Hague: Mouton.

Dixon, R. M. W. 1991. *A new approach to English grammar, on semantic principles.* New York: Oxford University Press.

Dougherty, Ray C. 1974. Generative Semantics methods: A Bloomfieldian counterrevolution. *International Journal of Dravidian Linguistics* 3:255–86.

Dougherty, Ray C. 1975. Reply to the critics of the Bloomfieldian counterrevolution. *International Journal of Dravidian Linguistics* 4:249–71.

Dougherty, Ray C. 1976a. Argument invention: The linguist's "feel" for science. In Wirth (1976:111–166).

Dougherty, Ray C. 1976b. Einstein and Chomsky on scientific methodology. *Linguistics* 170:5–14.

Dougherty, Ray C. 1976c. A methodological exorcism of semantic pseudo-problems. *Linguistics* 167:5–29.

Dowty, David R. 1972. On the syntax and semantics of the atomic predicate CAUSE. In Peranteau and others (1972:62–74).

Dowty, David R. 1979. *Word meaning in Montague grammar: The semantics of verbs and times in generative semantics and Montague's PTQ.* Dordrecht, Netherlands: Reidel.

Dubois-Charlier, Françoise. 1972. La sémantique générative—une nouvelle théorie linguistique? *Langages* 27:5–77.

Edgerton, Franklin. 1933. Review of Bloomfield (1933). *Journal of the American Oriental Society* 53:295–97.

Elgin, Suzette Haden. 1973. *What Is linguistics?* Englewood Cliffs, NJ: Prentice-Hall.

Elgin, Suzette Haden. 1979. *What Is linguistics?* Second edition. Englewood Cliffs, NJ: Prentice-Hall.

Emonds, Joseph E. 1970. *Root and structure-preserving transformations.* [MIT dissertation.] Bloomington: Indiana University Linguistic Club.

Empedocles. 1964. Thought. *Selections from early Greek philosophy.* Edited by Milton C. Nahm. New York: Appleton-Century-Crofts.

Eschliman, Herbert R., Robert C. Jones, and Tommy R. Burkett. 1966. *Generative English handbook.* Belmont, CA: Wadsworth.

Feyerabend, Paul K. 1978 [1975]. *Against method: Outline of an anarchistic theory of knowledge.* London: Verso

Fillmore, Charles J. 1966. A proposal concerning English prepositions. *Monograph Series on Language and Linguistics* 19:208–31.

Fillmore, Charles J. 1968 [1967]. The case for case. In Bach and Harms (1968:1–90).

Fillmore, Charles J. 1969 [1966]. Toward a modern theory of case. In Reibel and Schane (1969:361–75).

Fillmore, Charles J. 1972 [1969]. On generativity. In Peters (1972:1–20).

Fillmore, Charles J., and D. Terrence Langendoen, editors. 1971. *Studies in linguistic semantics.* New York: Holt, Rinehart and Winston.

Fodor, Janet Dean. 1980. *Semantics: Theories of meaning in generative grammar.* Cambridge, MA: Harvard University Press.

Fodor, Jerry A. 1970. Three reasons for not deriving *kill* from *cause to die*. *Linguistic Inquiry* 1:429–38.

Fodor, Jerry A. 1983. *The modularity of mind*. Cambridge, MA: MIT Press.

Fodor, Jerry A., Thomas G. Bever, and Merrill Garrett. 1974. *The psychology of language*. New York: McGraw-Hill.

Fodor, Jerry A., and Merrill Garrett [discussion by N. S. Sutherland, L. Jonathan Cohen, and others]. 1966. Some reflections on competence and performance. In Lyons and Wales (1966:135–79).

Fodor, Jerry A., and Jerrold J. Katz, editors. 1964. *The structure of language*. Englewood Cliffs, NJ: Prentice-Hall.

Francis, W. Nelson. 1963. The present state of grammar. *English Journal* 52:317–21.

Frantz, Donald. 1974. *Generative semantics: An introduction*. Bloomington: Indiana University Linguistic Club.

Frazier, Alexander, editor. 1967. *New directions in elementary English*. Champaign, IL: National Council of Teachers of English.

Fries, Charles C. 1940. *American English grammar:* English Monograph 10 of the National Council of Teachers of English. New York: Appleton-Century.

Fries, Charles C. 1952. *The structure of English*. New York: Harcourt, Brace, and Company.

Fries, Charles C. 1961. The Bloomfield 'school'. In Mohrmann, Sommerfelt, and Whatmough (1961:196–224).

Fromkin, Victoria A. 1991 [1989]. Language and brain: Redefining the goals and methodology of linguistics. In Kasher (1991:78–103).

Fujimura, Osamu, editor. 1973. *Three dimensions of linguistic theory*. Tokyo: Tokyo Institute for Advanced Studies of Language.

Galileo. 1954 [1638]. *Dialogues concerning two new sciences*. Translated by Henry Crew and Alfonso de Salvio. New York: Dover.

Gardner, B. T., and R. A. Gardner. 1969. Teaching sign language to a chimpanzee. *Science* 165:664–72.

Gardner, B. T., and R. A. Gardner. 1975. Evidence for sentence constituents in early utterances of child and chimpanzee. *Journal of Experimental Psychology* 104:244–67.

Gardner, Howard. 1985. *The mind's new science: A history of the cognitive revolution*. New York: Oxford University Press.

Garson, Barbara. 1966. *Macbird!* New York: Grove Press.

Gazdar, Gerald. 1977. *Implicature, presupposition and logical form*. Bloomington: Indiana University Linguistic Club.

Gazdar, Gerald. 1979. *Pragmatics: Implicature, presupposition, and logical form*. New York: Academic Press.

Gazdar, Gerald. 1981. On syntactic categories [and following discussion]. In Longuet-Higgins and others (1981:[53–70]).

Gazdar, Gerald. 1982. Reviews of Brame (1978; 1979). *Journal of Linguistics* 18:464–73.

Gazdar, Gerald, and Ewan Klein. 1978. Review of Keenan (1975b). *Language* 46:663–67.

Gazdar, Gerald, and Geoffrey K. Pullum. 1976. Truth-functional connectives in natural language. In Mufwene and others (1976:220–34).

Gazdar, Gerald, and others. 1985. *Generalized phrase structure grammar*. Cambridge: Cambridge University Press.

George, Alexander, ed. 1989. *Reflections on Chomsky*. Oxford: Basil Blackwell.

Givón, Talmy. 1979. *On understanding grammar*. New York: Academic Press.

Gleason, H. Allan. 1956. *An introduction to descriptive linguistics*. New York: Holt, Rinehart and Winston.

Gleason, H. Allan. 1961. *An introduction to descriptive linguistics.* 2d ed. New York: Holt, Rinehart and Winston.

Gleason, H. Allan. 1988. Theories in conflict: North American linguistics in the fifties and sixties. Manuscript.

Glucksberg, S., and J. H. Danks. 1969. Grammatical structure and recall: A function of the space in immediate memory or recall delay? *Perception and Psychophysics* 6:113–17.

Goldschmidt, Walter, editor. 1959. *The anthropology of Franz Boas: Essays of the centennial of his birth.* Memoir 89 of *The American Anthropologist* 61.5, part 2. San Francisco: American Anthropologist Association.

Goldsmith, John. 1987. George Lakoff stood . . . Paragraph on the back of the dustcover for Lakoff (1987).

Goldsmith, John. 1989. Review of van Riemsdijk and Williams (1986). *Language* 65:150–59.

Goldsmith, John, and Geoffrey J. Huck. 1991. Distribution et médiation dans le développement de la théorie linguistique. *Communications* 53:51–67.

Gordon, David, and George Lakoff. 1988 [1971]. Conversational postulates. In Schiller and others (1988).

Goulet, John. 1975. *Oh's profit.* New York: William Morrow.

Gragg, Gene B. 1972. Semi-indirect discourse and related nightmares. In Peranteau and others (1972:32–40).

Gray, Bennison [Barbara Bennison and Michael Gray]. 1974. Toward a semi-revolution in grammar. *Language Sciences* 29:1–12.

Gray, Bennison [Barbara Bennison and Michael Gray]. 1976. Counter-revolution in the hierarchy. *Forum Linguisticum* 1:38–50.

Gray, Bennison [Barbara Bennison and Michael Gray]. 1977. Now you see it—now you don't: Chomsky's *Reflections* [Review of Chomsky (1975b)]. *Forum Linguisticum* 2:65–74.

Green, Georgia. 1972. Some observations on the syntax and semantics of instrumental verbs. In Peranteau and others (1972:83–97).

Green, Georgia. 1973. A syntactic syncretism in English and French. In Kachru and others (1973:257–78).

Green, Georgia. 1974. *Semantics and syntactic irregularity.* Bloomington: Indiana University Press.

Green, Georgia. 1975. How to get people to do things with words: The whimperative question. In Cole and Morgan (1975:107–42).

Green, Georgia. 1981. Review of Napoli and Rando (1979). *Language* 57:703–7.

Green, Georgia. 1989. *Pragmatics and natural language understanding.* Hillsdale, NJ: Lawrence Erlbaum.

Greenberg, Joseph, editor. 1966. *Universals in language.* 2nd ed. Cambridge, MA: MIT Press. [1st edition, 1963.]

Greene, Judith. 1972. *Psycholinguistics: Chomsky and psychology.* Harmondsworth, England: Penguin.

Grinder, John. 1970. Super equi-NP deletion. In Chicago Linguistic Society (1970:297–317).

Grossman, Robin E., and others, editors. 1975a. *Papers from the eleventh regional meeting of the Chicago Linguistic Society.* Chicago: Chicago Linguistic Society.

Grossman, Robin E., and others, editors. 1975b. *Papers from the parasession on functionalism.* Chicago: Chicago Linguistic Society.

Gruber, Jeffrey S. 1976 [1965; 1967]. *Lexical structures in syntax and semantics* ["Studies in lexical relations," "Functions of the lexicon in formal descriptive grammars"]. New York: North-Holland.

Hacking, Ian. 1980. Review of Chomsky (1980b [1978]). *New York Review of Books* (23 October): 47–50.

Hagège, Claude. 1981 [1976]. *Critical reflections on generative grammar.* Translated by Robert A. Hall, Jr. Lake Bluff, IL: Jupiter Press.

Hale, Kenneth. 1976. Theoretical linguistics in relation to American Indian communities. In Chafe (1976:35–50).

Hall, Robert A., Jr. 1950. *Leave your language alone!* New York: Linguistica.

Hall, Robert A., Jr. 1968. *Essay on language.* New York: Chilton Books.

Hall, Robert A., Jr. 1969. Some recent studies on Port-Royal and Vaugelas. *Acta Linguistica* 12:207–33.

Hall, Robert A., Jr. 1987a [1965–85]. *Linguistics and pseudo-linguistics: Selected essays, 1965–1985.* Philadelphia: John Benjamins.

Hall, Robert A., Jr., editor. 1987b. *Leonard Bloomfield: Essays on his life and work.* Philadelphia: John Benjamins.

Hall, Robert A., Jr. 1990. *A life for language.* Studies in the History of the Language Sciences, vol. 55. Philadelphia: John Benjamins.

Halle, Morris. 1959a [1958]. Questions of linguistics. *Supplemento al Il nuovo cimento* 13, ser. 10:494–517.

Halle, Morris. 1959b. *The sound pattern of Russian.* The Hague: Mouton.

Halle, Morris, and others, editors. 1956. *For Roman Jakobson.* The Hague: Mouton.

Halle, Morris, Joan Bresnan, and George Miller, editors. 1978. *Linguistic theory and psychological reality.* Cambridge, MA: MIT Press.

Hankamer, Jorge. 1973. Unacceptable ambiguity. *Linguistic Inquiry* 4:17–68.

Harman, Gilbert. 1968. Review of Chomsky (1966a). *Philosophical Review* 77:229–35.

Harman, Gilbert. 1972. Deep structure as logical form. In Davidson and Harman (1972:30–45).

Harman, Gilbert, editor. 1974. *On Noam Chomsky: Critical essays.* New York: Anchor Books.

Harman, Gilbert. 1988. Cognitive science? In Hirst (1988:258–68).

Harris, R. Allen. 1989. Argumentation in *Syntactic Structures. Rhetoric Society Quarterly* 19:103–24.

Harris, R. Allen. 1990. The life and death of generative semantics. Ph.D. dissertation for Rensselaer Polytechnic Institute.

Harris, R. Allen. 1991. Review of George (1990). *Word* 42:327–35.

Harris, Zellig S. 1941. Review of Trubetskoy (1939). *Language* 17:345–49.

Harris, Zellig S. 1951 [1947]. *Structural linguistics.* [First published as *Methods in structural linguistics.*] Chicago: University of Chicago Press.

Harris, Zellig S. 1954. Distributional structure. *Word* 10:140–62.

Harris, Zellig S. 1957. Co-occurrence and transformation in linguistic structure. *Language* 33:283–340.

Harris, Zellig S. 1968. *The mathematics of language.* Dordrecht, Netherlands: Reidel.

Harris, Zellig S. 1970 [1952–65]. *Papers in structural and transformational linguistics.* Dordrecht, Netherlands: Reidel.

Harris, Zellig S. 1973. Review of Bloomfield (1970). *International Journal of American Linguistics* 39:252–55.

Hathaway, Baxter. 1962. Generative grammar: Toward unification and simplification. *English Journal* 51:94–99, 113.

Hathaway, Baxter. 1967. *A transformational syntax: The grammar of Modern English.* New York: Ronald Press.

Hawking, Stephen W. 1985 [1980]. Is the end in sight for theoretical physics? In Boslaugh (1985:131–50).

Hawking, Stephen W. 1988. *A brief history of time: From the big bang to black holes.* Toronto: Bantam Books.

Hayes, Curtis W. 1966. A study in prose styles: Edward Gibbon and Ernest Hemingway. *Texas Studies in Literature and Language* 4:371–86.

Hayes, Curtis W., Jacob Orenstein, and William W. Gage. 1977. *The ABCs of languages and linguistics: A practical primer to language sciences in today's world. The new, revised, and expanded edition.* Silver Spring, MD: The Institute of Modern Languages.

Hayes, John R., editor. 1970. *Cognition and the development of language.* New York: Wiley and Sons.

Heringer, James T. 1970. Research on quantifier-negative idiolects. In Chicago Linguistic Society (1970:287–95).

Hill, Archibald A., editor. 1953. *Report of the fourth annual round table meeting on linguistics and language teaching.* Washington: Georgetown University Press.

Hill, Archibald A. 1958. *Introduction to linguistic structure: From sound to sentence in English.* New York: Harcourt, Brace, and World.

Hill, Archibald A. 1961. Grammaticality. *Word* 17:1–10.

Hill, Archibald A., editor. 1962a [1956]. *The first Texas conference on problems of linguistic analysis in English.* Austin: University of Texas Press.

Hill, Archibald A., editor. 1962b [1957]. *The second Texas conference on problems of linguistic analysis in English.* Austin: University of Texas Press.

Hill, Archibald A., editor. 1962c [1958]. *The third Texas conference on problems of linguistic analysis in English.* Austin: University of Texas Press.

Hill, Archibald A., editor. 1969. *Linguistics today.* New York: Basic Books.

Hill, Archibald A. 1991. The Linguistic Society of America and North American linguistics, 1950–1968. *Historiographia Linguistica* 18:49–152.

Hintikka, Jaakko, and others, editors. 1973. *Approaches to natural language: Proceedings of 1970 Stanford Workshop on Grammar and Semantics.* Dordrecht, Netherlands: Reidel.

Hirst, William, editor. 1988. *The making of cognitive science: Essays in honor of George A. Miller.* Cambridge: Cambridge University Press.

Hobbes, Thomas. 1991 [1651]. *Leviathan.* Edited by Richard Tuck. Cambridge: Cambridge University Press.

Hockett, Charles F. 1940. Review of Boas (1938). *Language* 16:54–57.

Hockett, Charles F. 1948. A note on structure [a comment on Preston (1948)]. *International Journal of American Linguistics* 14:269–72.

Hockett, Charles F. 1954. Two models of grammatical description. *Word* 10:210–34.

Hockett, Charles F. 1955. *A manual of phonology.* Baltimore: Waverly Press.

Hockett, Charles F. 1958. *A course in modern linguistics.* New York: Macmillan.

Hockett, Charles F. 1965 [1964]. Sound change. *Language* 41:185–204.

Hockett, Charles F. 1967. Review of Lenneberg (1967). *Scientific American* 217:141–44.

Hockett, Charles F. 1968 [1966]. *The state of the art.* The Hague: Mouton.

Hockett, Charles F. 1980. Preserving the heritage. In Davis and O'Cain (1980:99–107).

Hockett, Charles F. 1987. *Refurbishing our foundations: Elementary linguistics from an advanced point of view.* Philadelphia: John Benjamins.

Hockney, D., and others, editors. 1973. *Contemporary research in philosophical logic and linguistic semantics.* Dordrecht, Netherlands: Reidel.

Hoenigswald, Henry M. 1986. Nineteenth-century linguistics on itself. In Bynon and Palmer (1986:172–88).

Hoffman, Abbie. 1971. *Woodstock nation.* New York: Pocket Books.

Holton, Gerald. 1988. *Thematic origins of scientific thought.* Rev. ed. Cambridge, MA: Harvard University Press. [1st ed., 1973.]

Works Cited   323

Horn, Laurence R. 1969. A presuppositional analysis of *only* and *even*. In Binnick and others (1969:98–107). Reprinted in Schiller and others (1988:162–70).

Horn, Laurence R. 1970. Ain't it hard (anymore). In Chicago Linguistic Society (1970:318–27).

Horn, Laurence R. 1988. Reprint of Horn (1969). In Schiller and others (1988:162–70).

Horn, Laurence R. 1989. *A natural history of negation.* Chicago: University of Chicago Press.

Horrocks, Geoffrey. 1987. *Generative grammar.* London: Longmans.

Householder, Fred W., Jr. 1958. Review of Hockett (1958). *Language* 35:503–27.

Householder, Fred W., Jr. 1962. Review of Lees (1960). *Word* 18:326–53.

Householder, Fred W., Jr. 1965. On some recent claims in phonological theory. *Journal of Linguistics* 1:13–34.

Householder, Fred W., Jr. 1970. Reviewer's reply. *Language Sciences* 10:35–36.

Huck, Geoffrey J., and John Goldsmith. Forthcoming. The deep structure debates: Generative semantics, interpretive semantics, and the modern orthodoxy.

Hughes, John P. 1962. *The science of language: An introduction to linguistics.* New York: Random House.

Humboldt, Wilhelm von. 1988 [1836]. *On language.* Translated by Peter Heath. Cambridge: Cambridge University Press.

Hunt, Kellog W. 1966. Recent measures in syntactic development. *Elementary English* 43:732–9.

Hunt, Kellog W. 1967 [1963]. How little sentences grow into big ones. In Frazier (1967:170–86).

Hymes, Dell, editor. 1974a. *Studies in the history of linguistics: Traditions and paradigms.* Bloomington: Indiana University Press.

Hymes, Dell. 1974b. Introduction: Traditions and paradigms. In Hymes (1974a:1–40).

Hymes, Dell, and John Fought. 1981 [1974]. *American structuralism.* The Hague: Mouton.

Ikeuchi, M. 1972. Adverbial clauses in noun phrases. *Studies in English Linguistics* 7:96–101.

Ingram, Jay. 1992. *Talk talk talk.* Toronto: Viking.

Jackendoff, Ray S. 1968. *An interpretive theory of pronouns and reflexives.* Bloomington: Indiana University Linguistic Club.

Jackendoff, Ray. S. 1971. Review of Robbins (1968). *Foundations of Language.* 7:138–42.

Jackendoff, Ray. S. 1972. *Semantic interpretation in generative grammar.* Cambridge, MA: MIT Press.

Jackendoff, Ray. S. 1975. Morphological and semantic regularities. *Language* 51:639–71.

Jackendoff, Ray. S. 1977. *X-bar syntax: A study of phrase structure.* Cambridge, MA: MIT Press.

Jackendoff, Ray S. 1984. *Semantics and cognition.* Cambridge, MA: MIT Press.

Jacobs, Roderick A., and Peter S. Rosenbaum, editors. 1970. *Readings in English transformational grammar.* Washington: Georgetown University Press.

Jakobson, Roman. 1959. Boas' view of grammatical meaning. In Goldschmidt (1959:139–144).

James, William. 1981 [1907]. *Pragmatism.* Indianapolis: Hackett Publishing.

Jazayery, Mohammad Ali, Edgar C. Polomé, and Werner Winter, editors. 1978. *Linguistic and literary studies in honor of Archibald A. Hill.* Four volumes. The Hague: Mouton.

Jespersen, Otto. 1922. *Language: Its nature, development, and origin.* London: Allen and Unwin.

Jespersen, Otto. 1937. *Analytic syntax.* London: Allen and Unwin. Most recent edition by University of Chicago Press, 1984.

Jespersen, Otto. 1949 [1909–1940]. *A Modern English grammar.* Seven volumes. London: Allen and Unwin.

Jespersen, Otto. 1954. Growth and structure of the English language. Two volumes. 9th edition. Oxford: Oxford University Press.

Johnson, David E., and Paul M. Postal. 1980. *Arc pair grammar.* Princeton, NJ: Princeton University Press.

Johnson-Laird, Philip N. 1987. Grammar and psychology. In Mogdil and Mogdil (1987a:147–56).

Joos, Martin. 1950. Description of language design. *Journal of the Acoustical Society of America* 22:701–8.

Joos, Martin, editor. 1957. *Readings in linguistics.* Washington: American Council of Learned Societies.

Joos, Martin, 1967. Bernard Bloch. *Language* 43:1–19.

Joos, Martin. 1986 [1976]. *Notes on the development of the Linguistic Society of America.* Ithaca, NY: Linguistica.

Kac, Michael B. 1975. Review of Jackendoff (1972). *Language Sciences* 36:23–31.

Kac, Michael B., editor. 1980. *Current syntactic theories.* Bloomington: Indiana University Linguistic Club.

Kachru, Braj, and others, editors. 1973. *Issues in linguistics: Papers in honor of Henry and Renee Kahane.* Urbana: University of Illinois Press.

Kampf, Louis. 1967. Review of Chomsky (1966a). *College English* 28:403–8.

Karttunen, Lauri. 1971. Definite descriptions with crossing reference. *Foundations of language* 7:157–82.

Kasher, Asa, editor. 1991. *The Chomskyan turn.* Oxford: Basil Blackwell.

Katz, Jerrold J. 1964. Semi-sentences. In Fodor and Katz (1964:400–36).

Katz, Jerrold J. 1966. *The philosophy of language.* New York: Harper and Row.

Katz, Jerrold J. 1970. Interpretive semantics vs. generative semantics. *Foundations of Language* 6:220–59.

Katz, Jerrold J. 1971. Generative Semantics is Interpretive Semantics. *Linguistic Inquiry* 2:313–31.

Katz, Jerrold J. 1972a. Interpretive semantics meets the zombies. *Foundations of Language* 549–96.

Katz, Jerrold J. 1972b. *Semantic theory.* New York: Harper and Row.

Katz, Jerrold J. 1976. Global rules and surface structure interpretation. In Bever and others (1976:415–25).

Katz, Jerrold J., and Thomas G. Bever. 1976 [1974]. The fall and rise of empiricism. In Bever and others (1976:11–64).

Katz, Jerrold J., and Jerry A. Fodor, editors. 1964a. *The structure of language.* Englewood Cliffs, NJ: Prentice-Hall.

Katz, Jerrold J., and Jerry A. Fodor. 1964b [1963]. The structure of a semantic theory. In Fodor and Katz (1964:479–518).

Katz, Jerrold J., and Paul M. Postal. 1964. *An integrated theory of linguistic descriptions.* Cambridge, MA: MIT Press.

Kay, Martin. 1970 [1967]. From semantics to syntax. In Bierwisch and Heidolph (1970:114–26).

Keenan, Edward L. 1975a. Some universals of passive in relational grammar. In Grossman and others (1975a:340–52).

Keenan, Edward L., editor. 1975b. *Formal semantics of natural language.* New York: Cambridge University Press.

Keenan, Edward L., and Bernard Comrie. 1977 [1972]. Non phrase accessibility and universal grammar. *Linguistic Inquiry* 8:63–99.

Kempson, Ruth. 1975. *Presupposition and the delimination of semantics.* Cambridge: Cambridge University Press.

Works Cited

Kesterton, Michael. 1993. Social studies. *The Globe and Mail* (11 February):A26.

Kiefer, Ferenc, editor. 1969. *Studies in syntax and semantics.* Dordrecht, Netherlands: Reidel.

Kiefer, Ferenc, and Nicolas Ruwet, editors. 1973. *Generative grammar in Europe.* Dordrecht: Reidel.

Kimball, John, editor. 1972. *Syntax and semantics.* Vol. 1. New York: Seminar Press.

Kimball, John. 1973. *The formal theory of grammar.* Englewood Cliffs, NJ: Prentice-Hall.

Kiparsky, Paul. 1974. From paleogrammarians to neogrammarians. In Hymes (1974a:331–45).

Kirsner, Robert S. 1991. Review of Rudzka-Ostyn (1988). *Studies in Language* 15:149–74.

Klima, Edward S. 1964 [1959]. Negation in English. In Fodor and Katz (1964:246–323).

Kneupper, Charles W., editor. 1989. *Rhetoric and ideology: Compositions and criticisms of power.* Arlington, TX: Rhetoric Society of America.

Koerner, E. F. Konrad, editor. 1975. *The transformational-generative paradigm and modern linguistic theory.* Philadelphia: John Benjamins.

Koerner, E. F. Konrad, editor. 1984. *Edward Sapir: Appraisals of his life and work.* Philadelphia: John Benjamins.

Koerner, E. F. Konrad. 1990. Wilhelm von Humboldt and North American ethnolinguistics: Boas (1984)—Hymes (1961). In Dinneen and Koerner (1990:111–29).

Koerner, E. F. Konrad, editor. 1991. *First person singular II.* Studies in the history of the language sciences, vol. 61. Philadelphia: John Benjamins.

Koerner, E. F. Konrad, and Matsuji Tajima, compilers. 1986. *Noam Chomsky: A personal bibliography.* Amsterdam Studies in the theory and history of linguistic science, series 5, volume 11. Philadelphia: John Benjamins.

Kretzmann, Norman. 1975. Transformationalism and the Port-Royal grammar. In Arnauld and Lancelot (1975 [1660]:176–95).

Krüger, Gustav. 1914 [1897–1911]. *Syntax der englischen Sprache.* Seven volumes. Dresden: C. A. Koch.

Kuhn, Thomas S. 1970. *The structure of scientific revolutions.* 2d ed. Chicago: University of Chicago Press. [1st ed., 1962.]

Kuiper Koenraad. 1975. Discussion of Ray C. Dougherty's Generative Semantics methods: A Bloomfieldian counterrevolution. *International Journal of Dravidian Linguistics* 4:159–61.

Kuroda, S.-Y. 1969 [1965]. Attachment transformations. In Reibel and Schane (1969:331–51).

Kuroda, S.-Y. 1976 [1967]. Linguistic harmony notes. In McCawley (1976a:227–28).

LaGaly, Michael W., and others. 1974. *Papers from the tenth regional Chicago Linguistic Society meeting.* Chicago: Chicago Linguistic Society.

Lakatos, Imre, and Alan Musgrave, editors. 1970. *Criticism and the growth of knowledge.* Cambridge: Cambridge University Press.

Lakoff, George. 1963. See Lakoff (1976a [1963]).

Lakoff, George. 1967. Letter to Noam Chomsky on "Remarks on Nominalization." Mimeograph.

Lakoff, George. 1968a. Pronouns and reference, Parts I and II. Bloomington: Indiana University Linguistic Club.

Lakoff, George. 1968b [1966]. Deep and surface grammar. Bloomington: Indiana University Linguistic Club.

Lakoff, George. 1969a. Empiricism without facts. *Foundations of language* 5:118–27.

Lakoff, George. 1969b. On derivational constraints. In Binnick and others (1969:117–39).

Lakoff, George. 1970a [1965]. *Irregularity in syntax* [On the nature of syntactic irregularity]. New York: Holt, Rinehart and Winston.

Lakoff, George. 1970b. Global rules. *Language* 46:627–40.

Lakoff, George. 1971a. Foreword. In Borkin and others (1968:i–v).

Lakoff, George. 1971b. On generative semantics. In Steinberg and Jakobovits (1971:232–96).

Lakoff, George. 1971c. Presupposition and relative well-formedness. In Steinberg and Jakobovits (1971:329–44).

Lakoff, George. 1971d. *Linguistik und natürliche Logik.* Translated by Udo Fries and Herald Mitterman. Frankfurt: Atheneum.

Lakoff, George. 1972a. Performative antimonies. *Foundations of Language* 7:569–72.

Lakoff, George. 1972b. Linguistics and natural logic. In Davidson and Harmon (1972:545–665).

Lakoff, George. 1972c. The arbitrary basis of transformational grammar. *Language* 48:76–87.

Lakoff, George. 1973a [1970]. Some thoughts on transderivational constraints. In Kachru and others (1973:442–52).

Lakoff, George. 1973b. Fuzzy grammar and the performance/competence terminology game. In Corum and others (1973:271–91).

Lakoff, George. 1973c. Deep language [letter to the editor]. *New York Review of Books* (8 February):33.

Lakoff, George. 1973d. Hedges: A study in meaning criteria and the logic of fuzzy concepts. In Hockney and others (1973:221–72) Also in Peranteau and others (1972:183–228).

Lakoff, George. 1974. Syntactic amalgams. In LaGaly and others (1974:321–34). Also in Schiller and others (1988:25–45).

Lakoff, George. 1975. Pragmatics in natural logic. In Keenan (1975b:253–86).

Lakoff, George. 1976a [1963]. Toward generative semantics. In McCawley (1976a:43–61).

Lakoff, George. 1976b [1968]. Pronouns and reference. In McCawley (1976a:275–335).

Lakoff, George. 1977a. Linguistic gestalts. In Beach and others (1977:236–87).

Lakoff, George. 1977b. Pragmatics in natural logic. In Rogers, Wall, and Murphy (1977:107–34).

Lakoff, George. 1987. *Women, fire, and dangerous things: What categories reveal about the mind.* Chicago: University of Chicago Press.

Lakoff, George. 1988 [1974]. Syntactic amalgams. In Schiller and others (1988:25–45).

Lakoff, George. 1992. Philosophical speculation and cognitive science: Comments on William Lycan's *Logical Form in Natural Language.* In Brentasi and others (1992:173–98).

Lakoff, George, and Mark Johnson. 1980. *Metaphors we live by.* Chicago: University of Chicago Press.

Lakoff, George, and John Robert [Háj] Ross. 1976 [1967]. Is deep structure necessary? In McCawley (1976a:159–64).

Lakoff, George, and Henry Thompson. 1975a. Introducing Cognitive Grammar. *Papers from the first annual meeting of the Berkeley Linguistics Society,* 295–313.

Lakoff, George, and Henry Thompson. 1975b. Dative questions in Cognitive Grammar. In Grossman and others (1975:337–50).

Lakoff, George, and Mark Turner. 1989. *More than cool reason: A field guide to the poetic metaphor.* Chicago: University of Chicago Press.

Lakoff, Robin Tolmach. 1968. *Abstract syntax and Latin complementation.* Research Monograph No. 49. Cambridge, MA: MIT Press.

Lakoff, Robin Tolmach. 1969a. A syntactic argument for negative transportation. In Binnick and others (1969:140–47).

Lakoff, Robin Tolmach. 1969b. Review: *Grammaire générale et raisonée. Language* 45:343–64.

Lakoff, Robin Tolmach. 1971a [1969]. If's, and's, and but's about conjunction. In Fillmore and Langendoen (1971:115–50).

Lakoff, Robin Tolmach. 1971b. Passive resistance. In Chicago Linguistic Society (1971:149–61).

Lakoff, Robin Tolmach. 1972a. The pragmatics of modality. In Peranteau and others (1972:229–46).

Lakoff, Robin Tolmach. 1972b. Language in context. *Language* 48:907–27.

Lakoff, Robin Tolmach. 1973a. The logic of politeness; or, minding your P's and Q's. In Corum and others (1973:292–305).

Lakoff, Robin Tolmach. 1973b. Questionable answers and answerable questions. In Kachru and others (1973:453–67).

Lakoff, Robin Tolmach. 1973c. Language and woman's place. *Language and Society* 2:45–80.

Lakoff, Robin Tolmach. 1977. What can you do with words: Politeness, pragmatics, and performatives. In Rogers, Wall, and Murphy (1977:78–106).

Lakoff, Robin Tolmach. 1989. The way we were; or, the real actual truth about generative semantics: A memoir. *Journal of Pragmatics* 13:939–88.

Lamb, Sidney. 1967. Review of Chomsky (1964b; 1965 [1964]). *American Anthropologist* 69:411–15.

Lancelot of Benwick [Robert I. Binnick], Morgan le Fay [Jerry Morgan], and The Green Knight [Georgia Green]. 1976 [1968]. Camelot, 1968. In McCawley (1976a:249–74).

Langacker, Ronald W. 1972. Review of Chafe (1970a). *Language* 48:134–61.

Langacker, Ronald W. 1973. Predicate raising: Some Uto-Aztecan evidence. In Kachru and others (1973:468–91).

Langacker, Ronald W. 1976a. Discussion [of Voegelin and Voegelin, 1976]. In Chafe (1976:99–104).

Langacker, Ronald W. 1976b. *Non-distinct arguments in Uto-Aztecan*. Berkeley: University of California Press.

Langacker, Ronald W., editor. 1977. *Studies in Uto-Aztecan 1: An overview of Uto-Aztecan grammar.* SIL Publications in Linguistics, vol. 56. Dallas: Summer Institute of Linguistics; Arlington: University of Texas.

Langacker, Ronald W., editor. 1979. *Studies in Uto-Aztecan 2: Modern Aztecan grammatical sketches.* SIL publications in linguistics, vol. 56. Dallas: Summer Institute of Linguistics; Arlington: University of Texas.

Langacker, Ronald W., editor. 1982. *Studies in Uto-Aztecan 3: Uto-Aztecan grammatical sketches.* SIL publications in linguistics, vol. 56. Dallas: Summer Institute of Linguistics; Arlington: University of Texas.

Langacker, Ronald W. 1986. An introduction to cognitive grammar. *Cognitive Science* 10:1–40.

Langacker, Ronald W. 1987. *Foundations of cognitive grammar 1: Theoretical prerequisites.* Stanford, CA: Stanford University Press.

Langacker, Ronald W. 1988a. Review of Lakoff (1987). *Language* 64:384–95.

Langacker, Ronald W. 1988b. An overview of cognitive grammar. In Rudzka-Ostyn (1988:3–48).

Langacker, Ronald W. 1990. *Concept, image, and symbol: The cognitive basis of grammar.* Berlin: Mouton de Gruyter.

Langacker, Ronald W. 1991. *Foundations of cognitive grammar 2: Descriptive application.* Stanford, CA: Stanford University Press.

Lawler, John. 1972. Generic to a fault. In Peranteau and others (1972:247–58).

Lawler, John. 1980. Remarks on [J. Ross on [G. Lakoff on cognitive grammar [and meta-phors]]]. In Kac (1980:51-61).

Leber, Justin. 1975. *Noam Chomsky.* Boston: Twayne.

Lees, Robert B. 1957. Review of Chomsky (1957a). *Language* 33:375-408.

Lees, Robert B. 1960. Review of Bolinger (1957). *Word* 16:119-25.

Lees, Robert B. 1962 [1960]. The grammatical basis of some semantic notions. In Choseed and Guss (1962:5-20).

Lees, Robert B. 1968 [1960; with prefaces dated 1962, 1964, and 1965]. *The grammar of English nominalizations.* 5th printing. Bloomington: Indiana University Press.

Lees, Robert B. 1970a [1967]. On very deep grammatical structure. In Jacobs and Rosenbaum (1970:134-44).

Lees, Robert B. 1970b [1967]. Problems in the grammatical analysis of English nominal compounds. In Bierwisch and Heidolph (1970:174-86).

Lehmann, Winfred. 1972. *Descriptive linguistics.* New York: Random House.

Leibniz, Gottfried Wilhelm. 1949 [1705]. *New essays concerning human understanding.* 3d ed. Translated by Alfred Gideon Langley. La Salle, IL: Open Court.

Lenneberg, Eric. 1967. *Biological foundations of language.* New York: Wiley and Sons.

Lepschy, Guilio. 1986. European linguistics in the twentieth century. In Bynon and Palmer (1986:189-201).

Lester, Mark. 1967. The value of transformational grammar in teaching composition. *College Composition and Communication* 18:227-31.

Levi, Judith N. 1978. *The syntax and semantics of complex nominals.* New York: Academic Press.

Levin, Samuel R. 1963. On automatic production of poetic sequences. *Texas studies in literature and language* 5:138-46.

Levin, Samuel R. 1965. Internal and external deviation in poetry. *Word* 21:225-37.

Levin, Samuel R. 1967. Poetry and grammaticalness. In Levin and Chatman (1967:224-30).

Levin, Samuel R., and Seymour Chatman, editors. 1967. *Essay on the language of literature.* Boston.

Levinson, Stephen C. 1983. *Pragmatics.* Cambridge: Cambridge University Press.

Liefrink, Frans. 1973. *Semantico-syntax.* London: Longman.

Lightfoot, David. 1980. Trace theory and explanation. In Moravscik and Wirth (1980:137-66).

Long, Ralph. 1958. Remarks made in Hill (1962c [1958]).

Longuet-Higgins, H. C., and others, organizers. 1981. *The psychological mechanisms of language: A joint symposium of the Royal Society and the British Academy.* London: Royal Society; British Academy.

Lounsbury, Floyd. 1956. A semantic analysis of Pawnee Kinship usage. *Language* 32:158-94.

Lounsbury, Floyd. 1964 [1962]. The structural analysis of kinship semantics. In Lunt (1964 [1962]:1073-93).

Luce, R. Duncan, and others, editors. 1963. *Handbook of mathematical psychology.* Three volumes. New York: Wiley and Sons.

Lunt, Horace G., editor. 1964 [1962]. *The proceedings of the ninth international congress of linguists.* The Hague: Mouton.

Luthy, M. J. 1977. Why transformational grammar fails in the classroom. *College Composition and Communication* 28:352-55.

Lyell, Sir Charles. 1870. *The geological evidences of the antiquity of man.* London: John Murray.

Lyons, John. 1970a. *Noam Chomsky.* Fontana Modern Masters Series. Cambridge: Cambridge University Press.

Lyons, John, editor. 1970b. *New horizons in linguistics.* Baltimore: Penguin Books.

Lyons, John. 1970c. Generative syntax. In Lyons (1970b).

Lyons, John. 1991. *Chomsky.* 3rd ed. (of *Noam Chomsky*). Fontana Modern Masters Series. Cambridge: Cambridge University Press.

Lyons, John, and R. J. Wales, editors. 1966. *Psycholinguistics papers: Proceedings of the 1966 Edinburgh Conference.* Edinburgh: Edinburgh University Press.

MacCorquodale, Kenneth. 1970. On Chomsky's review of Skinner's *Verbal behavior. Journal of the Experimental Analysis of Behavior* 13:83–99.

Maclay, Howard. 1971. Overview. In Steinberg and Jakobovits (1971:157–82).

MacKay, D. G., and T. G. Bever. 1967. In search of ambiguity. *Perception and Psychophysics* 2:193–200.

Maher, J. Peter. 1973a. Review of *Linguistic change and generative theory* [Stockwell and Macaulay (1972 [1969])]. *Language Sciences* 25:47–52.

Maher, J. Peter. 1973b. Repartee: Reply to Zwicky. *Language Sciences* 28:30–31.

Makkai, Adam, editor. 1975. *The first LACUS forum.* Washington: Hornbeam Press.

Mathews, G. H. 1965. *Hidatsa syntax.* The Hague: Mouton.

Mathews, P. H. 1986. Distributional syntax. In Bynon and Palmer (1986:245–79).

May, Robert C. 1977. The grammar of quantification [Ph.D. dissertation, MIT linguistics department]. Bloomington: Indiana University Linguistic Club.

May, Robert C. 1985. *Logical form.* Cambridge, MA: MIT Press.

Mayr, Ernst. 1988. *Toward a new philosophy of biology: Observations of an evolutionist.* Cambridge, MA: Harvard University Press.

McCawley, James D. See also: Quang Phuc Dong; Yuck Phoo.

McCawley, James D. 1968a. The role of semantics in a grammar. In Bach and Harms (1968:125–70).

McCawley, James D. 1968b [1965]. *The phonological component of a grammar of Japanese.* The Hague: Mouton.

McCawley, James D. 1970. Review: *Analytic syntax* (Jespersen, 1937). *Language* 46:442–49.

McCawley, James D. 1972. A program for logic. In Davidson and Harman (1972:498–544).

McCawley, James D. 1973a [1970]. Syntactic and logical arguments for semantic structures. In Fujimura (1973:260–376).

McCawley, James D. 1973b. Verbs of bitching. In Hockney and others (1973:313–32).

McCawley, James D. 1974a [1971]. Prelexical syntax. In Seuren (1974:29–42).

McCawley, James D. 1974b. On what is deep about Deep Structure. In Palermo and Weimer (1974:125–8).

McCawley, James D. 1974c. Review of *The sound pattern of English* (Chomsky and Halle, 1968). *International Journal of American Linguistics* 40:50–88.

McCawley, James D. 1975. Discussion of Ray C. Dougherty's "Generative Semantic methods: A Bloomfieldian counterrevolution." *International Journal of Dravidian Linguistics* 4:151–58.

McCawley, James D., editor. 1976a [1960–1970]. *Syntax and semantics 7: Notes from the linguistic underground.* New York: Academic Press.

McCawley, James D. 1976b [1964–1971]. *Grammar and meaning.* New York: Academic Press.

McCawley, James D. 1979 [1965–1967]. *Adverbs, vowels, and other objects of wonder.* Chicago: University of Chicago Press.

McCawley, James D. 1980a. A review of Newmeyer (1980a). *Linguistics* 18:911-30.

McCawley, James D. 1980b. An un-syntax. In Moravscik and Wirth (1980:167-93).

McCawley, James D. 1981. *Everything that linguists have always wanted to know about logic**. Chicago: University of Chicago Press.

McCawley, James D. 1982a. How far can you trust a linguist? In Simon and Scholes (1982:75-87).

McCawley, James D. 1982b [1973-1979] *Thirty million theories of grammar*. Chicago: University of Chicago Press.

McCawley, James D. 1988. *The syntactic phenomena of English*. Two volumes. Chicago: University of Chicago Press.

McCawley, James D. 1989. INFL, Spec, and other fabulous beasts. *Behavioral and brain sciences* 12:350-52.

McCloskey, James. 1988. Syntactic theory. In Newmeyer (1988b:18-59).

McDavid, Raven I. 1947. Pure and applied linguistics. *Studies in Linguistics* 5:27-32.

McDavid, Raven I. 1954. Review of Warfel (1952). *Studies in Linguistics* 12:30-32.

McMahan, Lee. 1963. Grammatical analysis as part of understanding a sentence. Unpublished Ph.D dissertation, Harvard University, Cambridge, MA.

McNeill, D. 1970. *The acquisition of language*. New York: Harper and Row.

McQuown, Norman A. 1952. Review of Harris (1951 [1947]). *Language* 28:495-504.

Mehler, Jacques. 1963. Some effects of grammatical transformations on recall of English sentences. *Journal of Verbal Learning and Verbal Behavior* 2:346-51.

Mehta, Ved. 1971. *John is easy to please*. London: Secker & Warburg.

Miller, George A., Eugene Galanter, and Karl H. Pribram. 1960. *Plans and the structure of behavior*. New York: Henry Holt and Company.

Miller, George A., and K. A. McKean. 1964. A chronometric study of some relations between sentences. *Quarterly Journal of Experimental Psychology* 16:297-308.

Miller, J. 1975. The parasitic growth of deep structures. *Foundations of Language*. 13:361-89.

Miller, Jonathan. 1983. *States of mind*. New York: Pantheon.

Mogdil, Sohan, and Celia Mogdil, editors. 1987a. *Noam Chomsky: Consensus and controversy*. Falmer International Master-Minds Challenged, vol. 3. New York: Falmer Press.

Mogdil, Sohan, and Celia Mogdil, editors, 1987b. *B. F. Skinner: Consensus and controversy*. Falmer International Master-Minds Challenged, vol. 5. New York: Falmer Press.

Mohrmann, Christine, Alf Sommerfelt, and Joshua Whatmough, eds. 1961. *Trends in European and American linguistics 1930-1960*. Antwerp: Spectrum.

Moore, Terrence, and Christine Carling. 1982. *Understanding language: Towards a post-Chomskyan linguistics*. London: Macmillan.

Moravcsik, Edith A., and Jessica R. Wirth, editors. 1980. *Current approaches to syntax*. Syntax and Semantics, vol. 13. New York: Academic Press.

Morgan, Jerry L. 1969a. On the treatment of presupposition in transformational grammar. In Binnick and others (1969:167-77).

Morgan, Jerry L. 1969b. On arguing about semantics. *Papers in linguistics* 1:49-70.

Morgan, Jerry L. 1973. Sentence fragments and the notion 'sentence.' In Kachru and others (1973:719-51).

Morgan, Jerry L. 1976 [1968 or 1969]. Cryptic note II and WAGS III. In McCawley (1976a:337-45).

Morgan, Jerry L. 1977. Conversational postulates revisisted. *Language* 53:277-84.

---

*but were ashamed to ask.

Morris, Charles. 1938. *Foundations of the theory of signs.* Chicago: University of Chicago Press.

Mufwene, Salikoko, and others, editors, 1976. *Papers from the twelfth regional Chicago Linguistic Society meeting.* Chicago: Chicago Linguistic Society.

Muntz, Jim. 1972. Reflections of the development of transformational theories. In Plötz (1972:251–74).

Murray, Stephen O. 1980. Gatekeepers and the "Chomskyan Revolution." *Journal of the History of the Behavioral Sciences* 16:73–88.

Murray, Stephen O. 1983. *Group formation in social science.* Current Inquiry into Language, Linguistics and Human Communication, vol. 44. Edmonton: Linguistic Research Institute.

Murray, Stephen O. 1991. The first quarter century of the Linguistic Society of America, 1924–49. *Historiographia Linguistica* 18:1–49.

Murray, Stephen O. Forthcoming. *American linguistic theorists and theory groups: A social history* [Second edition of *Group formation in social science*]. Philadelphia: John Benjamins.

Nagel, Ernest. 1961. *The structure of science.* New York: Harcourt, Brace, and World.

Nagel, Ernest, Patrick Suppes, and Alfred Tarski. 1962. *Logic, methodology and philosophy of science.* Stanford, CA: Stanford University Press.

Napoli, Donna Jo, and Emily Norwood Rando. 1979. *Syntactic argumentation.* Washington: Georgetown University Press.

Neisser, Ulric. 1988. Cognitive recollections. In Hirst (1988:81–88).

Neubauer, Paul. 1970. On the notion 'chopping rule.' In Chicago Linguistic Society (1970:400–407).

Neubauer, Paul. 1972. SUPER-EQUI revisited. In Peranteau and others (1972:287–93).

Newman, Paul. 1978. Review of *La grammaire generative* (Hagège 1981 [1976]). *Language* 54:925–9.

Newman, Stanley S. 1951. Review of Sapir (1949b). *International Journal of American Linguistics* 17:180–86. Reprinted in Koerner (1984:59–65).

Newmeyer, Frederick J. 1970. On the alleged boundary between syntax and semantics. *Foundations of Language* 6:178–86.

Newmeyer, Frederick J. 1971. A problem with the verb-initial hypothesis. *Papers in Linguistics* 4:390–94.

Newmeyer, Frederick J. 1979. Review of Levi (1978). *Language* 55:396–407.

Newmeyer, Frederick J. 1980a. *Linguistic theory in America.* New York: Academic Press.

Newmeyer, Frederick J. 1980b. A reply to James McCawley's review of *Linguistic theory in America. Linguistics* 18:931–37.

Newmeyer, Frederick J. 1986a. *Linguistic theory in America.* 2d ed. New York: Academic Press.

Newmeyer, Frederick J. 1986b. *The politics of linguistics.* Chicago: Chicago University Press.

Newmeyer, Frederick J. 1988a. Bloomfield, Jakobson, Chomsky, and the roots of generative grammar. Manuscript of a paper delivered at the First Annual Meeting of the North American Association for the History of the Language Sciences.

Newmeyer, Frederick J., editor. 1988b. *Linguistics: The Cambridge survey.* Two volumes. London: Cambridge University Press.

Newmeyer, Frederick J. 1990. Competence vs. performance; theoretical vs. applied: The development and interplay of two dichotomies in modern linguistics. In Dinneen and Koerner (1990:167–82).

Newmeyer, Frederick J. 1991 [1989]. Rules and principles in the historical development of generative syntax. In Kasher (1991:200–30).

Newmeyer, Frederick J., and Joseph Emonds. 1971. The linguist in American society. In Chicago Linguistic Society (1971:285–303).

Newton, Sir Isaac. 1960 [1686]. *Mathematical principles of natural philosophy.* Translated by Andrew Motte [1729], revised by Florian Cajori [1934]. Berkeley, CA: University of California Press.

Nida, Eugene A. 1951. A system for the description of semantic elements. *Word* 7:1–14.

Nida, Eugene. 1960 [1943]. *A synopsis of English syntax.* Norman, OK: The Summer Institute of Linguistics.

Nida, Eugene A. 1975. *Componential analysis of meaning.* The Hague: Mouton.

O'Donnell, W. R. 1974. On generative gymnastics. *Archivum linguistica* (new series) 5:53–82.

Ohmann, Richard. 1964. Generative grammars and the concept of literary style. *Word* 20:423–39.

Ohmann, Richard. 1966. Literature as sentences. *College English* 27:261–67.

O'Neill, William L. 1971. *Coming apart: An informal history of America in the 1960's.* Chicago: Quadrangle Books.

Onions, C. T. 1911. *An advanced English syntax.* London: Swan Sonnenshein and Company.

Ortegay Gasset, Jose. 1959. The difficulty of reading. *Diogenes* 28:1–17.

Otero, Carlos P. 1986. Dissertations written under the supervision of Noam Chomsky. In Koerner and Tajima (1986:183–204).

Otero, Carlos P. 1988. The third emancipatory phase of history [Introduction]. In Chomsky (1988c:22–81).

Palermo, David, and Walter Weimar, editors. 1974. *Cognition and the symbolic processes.* Hillsdale, NJ: Lawrence Erlbaum.

Papert, Seymour. 1988. One AI or many? *Daedalus* 117:1–14.

Parret, Herman. 1974 [1972–1973]. *Discussing language.* The Hague: Mouton.

Partee, Barbara Hall. 1971 [1969]. On the requirement that transformations preserve meaning. In Fillmore and Langendoen (1971:1–22).

Partee, Barbara Hall, editor. 1976. *Montague grammar.* New York: Academic Press.

Passmore, John. 1985. *Recent philosophers.* La Salle, IL: Open Court.

Patterson, Francine. 1978. Conversations with a gorilla. *National Geographic* (October):438–65.

Patterson, Francine, and Eugene Linden. 1981. *The education of Koko.* New York: Holt, Rinehart.

Paul, Hermann. 1891 [1886]. *Principles of the history of language.* Translated by H. A. Strong. London: Longmans, Green, and Company.

Peranteau, Paul M., and others, editors. 1972. *Papers from the eighth regional Chicago Linguistic Society meeting.* Chicago: Chicago Linguistic Society.

Percival, W. Keith. 1971. Review of Salus (1969). *Language* 47:181–84.

Perlmutter, David M. 1971 [1968]. *Deep and surface structure constraints in syntax.* New York: Holt, Rinehart and Winston.

Perlmutter, David M. 1980. Relational grammar. In Moravcsik and Wirth (1980:195–230).

Perlmutter, David M., and C. Rosen, editors. 1984. *Studies in relational grammar.* Vol. 2. Chicago: University of Chicago Press.

Peters, Stanley, editor. 1972. *Goals of linguistic theory.* Englewood Cliffs, NJ: Prentice-Hall.

Peters, Stanley, and Robert Ritchie. 1969. A note on the universal base hypothesis. *Journal of Linguistics* 5:150–52.

Peters, Stanley, and Robert Ritchie. 1971. On restricting the base component of transformational grammar. *Information and Control* 18:493–501.

Peters, Stanley, and Robert Ritchie. 1973a. On the generative power of transformational grammars. *Information Sciences* 6:49–83.

Peters, Stanley, and Robert Ritchie. 1973b. Nonfiltering and local-filtering transformational grammars. In Hintikka and others (1973:180–94).

Piattelli-Palmarini, M., editor. 1983. *Language and learning: The debate between Jean Piaget and Noam Chomsky.* London: Routledge and Kegan Paul.

Plötz, Senta, editor. 1972. *Transformational analysis: The transformational theory of Zellig Harris and its development.* Frankfurt: Atheneum Verlag.

Pop, Fom [Thomas Priestly]. 1970. A concise history of modern art. In Zwicky and others (1970:121–24).

Popper, Sir Karl. 1970 [1965]. Normal science and its dangers. In Lakatos and Musgrave (1970:51–9).

Postal, Paul M. 1964. *Constituent structure.* Publication 30 of the Indiana University Research Center in Anthropology, Folklore, and Linguistics. Bloomington: Indiana University Press.

Postal, Paul M. 1966a. On so-called "pronouns" of English. In Dinneen (1966:177–206).

Postal, Paul M. 1966b. Review of Martinet (1964). *Foundations of Language* 2:151–77.

Postal, Paul M. 1966c. Review of Dixon (1963). *Language* 42:84–92.

Postal, Paul M. 1968 [1965]. *Aspects of phonological theory.* New York: Harper and Row.

Postal, Paul M. 1970. On coreferential complement subject deletion. *Linguistic Inquiry* 1:439–500.

Postal, Paul M. 1971a [1968]. *Cross-over phenomena.* New York: Holt, Rinehart and Winston.

Postal, Paul M. 1971b [1969]. On the surface verb *remind.* In Fillmore and Langendoen (1971:180–270).

Postal, Paul M. 1972a [1969]. The best theory. In Peters (1972:131–70).

Postal, Paul M. 1972b [1970]. A global constraint on pronominalization. *Linguistic Inquiry* 3:36–58.

Postal, Paul M. 1972c. On some rules that are not successive cyclic. *Linguistic Inquiry* 3:211–22.

Postal, Paul M. 1972d. A few factive facts. *Linguistic Inquiry* 3:396–400.

Postal, Paul. 1972e. A remark on the verb-initial hypothesis. *Papers in Linguistics* 5:124–37.

Postal, Paul M. 1974. *On raising: One rule of grammar and its theoretical implications.* Cambridge, MA: MIT Press.

Postal, Paul M. 1976 [1967–1970]. Linguistic anarchy notes. In McCawley (1976a:201–225).

Postal, Paul M. 1988a [1969]. Anaphoric islands. In Schiller and others (1988:67–94).

Postal, Paul M. 1988b. Topic . . . comment: Advances in linguistic rhetoric. *Natural Language and Linguistic Theory* 6:129–37.

Postal, Paul M., and Geoffrey Pullum. 1978. Traces and the description of English complementizer contraction. *Linguistic Inquiry* 9:1–30.

Poutsma, H. 1914. *A grammar of late Modern English.* 3 vols. Groningen, Netherlands: P. Noordhoff.

Premack, A. J., and D. Premack. 1972. Teaching language to an ape. *Scientific American* 227:92–99.

Preston, W. D. 1948. Review of de Goeje (1946). *International Journal of American Linguistics* 14:131–34.

Prideaux, Gary D. 1967. Review of Chomsky (1966a). *Canadian Journal of Linguistics* 13:50–51.

Prideaux, Gary D. 1985. *Psycholinguistics: The Experimental Study of Language.* New York: Guilford Press.

Prideaux, Gary D., and William J. Baker. 1976. The recognition of ambiguity. *Human Communication* 4:51–58.

Propp, Vladimir. 1968 [1928]. *Morphology of the folktale.* 2nd edition. Translated by Laurence Scott. Austin: University of Texas Press.

Pullum, Geoffrey K. 1987. Topic . . . comment: Trench-mouth comes to Trumpington Street. *Natural Language and Linguistic Theory* 5:139–47.

Pullum, Geoffrey K. 1988. Topic . . . comment: Citation etiquette beyond thunderdome. *Natural Language and Linguistic theory* 6:580–88.

Pullum, Geoffrey K. 1989. Topic . . . comment: Formal linguistics meets the boojum. *Natural Language and Linguistic Theory* 7:137–43.

Pullum, Geoffrey K. 1991 [1983–89]. *The great Eskimo vocabulary hoax and other irreverent essays on the study of language.* Chicago: University of Chicago Press.

Pullum, Geoffrey K., and Paul M. Postal. 1979. On an inadequate defence of "Trace Theory." *Linguistic Inquiry* 10:689–706.

Pullum, Geoffrey K., and Deirdre Wilson. 1977. Autonomous syntax and the analysis of auxiliaries. *Language* 53:741–88.

Putnam, Hilary. 1975 [1960–1975]. *Mind, language, and reality:* Volume 2 of *Philosophical papers.* Cambridge: Cambridge University Press.

Putnam, Hilary. 1983. What is innate and why. In Piattelli-Palmarini (1983:287–309).

Quang Phuc Dong [James D. McCawley]. 1971a [1968]. English sentences without overt grammatical subjects. In Zwicky and others (1971:3–10).

Quang Phuc Dong [James D. McCawley]. 1971b. A note on conjoined noun phrases. In Zwicky and others (1971:11–18).

Quang Phuc Dong [James D. McCawley]. 1971c. The applicability of transformations to idioms. In Chicago Linguistic Society (1971:198–205). Reprinted in Schiller and others (1988:95–100).

Quine, Willard Van Orman. 1969. Replies. In Davidson and Hintikka (1969:292–352).

Quine, Willard Van Orman. 1985. *The time of my life.* Cambridge, MA: MIT Press.

Quintilian. 1891. *Institutes of Oratory.* Translated by John Selby Watson. 2 vols. London: George Bell and Sons.

Radford, Andrew. 1981. *Transformational syntax: A student's guide to Chomsky's extended standard theory.* Cambridge: Cambridge University Press.

Raskin, Victor. 1975. Review of Steinberg and Jakobovits (1971). *Foundations of language* 13:457–67.

Raup, David M. 1986. *The Nemesis affair: A story of the death of dinosaurs and the ways of science.* New York: W. W. Norton.

Reibel, David A., and Sanford A. Schane. 1969. *Modern studies in English.* Englewood Cliffs, NJ: Prentice-Hall.

Reich, Charles A. 1970. *The greening of America: How the youth culture is trying to make America livable.* New York: Random House.

Rieber, Robert W. 1983. *Dialogues on the psychology of language and thought.* New York: Plenum Press.

Riemsdijk, Henk van, and Edwin Williams. 1986. *Introduction to the theory of grammar.* Dordrecht, Netherlands: Foris.

Roazen, Paul. 1986 [1969]. *Brother animal: The story of Freud and Tausk.* New York: New York University Press.

Robbins, Beverly L. 1968. *The definite article in English transformations.* The Hague: Mouton.

Roberts, Paul. 1962. *English sentences.* New York: Harcourt, Brace, and World.

Roberts, Paul. 1963. Linguistics and the teaching of composition. *English Journal* 52:331–335.

Roberts, Paul. 1964. *English syntax: A book of programmed lessons: An introduction to transformational grammar.* New York: Harcourt, Brace, and World.

Roberts, Paul. 1967. *Modern Grammar.* New York: Harcourt, Brace, and World.

Robertson, Priscilla. 1952. *Revolutions of 1848: A social history.* Princeton, NJ: Princeton University Press.

Robins, R. H. 1967. *A short history of linguistics.* Bloomington: Indiana University Press.

Robinson, I. 1975. *The new grammarians' funeral.* Cambridge: Cambridge University Press.

Rogers, Andy. 1971. Three kinds of physical perception verbs. In Chicago Linguistic Society (1971:206–22).

Rogers, Andy, Robert Wall, and J. P. Murphy, editors. 1977. *Proceedings of the Texas conference on performatives, presupposition, and implicature.* Arlington, VA: Center for Applied Linguistics.

Rogovin, Syrrell. 1965. *Modern English sentence structure.* New York: Random House.

Ronat, Mitsou. 1972 [1970]. A propos du verbe *remind. Studi di Linguistica Teorica ed Applicata* 2:241–67.

Rosch, Eleanor, and Barbara B. Lloyd, editors. 1978 [1976]. *Cognition and categorization.* Hillsdale, N.J.: Lawrence Erlbaum Associates.

Ross, John Robert [Háj]. 1969a. [1966]. Adjectives as noun phrases. In Reibel and Schane (1969:352–60).

Ross, John Robert [Háj]. 1969b [1967]. Auxiliaries as main verbs. *Studies in Philosophical Linguistics* 1:77–102.

Ross, John Robert [Háj]. 1970a [1967]. Gapping and the order of constituents. In Bierwisch and Heidolph (1970:249–59).

Ross, John Robert [Háj]. 1970b [1968]. On declarative sentences. In Jacobs and Rosenbaum (1970:222–72).

Ross, John Robert [Háj]. 1972a. Doubl-ing. In Kimball (1972:157–86).

Ross, John Robert [Háj]. 1972b. Parentage. *Foundations of language* 7:573.

Ross, John Robert [Háj]. 1972c. The category squish: Endstation Hauptwork. In Peranteau and others (1972:316–28).

Ross, John Robert [Háj]. 1972d. Act. In Davidson and Harman (1972:70–126).

Ross, John Robert [Háj]. 1973a. Slifting. In Gross, Halle, and Schützenberger (1973:133–72).

Ross, John Robert [Háj]. 1973b. The penthouse principle and the order of constituents. In Corum and others (1973b:397–422).

Ross, John Robert [Háj]. 1973c. Nouniness. In Fujimura (1973).

Ross, John Robert [Háj]. 1974a. There, there, (there, (there, (there . . .))). In LaGaly and others (1974:569–87).

Ross, John Robert [Háj]. 1974b. Three batons for cognitive psychology. In Palermo and Weimer (1974:63–124).

Ross, John Robert [Háj]. 1975. Clausematiness. In Keenan (1975a:422–75).

Ross, John Robert [Háj]. 1986 [1967]. *Infinite syntax!* [Constraints on variables in syntax]. Norwood: ABLEX Publishing.

Ross, John Robert [Háj]. 1991. Toward a _____ linguistics. Manuscript.

Ross, John Robert [Háj], and George Lakoff. 1967. Stative adjectives and verbs. NSF report #17. Aiken Computational Laboratory, Harvard University, Cambridge, MA.

Roszak, Theodore. 1969. *The making of a counter culture: Reflections on the technocratic society and its youthful opposition.* New York: Doubleday and Company.

Rubin, Jerry. 1971. *We are everywhere.* New York: Harper and Row.

Rudwick, Martin J. S. 1985. *The great Devonian controversy: The shaping of scientific knowledge among gentlemanly specialists.* Chicago: University of Chicago Press.

Rudzka-Ostyn, Brygida, editor. 1988. *Topics in cognitive linguistics.* Philadelphia: John Benjamins.

Russell, Bertrand. 1967 [1912]. *The problems of philosophy.* New York: Oxford University Press.

Russell, Bertrand, and Alfred North Whitehead. 1925. *Principia mathematica.* Cambridge: Cambridge University Press.

Ruwet, Nicolas. 1991 [1982-89]. *Syntax and human experience.* Edited and translated by John Goldsmith. Chicago: University of Chicago Press.

Rynin, David 1957. Vindication of L*G*C*L P*S*T*V*SM. *Proceedings and Addresses of the American Philosophy Association* 30:45-67.

Sadock, Jerrold M. 1969. Hypersentences. *Papers in Linguistics* 1:283-370.

Sadock, Jerrold M. 1970. Whimperatives. In Sadock and Vanek (1970:223-38).

Sadock, Jerrold M. 1971. Quelaratives. In Chicago Linguistic Society (1971:223-31).

Sadock, Jerrold M. 1972. Speech act idioms. In Peranteau and others (1972:329-39).

Sadock, Jerrold M. 1974a. *Towards a linguistic theory of speech acts.* New York: Academic Press.

Sadock, Jerrold M. 1974b. Read at your own risk: syntactic and semantic horrors you can find in your medicine chest. In LaGaly and others (1974:599-607), Schiller and others (1988:202-8).

Sadock, Jerrold M. 1975. The soft, interpretive underbelly of generative semantics. In Cole and Morgan (1975:283-96).

Sadock, Jerrold M. 1976. On significant generalizations: Notes on the Hallean syllogism. In Wirth (1976:85-94).

Sadock, Jerrold M. 1985a [1979]. On the performadox, or A semantic defence of the performative hypothesis. *University of Chicago working papers in linguistics* 1:160-69.

Sadock, Jerrold M. 1985b. Autolexical syntax. *Natural Languages and linguistic theory* 3:379-439.

Sadock, Jerrold M. 1988. Speech act distinctions in grammar. In Newmeyer (1988b. 2:183-97).

Sadock, Jerrold M. 1991. *Autolexical syntax: A theory of parallel grammatical representations.* Chicago: University of Chicago Press.

Sadock, Jerrold M., and Anthony L. Vanek, editors. 1970. *Studies presented to Robert B. Lees by his students.* Edmonton: Linguistic Research Institute.

Salkie, Raphael. 1990. *The Chomsky update: Linguistics and politics.* Boston: Unwin Hyman.

Salmon, Vivian. 1969. Review of Chomsky (1966a). *Journal of Linguistics* 12:177-88.

Salus, Peter. 1969. *Linguistics.* New York: Bobbs-Merrill.

Sampson, Geoffrey. 1975. *The form of language.* London: Weidenfield and Nicolson.

Sampson, Geoffrey. 1976. Review of Cohen (1974). *Journal of Linguistics* 12:177-88.

Sapir, Edward. 1922. The Takelma language of Southwestern Oregon. *Handbook of American Indian Languages.* Edited by Franz Boas. Washington: The Bureau of American Ethnography. Bulletin 40, part II, 3-296.

Sapir, Edward. 1929. The status of linguistics as a science. *Language* 5:207-14.

Sapir, Edward. 1949a [1921]. *Language: An introduction to the study of speech.* New York: Harcourt Brace.

Sapir, Edward. 1949b [1907-1939]. *Selected writings.* Edited by David G. Mandelbaum. Berkeley: University of California Press.

Saporta, Sol, editor. 1961. *Psycholinguistics: A book of readings.* New York: Holt, Rinehart and Winston.

Saporta, Sol. 1965. Review of *Psychology, Study of a science. Language* 41:95–100.

de Saussure, Ferdinand. 1966 [1915]. *Course in general linguistics.* Translated by Wade Baskin. New York: The Philosophical Library.

Savin, H. B., and E. Perchonock. 1965. Grammatical structure and immediate recall of English sentences. *Journal of Verbal Learning and Verbal Behavior* 4:348–53.

Schachter, Paul. 1962. Review of Lees (1960). [In this bibliography as "Lees (1968 [1960])"] *International Journal of American Linguistics* 28:134–46.

Schachter, Paul. 1964 [1962]. Kernel and non-kernel sentences in transformational grammar. In Lunt (1964 [1962]:692–97).

Schiller, Eric, and others. 1988 [1968–1975]. *The best of CLS: A selection of out-of-print papers from 1968 to 1975.* Chicago: Chicago Linguistic Society.

Schmerling, Susan. 1971. Presupposition and the notion of normal stress. In Chicago Linguistic Society (1971:242–53).

Searle, John R. 1972. Chomsky's revolution. *New York Review of Books* (29 June):16–24.

Sebeok, Thomas A., editor. 1966. *Current trends in linguistics 3: Theoretical foundations.* The Hague: Mouton.

Sebeok, Thomas. 1982. The not so sedulous ape. *Times Literary Supplement* (10 Sept.):976.

Selkirk, Elizabeth. 1972. The phrase phonology of English and French. Ph.D. dissertation for MIT.

Sells, Peter, 1985. *Lectures on contemporary syntactic theories.* Stanford, CA: Center for the Study of Language and Information, Stanford University.

Seuren, Pieter A. M. 1973 [1971]. The comparative. In Kiefer and Ruwet (1973:528–64).

Seuren, Pieter A. M., editor. 1974. *Semantic syntax.* Oxford: Oxford University Press.

Seuren, Pieter A. M. 1975. Referential constraints on lexical items. In Keenan (1975a:84–98).

Seuren, Pieter A. M. 1985. *Discourse semantics.* New York: Basil Blackwell.

Sgall, Petr, E. Hajičová, and E. Benešová. 1973. *Topic, focus, and generative semantics.* Kronberg: Atheneüm.

Shannon, Claude E., and Warren Weaver. 1949. *The mathematical theory of communication.* Urbana: University of Illinois Press.

Shenker, Israel. 1971. "Chomsky is difficult to please." "Chomsky is easy to please." "Chomsky is certain to please." *Horizon* 13.2 (Spring):105–9.

Shenker, Israel. 1972. Former Chomsky disciples hurl harsh words at the master. *New York Times* (10 September):70.

Simon, Thomas W., and Robert T. Scholes, editors. 1982. *Language, mind, and brain.* Hillsdale, N.J.: Lawrence Erlbaum.

Simpson, J.M.Y. 1979. *A first course in linguistics.* Edinburgh: Edinburgh University Press.

Sinha, Anil C. 1977a. Review of Southworth and Daswani (1974). *Language Sciences* 48:32–36.

Sinha, Anil C. 1977b. Some issues in semantics [Review of Seuren (1974)]. *Semiotica* 20:271–356.

Skinner, B. F. 1957. *Verbal behavior.* New York: Appleton-Century-Crofts.

Skinner, B. F. 1987. Controversy? In Mogdil and Mogdil (1987b:11–12).

Sledd, James H. 1955. A review of Trager and Smith (1957 [1951]) and Fries (1952). *Language* 31:312–45.

Sledd, James H. 1959. *A short introduction to English grammar.* Chicago: University of Chicago Press.

Sledd, James H. 1962 [1958]. Prufrock among the grammarians. In Hill (1962c [1958]).

Slobin, Dan J. 1966. Grammatical transformations and sentence comprehension in childhood and adulthood. *Journal of Verbal Learning and Verbal Behavior* 5:219–27.

Smith, Neil V. 1989. *The twitter machine: Reflections on language.* Oxford: Basil Blackwell.

Smith, Neil V., and Deirdre Wilson. 1979. *Modern linguistics: The results of Chomsky's revolution.* Bloomington: Indiana University Press.

Southworth, Franklin C., and Chandler J. Daswani. 1974. *Foundations of linguistics.* New York: Free Press.

Spencer, Herbert. 1865. *Essays: Moral, political, and aesthetic.* New York: D. Appleton.

Stark, Bruce R. 1972. The Bloomfieldian model. *Lingua* 30:385–421.

Steinberg, Danny D., and Leon A. Jakobovits. 1971. *Semantics: An interdisciplinary reader in philosophy, linguistics and psychology.* Cambridge: Cambridge University Press.

Steiner, George. 1971 [1969–1970]. *Extraterritorial papers on literature and the language revolution.* New York: Atheneüm.

Steinmann, Martin, Jr., editor. 1967. *New rhetorics.* New York: Charles Scribner's Sons.

Stewart, Ian. 1990. The symplectic revolution. *Sciences* 30 (May/June):29–36.

Stockwell, Robert P. 1977. *Foundations of syntactic theory.* Englewood Cliffs, NJ: Prentice-Hall.

Stockwell, Robert P., and Ronald K. S. Macaulay, eds. 1972 [1969]. *Linguistic change and generative theory: Essays from the UCLA Conference on Historical Linguistics in the Perspective of Transformational Theory.* Bloomington: Indiana University Press.

Stout, Carol. 1973. Problems of a Chomskyan analysis of Zuni transitivity. *International Journal of American Linguistics* 39:207–23.

Stugrin, M. 1979. Sentence-combining, conceptual sophistication and precision in technical exposition. *Technical Writing Teacher* 7:28–34.

Swadesh, Morris. 1934. The phonemic principle. *Language* 10:117–29.

Swadesh, Morris. 1948. On linguistic mechanism. *Science and Society* 12:254–59.

Teeter, Karl V. 1969. Leonard Bloomfield's linguistics. *Language Sciences* 7:1–6.

Terrace, Herbert S. 1979. *Nim: A chimpanzee who learned sign language.* New York: Knopf.

Thomas, Owen. 1965. *Transformational grammar and the teacher of English.* New York: Holt, Rinehart and Winston.

Thorne, James P. 1965. Stylistics and generative grammars. *Journal of Linguistics* 1:49–59.

Trager, George L. 1968. Review of Hockett (1968). *Studies in Linguistics* 20:77–84.

Trager, George L., and Henry Lee Smith. 1957 [1951]. *An outline of English structure.* 3d printing. Washington: American Council of Learned Societies.

Trubetzkoy, N. 1939. *Grundzüge der Phonologie.* Prague: Cercle linguistique de Prague.

Van Riemsdijk. See Riemsdijk, Henk van.

Vater, Heinz. 1971. Linguistics in West Germany. *Language sciences* 16:6–24.

Vechtman-Veth, A.C.E. 1942. *A syntax of living English.* 2nd edition. Gröningen: P. Noordhoff. First edition in 1928.

Voegelin, Carl F. 1958. Review of Chomsky (1957a). *International Journal of American Linguistics* 24:229–31.

Voegelin, Carl F., and Florence M. Voeglin. 1963. On the history of structuralizing in 20th century America. *Anthropological Linguistics* 5:12–37.

Voegelin, Carl F., and Florence M. Voegelin. 1976. Some recent (and not so recent) attempts to interpret semantics of native languages in North America. In Chafe (1976:75–98).

Vroman, William Viera. 1976. Predicate raising and the syntax-morphology-semantics cycle: Latin and Portuguese. Unpublished Ph.D. dissertation, University of Michigan, Ann Arbor.

Wall, Robert. 1970. On the notion 'derivational constraint in grammar' or, The Turing machine doesn't stop here anymore (if it ever will). In Zwicky and others (1970:163–70).

Wanner, Eric. 1974. *On remembering, forgetting, and understanding sentences: A study of the deep structure hypothesis.* The Hague: Mouton.

Wanner, Eric. 1988. Psychology and linguistics in the sixties. In Hirst (1988:143–52).

Wardaugh, Ronald, 1977. *Introduction to linguistics.* 2nd ed. New York: McGraw-Hill.

Warfel, Harry R. 1952. *Who killed grammar?* Gainesville: University of Florida Press.

Wasow, Thomas. 1976. McCawley on generative semantics. *Linguistic Analysis* 2:279–301.

Wasow, Thomas. 1985. Postscript. In Sells (1985:193–205).

Watson, James D. 1968. *The double helix: A personal account of the discovery of the structure of DNA.* New York: Mentor.

Waugh, Auberon. 1988. From Oxymoron to boiled egg [a review of Chomsky (1987)]. *The Independent* (26 March).

Weigel, John A. 1977. *B. F. Skinner.* Boston: Twayne Publishers.

Weinreich, Uriel. 1966 [1964]. *Explorations in semantic theory.* Current trends in linguistics, vol. 3. The Hague: Mouton.

Weinreich, Uriel. 1968. On arguing with Mr. Katz: A rejoinder. *Foundations of language* 3:284–87.

Weiss, Alfred. 1925. A set of postulates for psychology. *Psychological Review* 32:83–87.

Wells, Rulon. 1947a. De Saussure's system of linguistics. *Word* 3:1–31.

Wells, Rulon. 1947b. Immediate constituents. *Language* 23:81–117.

Wells, Rulon. 1963. Some neglected opportunities in descriptive linguistics. *Anthropological Linguistics* 5:38–49.

Wendt, G. 1911. *Syntax des heutigen Englisch.* 2 vols. Heidelberg: Carl Winters Universitätsbuchhandlung.

Whately, Richard. 1963 [1846]. *Elements of rhetoric.* Edited by Douglas Ehninger. Carbondale: University of Southern Illinois Press.

Whewell, William. 1837. *The history of the inductive sciences.* 3 vols. London: Frank Cass & Co.

Whitehead, Alfred North. 1929. *Process and reality.* Cambridge: Cambridge University Press.

Whitney, William D. 1910 [1867]. *Language and the study of language.* New York: Charles Scribner's Sons.

Whorf, Benjamin Lee. 1956 [1927–1941]. *Language, thought, and reality.* Edited by John B. Carroll. Cambridge, MA: MIT Press.

Wierzbicka, Anna. 1972. *Semantic primitives.* Frankfurt: Atheneüm Verlag.

Wierzbicka, Anna. 1976 [1967]. Mind and body. In McCawley (1976a:129–58).

Wirth, Jessica R., editor. 1976. *Assessing linguistic arguments.* Washington: Hemisphere Publishing.

Wittgenstein, Ludwig. 1961 [1921]. *Tractatus Logico-Philosophicus.* Translated by D. F. Pears and B. F. McGuiness. London: Routledge and Kegan Paul.

Woodhouse, A.S.P. 1952. The nature and function of the humanities. *Transactions of the Royal Society of Canada* 46.3 Section 2 (June, 1952).

Woodworth, Elisabeth Delorme, and Robert J. DiPeitro, editors. 1962. *Report of the thirteenth annual round table meeting on linguistics and language studies.* Washington: Georgetown University Press.

Yamanashi, Masa-Aki. 1972. Lexical decomposition and implied proposition. In Peranteau and others (1972:242–53).

Yuck Foo [James D. McCawley]. 1971. A selectional restriction involving pronoun choice. In Zwicky and others (1971:19–22).

Zadeh, Lofti Asker. 1965. Fuzzy sets. *Information and Control* 8:338–53.

Zadeh, Lofti Asker. 1987. *Fuzzy sets and applications: Selected papers.* Edited by R. R. Yager. New York: Wiley.

Zimmer, Karl E. 1968. Review of Chomsky (1966a). *International Journal of American Linguistics* 34:290–303.

Zwicky, Arnold M. See also: Craft, Ebbing.

Zwicky, Arnold M. 1976. Well, this rock and roll has to stop. Junior's head is hard as a rock. In Mufwene and others (1976:676–97).

Zwicky, Arnold M., Peter H. Salus, Robert I. Binnick, and Anthony L. Vanek, editors. 1970. *Studies out in left field: Defamatory essays presented to James D. McCawley.* Current Inquiry into language and linguistics, vol. 4. Edmonton: Linguistic Research Institute. Reprinted by John Benjamins, Philadelphia, 1992.

Index